ATOMIC UNITS

Quantity	Symbol or Expression	CGS Equivalent	Important Related Properties
Mass	m_e	9.10939×10^{-28} g	m_e = electron mass
Charge	e	4.803207×10^{-10} stat C	$-e$ = electron charge
Angular momentum	$\hbar$	1.05457×10^{-27} erg s	
Length (Bohr)	$a_0 = \hbar^2/m_e e^2$	0.5291772×10^{-8} cm	a_0 = radius of lowest energy Bohr orbit
Energy (Hartree)	$E_h = e^2/a_0$	4.35975×10^{-11} erg	$\frac{1}{2}E_h$ = ionization energy of hydrogen
Time	$\tau_0 = \hbar^3/m_e e^4$	2.41887×10^{-17} s	$2\pi\tau_0$ = orbital period of lowest Bohr orbit
Frequency	$m_e e^4/\hbar^3$	4.13416×10^{16} s^{-1}	
Velocity	$e^2/\hbar$	2.18770×10^8 cm/s	$e^2/\hbar$ = velocity of electron in lowest Bohr orbit
Force	e^2/a_0^2	8.23873×10^{-3} dyne	
Electric field	e/a_0^2	1.71526×10^7 stat C/cm^2 (5.14221×10^9 V/cm)	e/a_0^2 = field experienced by electron in lowest Bohr orbit
Electric potential	e/a_0	9.07675×10^{-2} stat C/cm (27.2114 V)	
Fine structure constant	$\alpha = e^2/\hbar c$	1/137.036	
Power	$m_e^2 e^8/\hbar^5$	1.80239×10^6 erg/s (0.180239 W)	
Magnetic moment	$\beta_e = e\hbar/2m_e c$	0.92740×10^{-20} erg/gauss	β_e = magnetic moment of orbital motion in lowest Bohr orbit
Magnetic field	$e\hbar/m_e c a_0^3$	1.25168×10^5 gauss	Magnetic field at nucleus due to orbital motion in first Bohr orbit

Source: G.C. Schatz and M.A. Ratner, *Quantum Mechanics in Chemistry* (Prentice-Hall, Englewood Cliffs, 1993).

This book provides a detailed presentation of modern quantum theories for treating the reaction dynamics of small molecular systems. Its main focus is on the recent development of successful quantum dynamics theories and computational methods for studying the molecular reactive scattering process, with specific applications given in detail for a number of benchmark chemical reaction systems in the gas phase and the gas surface.

In contrast to traditional books on collisions in physics focusing on abstract theory for nonreactive scattering, this book deals with both the development and application of modern reactive or rearrangement scattering theory. It is written in a fashion in which the development of reactive scattering theory is closely coupled with its computational aspects for practical applications to realistic molecular reactions. The topics discussed include methods for calculating the rovibrational states of molecules, fundamental quantum theory for scattering (nonreactive and reactive), modern time-independent computational methods for reactive scattering, general time-dependent wave packet methods for reactive scattering, dynamics theory of chemical reactions, dynamics of molecular fragmentation, and semiclassical description of quantum mechanics. Some useful appendices are also provided.

The book is intended for the reader to not only understand molecular reaction dynamics from the fundamental scattering theory, but also to utilize the described computational methodologies in their practical applications. It will be of benefit to graduate students and researchers in the field of chemical physics.

JOHN ZENG HUI ZHANG is Professor of Chemistry in the Department of Chemistry at New York University. He was conferred his BSc in physics in 1982 from the East China Normal University in Shanghai, China. In 1987, he received his PhD in chemical physics from the University of Houston in Texas, USA. He is also a recipient of the National Science Foundation Presidential Faculty Fellowship, the Alfred P. Sloan Research Fellowship, and the Dreyfus Teacher-Scholar Award.

Theory and Application of Quantum Molecular Dynamics

John Zeng Hui Zhang

Department of Chemistry
New York University

World Scientific
Singapore • New Jersey • London • Hong Kong

Published by

World Scientific Publishing Co. Pte. Ltd.

P O Box 128, Farrer Road, Singapore 912805

USA office: Suite 1B, 1060 Main Street, River Edge, NJ 07661

UK office: 57 Shelton Street, Covent Garden, London WC2H 9HE

Library of Congress Cataloging-in-Publication Data

Zhang, John Z. H.

Theory and application of quantum molecular dynamics / John Zeng Hui Zhang.

p. cm.

Includes bibliographical references (p.) and index.

ISBN 9810233884 (alk. paper)

1. Quantum chemistry. 2. Molecular dynamics. 3. Chemical kinetics. I. Title.

QD462.Z43 1999

541.2'8--dc21 98-38058

CIP

British Library Cataloguing-in-Publication Data

A catalogue record for this book is available from the British Library.

Printed in Singapore by Uto-Print

Preface

The writing of this book is motivated by recent advances in the field of theoretical reaction dynamics since the mid-1980's. Prior to the mid-1980's, the field of reactive scattering was dominated by model theoretical studies of collinear atom-diatom reactive scattering and approximate three dimensional atom-diatom reactive scattering including only two effective degrees of freedom. The post mid-1980's period has been marked by rapid progress in the development of rigorous quantum mechanical methods for computation of molecular reactive scattering in the gas-phase. The rapid development in computational methodology was motivated in part by the availability of new and faster computers to theoretical chemists. The new theoretical and computational development has quickly eclipsed collinear atom-diatom scattering calculations and made exact 3D atom-diatom reactive scattering calculations routine applications in the gas-phase. For the first time exact quantum dynamics calculations have enabled theoretical chemists to make accurate dynamics predictions for some simple triatomic reactions. Such exact quantum dynamics methods provide rigorous theoretical tools for analyzing reaction dynamics at an unprecedented level of detail and accuracy. Exact dynamics calculations can be regarded as "simulated experiments", especially when coupled with *ab initio* quantum chemistry calculations. These theoretical developments have bridged a huge gap between theory and experiment in chemical reaction dynamics. Some of the progress has been written up in a number of reviews. Since the early nineties, new progress has been made in further development of theoretical methods to extend rigorous quantum dynamical calculations to chemical reactions beyond atom-diatom systems. In particular the time-dependent approach has spear-headed the computational advance for rigorous quantum dynamics calculations for tetraatomic reactions. It has proven to be a promising powerful tool for dynamics study of complex chemical systems.

Although this book is motivated by recent theoretical advances, it is not written simply to describe the progress in the field of chemical reaction dynamics. Rather, the book is intended to provide a balanced presentation of both fundamental dynamics theories and their applications to realistic chemical dynamics problems. From my personal knowledge and experience, there is a real need for a book that can combine both the abstract theory on quantum scattering and practical methodologies that are tailored for numerical applications to realistic molecular systems. Currently, there are basically two categories of books that are being read by graduate students and researchers in the field. On the one side are books on abstract scattering theory mainly written by and for theoretical physicists. These books are usually quite abstract and mainly describe theories on elastic scattering with few, if any, discussions of treatment of reactive or arrangement scattering problems that are relevant to reaction dynamics. On the other side are monographs and proceedings on very specific topics in quantum reactive scattering that are mainly tailored for those specialists who are already experts in the field.

In view of the above considerations, the contents of this book are specially organized to contain three main parts: (1) fundamental theory of scattering (2) development of practical computational methodologies based on fundamental theory and (3) applications of theory to specific benchmark molecular systems. In particular, the section dealing with scattering theory in the book is written in a coherent fashion. The central theme of the theory is based on the time-dependent picture in order to help the reader visualize the dynamical process and gain a physically intuitive picture. This is in contrast to most traditional books on scattering theory that tend to focus on the abstract time-independent theory of scattering. I believe that the time-dependent theme will make the topic of scattering theory much easier to understand and grasp by graduate students and anyone who is new to the field. The presentation of the theory in this book is then generally followed by applicable computational methods, and in addition, efforts are made to show specific applications of theories or methods to chemical problems. Besides the main topics in quantum dynamics, the book also has other useful contents including an introductory chapter on Hartree-Fock theory for electron structure and a chapter on the basic theoretical treatment of separation of electronic and nuclear degrees of freedom. There is also an appendix at the end of the book to provide the reader with many useful mathematical formulas and relations that are frequently encountered in theoretical studies and computation of quantum molecular dynamics problems. Taken as a whole, this book is thus intended to be a comprehensive treatment of quantum scattering theory and its application in molecular reaction dynamics.

This book is organized as follows; Chapter 1 describes the basic theory and treatment for molecular systems with both electronic and nuclear degrees of freedom. Chapter 2 introduces the basic theory of the Hartree-Fock equations for solving electronic problems. This is followed by a description of theoretical methodologies to solve general bound state problems for molecules in Chapter 3. The fundamental theory of quantum scattering is presented in detail in Chapter 4. Chapter 5 focuses on the development of time-independent computational methodologies to solve gas phase reactive scattering problems with specific applications to prototype atom-diatom reactive scattering. Chapter 6 presents the time-dependent wavepacket approach for solving large scale quantum dynamics problems with specific applications to benchmark tetraatomic reactions. Chapter 7 describes various equivalent forms of state-to-state S matrix elements and their applications. It also presents a general reactant-product decoupling (RPD) approach to computing state-to-state S matrix elements for large systems. The theory and phenomena of dynamical resonances and relation of the exact quantum mechanical rate constant to the transition state theory are given in Chapter 8. Chapter 9 introduces basic theory for molecular interactions with the radiation field and presents methodologies for calculating fragmentation dynamics of molecules. Theoretical models with applications to molecular reactions on rigid solid surfaces are presented in Chapter 10. Finally, Chapter 11 presents general theories of semiclassical approximation in quantum mechanics.

This book is written as a self-contained and comprehensive yet very compact book that is suitable for a variety of readers including graduate students, researchers and/or other professionals who are interested in the theory and applications in the field of molecular reaction dynamics.

Many people have contributed to my writing of this book. I especially thank Prof. Donald J. Kouri and Prof. William H. Miller from whom I have learned many of the theories presented in the book while I was a graduate student and postdoctoral fellow. Many of the research topics and specific applications presented in the book are results of years of research work done by my former students and postdoctoral fellows. Here I would like to thank Dr. Donghui Zhang, Dr. Jiqiang Dai, Dr. Wei Zhu for their contributions to the research topics presented in the book. I thank my colleague Prof. Henry Brenner for careful reading of the manuscript and for correcting errors. The writing of this book began while I was on sabbatical for one semester at Hong Kong University of Science and Technology in 1997.

Contents

Theory and Application of Quantum Molecular Dynamics

Chapter 1

Separation of Electronic and Nuclear Motions

1.1 Adiabatic Representation

For a general molecular system consisting of electrons and nuclei, the Hamiltonian can be written as

$$H(r,R) = T_N + H_e(r) + V_{eN}(r,R) \tag{1.1}$$

where T_N represents nuclear kinetic energy operators, H_e is the electronic Hamiltonian and $V_{eN}(r,R)$ includes all the electron–nuclear and nuclear–nuclear interactions. The electronic Hamiltonian H_e can be written as

$$H_e(r) = T_e + V_{ee} \tag{1.2}$$

where T_e represents electron kinetic energy operators and V_{ee} includes all electron–electron interactions. Here, we use r and R as collective indexes to denote, respectively, the coordinates of the electrons and nuclei, For a system with N nuclei and n electrons, various terms are explicitly given by

$$T_N = \sum_k^N \left(-\frac{\hbar^2}{2M_k}\right) \nabla_k^2 \tag{1.3}$$

$$T_e = \sum_i^n \left(-\frac{\hbar^2}{2m_e}\right) \nabla_i^2 \tag{1.4}$$

$$V_{ee} = \frac{1}{2}\sum_{i\neq j}^{n}\frac{1}{r_{ij}} \tag{1.5}$$

$$V_{eN} = -\sum_{k}^{N}\sum_{i}^{n}\frac{Z_k}{|\mathbf{R}_k - \mathbf{r}_i|} + \frac{1}{2}\sum_{k\neq k'}^{N}\frac{Z_k Z_{k'}}{R_{kk'}} \tag{1.6}$$

where M_k and Z_k are, respectively, the mass and charge of kth nucleus, and m_e is the electron mass.

Now, if all the nuclei were fixed in space (R fixed), then the motions of electrons would be governed by the following Hamiltonian equation,

$$\left[H_e(r) + V_{eN}(r,R)\right]\phi_n(r,R) = \epsilon_n(R)\phi_n(r,R) \tag{1.7}$$

where $\phi_n(r,R)$ and $\epsilon_n(R)$ are called ***adiabatic*** eigenfunctions and eigenvalues of the electrons with the fixed nuclear coordinates R as parameters. Since the adiabatic eigenfunctions $\phi_n(r,R)$ form a complete orthonormal set, the molecular wavefunction $\Psi(r,R)$ satisfying the Schrödinger equation

$$H(r,R)\Psi(r,R) = E\Psi(r,R) \tag{1.8}$$

can be expanded in the adiabatic basis $\phi_n(r,R)$

$$\Psi(r,R) = \sum_n \chi_n(R)\phi_n(r,R) \tag{1.9}$$

where $\chi_n(R)$ is the corresponding nuclear wavefunction in the ***adiabatic representation.*** By substituting the expansion in Eq. (1.9) into Eq. (1.8), we obtain, after integrating over the electron coordinates, the coupled equations

$$\left[T(R) + \epsilon_m(R)\right]\chi_m(R) + \sum_n \Lambda_{mn}(R)\chi_n(R) = E\chi_m(R) \tag{1.10}$$

Here $\Lambda_{mn}(R)$ is the nonadiabatic coupling matrix operator which arises from the action of the nuclear kinetic energy operator $T(R)$ on the electron wavefunction $\phi_n(r,R)$. It can be shown easily that Λ_{mn} is given by

$$\Lambda_{mn}(R) = -\hbar^2\sum_i \frac{1}{M_i}\left(A^i_{mn}\frac{\partial}{\partial R_i} + \frac{1}{2}B^i_{mn}\right) \tag{1.11}$$

where the matrices are defined as

$$A^i_{mn} = <\phi_m|\frac{\partial}{\partial R_i}|\phi_n> = \int \phi^*_m \frac{\partial}{\partial R_i}\phi_n \, dr \tag{1.12}$$

$$B^i_{mn} = <\phi_m|\frac{\partial^2}{\partial R_i^2}|\phi_n> = \int \phi^*_m \frac{\partial^2}{\partial R_i^2}\phi_n \; dr \tag{1.13}$$

Equation (1.10) can be written in matrix form

$$(\mathbf{T}+\mathbf{V})\mathbf{X}(R) = E\mathbf{X}(R) \tag{1.14}$$

where the diagonal matrix

$$\mathbf{V}_{mn}(Q) = \epsilon_m(R)\delta_{mn} \tag{1.15}$$

is called the ***adiabatic potential*** and the nondiagonal kinetic matrix is given by

$$\mathbf{T}_{mn}(R) = T(R)\delta_{mn} + \Lambda_{mn}(R) \tag{1.16}$$

Thus in the adiabatic representation, the nuclear potential operator in the Schrödinger equation is diagonal while the kinetic energy operator is not.

1.2 Born-Oppenheimer Approximation

Equation (1.10) solves rigorously the coupled channel Schrödinger equation for the nuclear wavefunction in the adiabatic representation. The nondiabatic coupling between different adiabatic states is given by the nonadiabatic operator of (1.11) which is responsible for nonadiabatic transitions. The direct calculation of the nonadiabatic coupling matrix is usually a very difficult task in quantum chemistry. In addition, the coupled equation (1.10) is difficult to solve. However, what makes the adiabatic representation so powerful is the use of adiabatic approximation in which the off-diagonal couplings Λ_{mn} $(m \neq n)$ are discarded. This approximation is based on the rationale that the nuclear mass is much larger than the electron mass, and therefore the nuclei move much slower than the electrons. Thus the nuclear kinetic energies are generally much smaller than those of electrons and consequently the nonadiabatic coupling matrices A^i_{mn} and B^i_{mn}, which result from nuclear motions, are generally small.

If we neglect the nonadiabatic coupling, which is equivalent to retaining just a single term in the adiabatic expansion of the wavefunction

$$\Psi(r,R) = \chi_n(R)\phi_n(r,R) \tag{1.17}$$

we obtain the adiabatic approximation for the nuclear wavefunction

$$H^{ad}_n\chi_n(R) = E\chi_n(R). \tag{1.18}$$

where the adiabatic Hamiltonian is defined as

$$H_n^{ad} = T_N + \epsilon_n(R) + \Lambda_{nn}(R) \tag{1.19}$$

Here $\epsilon_n(R)$ is the adiabatic potential for the nuclear motion defined in Eq. (1.7). A more rigorous way to state the validity condition of the adiabatic approximation is that the nuclear kinetic energy should be much smaller than the energy gaps between (adiabatic) electronic states.

Since the electronic eigenfunction $\phi_n(r, R)$ is indeterminate to a phase factor of R, viz., $\exp[if(R)]$, a common practice is to choose $\phi_n(r, R)$ to be real. In this case, the function $A_{nn}^{(i)}(R)$ in Eq. (1.12) vanishes and therefore the diagonal operator $\Lambda_{nn}(R)$ does not include differential operators. This follows directly by carrying out partial integration in Eq. (1.12)

$$\begin{aligned} A_{nn}^i &= <\phi_n|\frac{\partial}{\partial R_i}\phi_n> \\ &= \frac{\partial}{\partial R_i} <\phi_n|\phi_n> - <\frac{\partial}{\partial R_i}\phi_n|\phi_n> \\ &= -A_{nn}^i \end{aligned} \tag{1.20}$$

where the reality condition $\phi_n = \phi_n^*$ and the normalization condition $<\phi_n|\phi_n>=1$ have been used in the above equation. It should be noted here that the Dirac inner product notion of $<\phi_n|\phi_n>= 1$ for electrons includes only the integration over the electronic coordinates r, with the nuclear coordinates R being just parameters of ϕ_n.

In most situations, the dependence of $B_{nn}(R)$ on nuclear coordinates R is relatively weak compared to that of the adiabatic potential $\epsilon_n(R)$.. Thus the term $\Lambda_{nn}(R)$ is often neglected in the adiabatic approximation and one obtains the familiar Born-Oppenheimer approximation

$$\left[T_N + V_n(R)\right]\chi_n(R) = E\chi_n(R) \tag{1.21}$$

where $\epsilon_n(R)$ is replaced by the more familiar notation $V_n(R)$. Although the term adiabatic approximation is often used interchangeably with the Born-Oppenheimer approximation, it is important to bear in mind the difference between the proper adiabatic approximation of (1.18) and the Born-Oppenheimer approximation of (1.21). Thus in the adiabatic or Born-Oppenheimer approximation, one achieves a complete separation of electronic motion from that of nuclei; one first solves for electronic eigenvalues $\epsilon(R)$ at given nuclear geometries and then solves the nuclear dynamics problem using $\epsilon(R)$ as the potential for the nuclei.

The physical meaning of the adiabatic approximation is clear: the slow nuclear motion only leads to the deformation of the electronic states but not to transitions between different electronic states. The electronic wavefunction deforms instantaneously to adjust to the slow motion of nuclei. The general criterion for the validity of the adiabatic approximation is that the nuclear kinetic energy be small relative to the energy gaps between electronic states such that the nuclear motion does not cause transitions between electronic states, but only distortions of electronic states. This argument can be deduced from the nonadiabatic coupling terms in Eq. (1.10).

1.3 Hellmann-Feynman Theory

The Hellmann-Feynman theory provides a convenient means to compute Ehrenfest forces on the nuclei for a molecular system in the adiabatic or Born-Oppenheimer approximation. The theory is proved as follows. From the adiabatic equation of (1.7) for electrons, the adiabatic eigenvalue $\epsilon(R)$ can be expressed as the expectation value of the Hamiltonian

$$\epsilon(R) = <\phi(R)|H(R)|\phi(R)> \tag{1.22}$$

where the Hamiltonian for electronic motion is defined as

$$H(R) = H_e(r) + V_{eN}(r, R) \tag{1.23}$$

Here the subscript that labels the adiabatic eigenvalue and eigenfunction has been removed for clarity and the nuclear coordinate R is simply a parameter. Differentiating Eq. (1.22) with respect to the nuclear coordinate R gives

$$\begin{aligned}\frac{\partial \epsilon(R)}{\partial R} &= <\frac{\partial \phi(R)}{\partial R}|H(R)|\phi(R)> + <\phi(R)|H(R)|\frac{\partial \phi(R)}{\partial R}> \\ &\quad + <\phi(R)|\frac{\partial V_{eN}}{\partial R}|\phi(R)> \end{aligned} \tag{1.24}$$

Using the fact that $|\phi(R)>$ is an eigenstate of the Hamiltonian $H(R)$ with eigenvalue $\epsilon(R)$, the above equation can be simplified to yield

$$\frac{\partial \epsilon(R)}{\partial R} = \epsilon(R)\frac{\partial}{\partial R} <\phi(R)|\phi(R)> + <\phi(R)|\frac{\partial V_{eN}}{\partial R}|\phi(R)> \tag{1.25}$$

Because $|\phi>$ is normalized, the first term vanishes and we obtain the final relation

$$\frac{\partial \epsilon(R)}{\partial R} = <\phi(R)|\frac{\partial V_{eN}}{\partial R}|\phi(R)> \tag{1.26}$$

or in vector notation

$$\nabla_{\mathbf{R}}\epsilon(\mathbf{R}) = <\phi(\mathbf{R})|\nabla_{\mathbf{R}} V_{eN}|\phi(\mathbf{R})> \tag{1.27}$$

Thus the Ehrenfest force acting on the ith nucleus can be calculated via the Hellmann-Feynman theory

$$\mathbf{F_i} = -<\phi|\nabla_i V_{eN}|\phi> = -\nabla_i \epsilon(\mathbf{R}) \tag{1.28}$$

Equation (1.28) states that the force acting on the ith nucleus is simply the negative gradient of the adiabatic eigenvalue or potential $\epsilon(\mathbf{R})$ with respect to the ith nuclear coordinate. Thus, the force on the nucleus can be obtained by simply calculating the gradient of $\epsilon(\mathbf{R})$ without the need of averaging over the adiabatic electronic wavefunction ϕ. This is a significant simplification because no electronic wavefunction is needed in the numerical calculation of forces for nuclei! For simplicity, we left out the nuclear–nuclear interaction V_{NN} in the above derivation. The forces due to V_{NN} can be trivially added to the right side of Eq. (1.28).

1.4 Diabatic Representation

Although the adiabatic representation is widely used for molecular applications, it is not particularly convenient to solve nonadiabatic problems in numerical calculations because the nonadiabatic coupling matrix is difficult to compute directly. Thus in solving nonadiabatic problems, one often starts from the *diabatic* representation . In the diabatic representation, one chooses the electronic wavefunction calculated for a fixed reference nuclear configuration R_0 by solving the Schrödinger equation

$$\left[H(r) + V_{eN}(r, R_0)\right]\phi_n(r, R_0) = \epsilon_n(R_0)\phi_n(r, R_0) \tag{1.29}$$

where the nuclear configuration R_0 is chosen at a fixed reference value regardless of the actual spatial positions of the nuclei. By using $\phi_n(r, R_0)$ as basis set, the molecular wavefunction can be expanded as

$$\Psi(r, R) = \sum_n \chi_n^0(R)\phi_n(r, R_0) \tag{1.30}$$

Substituting the expansion of (1.30) into the Schrödinger Eq. (1.8) and integrating over the electronic wavefunction, one obtains the coupled equation for the nuclear wavefunction in the diabatic representation

$$T_N\chi_m^0(R) + \sum_n U_{mn}(R)\chi_n^0(R) = E\chi_m^0(R). \tag{1.31}$$

Here the nondiagonal coupling U_{mn} arises from the electron–nuclear interaction $V_{eN}(r, R)$ and is given by

$$\begin{aligned} U_{mn}(R) &= <\phi_m|H_e + V_{eN}(R)|\phi_n> \\ &= \epsilon_m(R_0)\delta_{mn} + <\phi_m|V_{eN}(R) - V_{eN}(R_0)|\phi_n> \\ &= \epsilon_m(R_0)\delta_{mn} + \int \phi_m^*(r, R_0)[V_{eN}(r, R) - V_{eN}(r, R_0)] \\ &\quad \times \phi_n(r, R_0)dr \end{aligned} \tag{1.32}$$

Equation (1.31) can be written in matrix form as

$$(\mathbf{T} + \mathbf{U})\mathrm{X}^0(R) = E\mathrm{X}^0(R) \tag{1.33}$$

where the kinetic energy operator is diagonal

$$\mathbf{T}_{mn}(R) = T_N\delta_{mn} \tag{1.34}$$

but the potential energy operator is nondiagonal with its matrix element given by Eq. (1.32). If the nondiagonal coupling can be neglected, we arrive at the diabatic approximation

$$[T_N + V_m^d(R)]\chi_m^0(R) = E\chi_m^0(R). \tag{1.35}$$

where the *diabatic potential* is given by $V_m^d(R) = U_{mm}(R)$.

Although the diabatic approximation is mathematically simpler because one only needs to carry out a calculation for the electronic wavefunction at a single fixed nuclear coordinate, it is less useful than the adiabatic approximation in practical situations in chemistry. This is simply explained by the conditions of validity of both approximations. In the adiabatic representation, the nonadiabatic coupling is caused by the nuclear kinetic energy operator or nuclear motion which acts like a small perturbation. Thus the condition for the validity of the adiabatic approximation is that the nuclear kinetic energy be relatively small compared to energy gaps between the adiabatic electronic states. This is not too difficult to achieve because of the large mass differential between the electrons and nuclei. A crude estimation gives a rough ratio of $M/m_e \geq 1800$ where m_e and M are, respectively, the electron and nuclear mass. Another way to understand this is from the time-dependent point of view in that the electrons can quickly adapt themselves to the new configuration of the nuclei if the latter move slowly enough. Thus if the nuclei are not moving too fast (having too much kinetic energy in comparison to the energy gaps between the adiabatic states), the adiabatic approximation should be a reasonably good approximation.

On the other hand, the validity condition of the diabatic approximation is quite the opposite. In the diabatic representation, the coupling of electronic diabatic states is caused by the electron–nuclear interaction potential $V_{eN}(r, R)$. Thus the validity of the diabatic approximation requires that this interaction be small compared to the nuclear kinetic energy as can be seen from Eq. (1.35). Again using the time-dependent point of view, this condition is satisfied if the nuclei move very fast because in this case the electrons do not have sufficient time to adjust to the nuclear motion and their wavefunction will remain the same as at R_0. To summarize, we can think of the adiabatic approximation as the *low* kinetic energy limit of the nuclear motion, while the diabatic approximation as the *high* kinetic energy limit of the nuclear motion.

1.5 Transformation Between Representations

When nonadiabatic processes are involved and more than one adiabatic electronic state is coupled, it is often more convenient to use the diabatic representation to solve for the nuclear wavefunction. This is because in the diabatic representation, only the electron wavefunctions at fixed nuclear configurations are needed in order to construct the coupling matrix and no derivative coupling is required. On the other hand, it is often desirable to use the adiabatic wavefunction for practical purposes, As a result, one often solves nonadiabatic problems in the diabatic representation first and then transforms the diabatic nuclear wavefunction to the adiabatic ones. Since the adiabatic and diabatic representations are equivalent provided both sets are complete, they are related to each other through a unitary transformation.

If the electronic basis set is sufficiently complete, both the adiabatic and diabatic expansions of the molecular wavefunction should be equivalent, viz,

$$\begin{aligned} \Psi(r, R) &= \sum_n \chi_n(R)\phi_n(r|R) \\ &= \sum_n \chi_n^0(R)\phi_n(r|R_0) \end{aligned} \tag{1.36}$$

By integrating over the electronic coordinates, we obtain the transformation relation between nuclear wavefunctions in the two representations

$$\chi_m(R) = \sum_m D_{mn}(R|R_0)\chi_n^0(R) \tag{1.37}$$

where the transformation matrix is given by

$$D_{mn}(R|R_0) = <\phi_m(R)|\phi_n(R_0)> = \int \phi_m^*(r|R)\phi_n(r|R_0)dr \tag{1.38}$$

However, direct calculation of Eq. (1.38) to generate the transformation matrix $D_{mn}(R|R_0)$ is difficult because it requires the calculation of adiabatic eigenfunction $\phi_n(r, R)$. In order to avoid this, one can start from the diabatic representation of Eq. (1.33) by constructing the diabatic coupling potential matrix $\mathbf{U}(R)$ which is much easier to calculate than the adiabatic coupling terms. The diabatic potential matrix $\mathbf{U}(R)$ can then be diagonalized to yield the diagonal potential which is just the adiabatic potential $\mathbf{V}(R)$

$$\mathbf{V}(R) = \mathbf{D}^\dagger(R|R_0)\mathbf{U}(R)\mathbf{D}(R|R_0) \tag{1.39}$$

where $\mathbf{D}$ is just the transformation matrix between diabatic and adiabatic representations defined in Eq. (1.38). The adiabatic nuclear wavefunction is then given by a simple transformation in Eq. (1.37). Thus one can simply solve for the diabatic wavefunction $\chi^0(R)$ first and carry out a simple transformation to obtain the adiabatic wavefunction $\chi(R)$ when desired. This approach is much simpler than directly solving nonadiabatic problems in the adiabatic representation.

As discussed above, the adiabatic approximation is valid around the nuclear configuration R where the nuclear kinetic energy is small relative to the energy gaps between the neighboring electronic states. If the electronic energy gap is small or states even cross at R, the adiabatic approximation breaks down in this region. As is well known in perturbation theory, when two or more states are degenerate or nearly degenerate, the standard perturbation theory breaks down and one needs to use perturbation theory for degenerate states. To use the simple analogy with degenerate perturbation theory, we can imagine that the "original" adiabatic states that are degenerate or nearly degenerate are no longer valid representations of the electronic motion. Instead, we need to form linear combinations of these degenerate states and use the nuclear kinetic operator (treated as a perturbation) to lift the degeneracy by solving the secular equation. There are many cases of nonadiabatic couplings that are extremely important in molecular processes and the study of nonadiabatic processes is currently an active area of research.

1.6 Crossing of Adiabatic Potentials

The adiabatic potential curves $V_n(R)$ can sometimes cross or come near each other at some nuclear configurations. This corresponds to the case of degeneracy or quasi degeneracy of electronic states. In order for adiabatic potentials to cross, certain conditions must be satisfied. Usually, the crossing of adiabatic potentials that belong to the same electronic symmetry can only occur at nuclear configurations that correspond to certain symmetries of the molecular configuration. This does not apply to adiabatic states that have different symmetries. The following is a simple heuristic derivation of the condition for the crossing of two adiabatic potential curves which are of the same symmetry.

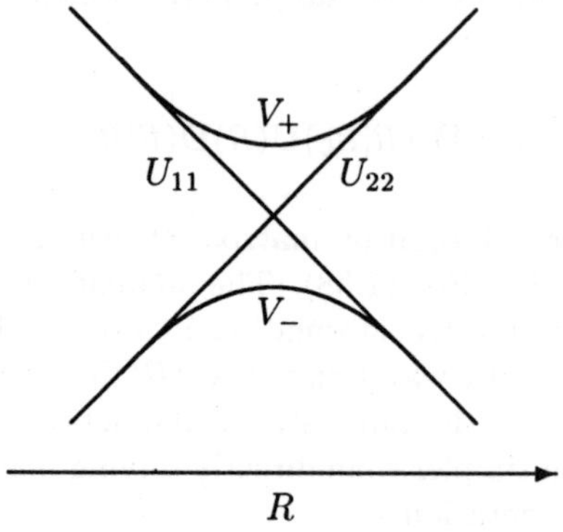

Figure 1.1: Adiabatic and diabatic potentials in a two state model.

For a coupled two state problem, the diabatic potential matrix can be written as a 2×2 hermitian matrix

$$\mathbf{U} = \begin{bmatrix} U_{11} & U_{12} \\ U_{21} & U_{22} \end{bmatrix} \tag{1.40}$$

By diagonalizing the matrix operator $\mathbf{U}$, we obtain the adiabatic potentials

$$V_{\pm} = \frac{U_{11}+U_{22}}{2} \pm \frac{1}{2}\sqrt{(U_{11}-U_{22})^2 + 4|U_{12}|^2} \tag{1.41}$$

and the gap between the two adiabatic potentials is given by

$$\Delta V = V_+ - V_- = \sqrt{(U_{11}-U_{22})^2 + 4|U_{12}|^2} \tag{1.42}$$

The relationship of the potentials is illustrated in Fig. 1.1.

For two adiabatic potentials to cross, the two positive terms in Eq. (1.42) must satisfy the following equations simultaneously

$$\begin{cases} U_{11}(R) = U_{22}(R) \\ U_{12}(R) = 0 \end{cases} \tag{1.43}$$

This is the general condition for the crossing of two adiabatic potential curves with the same symmetry. For a molecular system with N internal nuclear degrees of freedom, the crossing can occur on an $N-2$ dimensional hypersurface. For diatomic molecules, there is only one internal nuclear degree of freedom and thus the adiabatic potentials cannot cross. For triatomic molecules, there are three internal degrees of freedom and the crossing can occur along a curve.

Further reading

References [1–4] are good sources for further reading on the topics of this chapter.

Chapter 2

Ab initio Theory for Electrons

2.1 Hartree-Fock Theory

2.1.1 Hartree-Fock Equation

To solve chemical dynamics problems, one first needs to solve the electron problem. In dynamical models based on the Born-Oppenheimer approximation, one needs to solve for the adiabatic eigenvalues $\epsilon(R)$ which form the potential for nuclear motion. For a molecular system consisting of M nuclei and N electrons, the electronic Hamiltonian for a fixed nuclear configuration is the sum of one-electron terms $h(i)$ and electron–electron interactions (in atomic units)

$$H = \sum_{1}^{N} h(i) + \frac{1}{2} \sum_{i \neq j} \frac{1}{r_{ij}} \tag{2.1}$$

where the one-electron term is given by

$$h(i) = -\frac{1}{2}\nabla_i^2 + \sum_{\alpha}^{M} \frac{Z_\alpha}{r_{i\alpha}} \tag{2.2}$$

Here Z_α is the electric charge of the αth nucleus, $r_{i\alpha} = |\mathbf{r}_i - \mathbf{R}_\alpha|$, and $r_{ij} = |\mathbf{r}_i - \mathbf{r}_j|$. The electron wavefunction Φ_n satisfies the Schrödinger equation

$$\left[\sum_{i}^{N} h(i) + \frac{1}{2}\sum_{i\neq j}\frac{1}{r_{ij}}\right]\Phi_n(r) = \epsilon_n\Phi_n(r) \tag{2.3}$$

where the dependence of ϵ and Φ_n on nuclear coordinates R has been omitted for clarity, and also the nuclear–nuclear interactions have been excluded.

The Hartree-Fock wavefunction

Since solving Eq. (2.3) is a complex many body problem, it is not feasible at present to solve it exactly. It is necessary to introduce various approximations to solve Eq. (2.3). One of the most important approximations in solving the electron problem is the Hartree-Fock approximation. In the Hartree-Fock treatment, the electron wavefunction Φ is approximated by the symmetrized Hartree product of one-electron *spin orbitals*

$$\Phi = \prod_{k=1}^{N}\psi_k(k) = \psi_1(1)\psi_2(2)\cdots\psi_N(N) \tag{2.4}$$

where the spin orbital $\psi_k(\mathbf{x})$ is defined as the product of the *spatial* orbital (wavefunction) $\phi_k(\mathbf{r})$ and the *spin* wavefunction χ_s of the kth electron, viz.,

$$\psi_k = \phi_k\chi_s \tag{2.5}$$

The spin orbital is usually expressed by a two-component tensor representing two spin states ψ_+ and ψ_-. These spin orbitals are chosen to be orthonormal

$$<\psi_k|\psi_{k'}>= \delta_{kk'} \tag{2.6}$$

In order to take into account the antisymmetry property of electron, the direct product wavefunction is properly symmetrized by an antisymmetry operator $\mathcal{A}$

$$\Phi_A = \mathcal{A}\prod_{k=1}^{N}\psi_k(k) \tag{2.7}$$

where the antisymmetry operator is defined as

$$\mathcal{A} = \frac{1}{\sqrt{N!}}\sum(-1)^p\hat{p} \tag{2.8}$$

Here $\hat{p}$ is a permutation operator and the summation is over all electron permutations (total of $N!$ terms). The sign in front of the operator $\hat{p}$ is positive for an even permutation and negative for an odd permutation. Equation (2.7) can be written in the form of a determinant called the Slater determinant

$$\Phi_A = \frac{1}{\sqrt{N!}} \begin{vmatrix} \psi_1(1) & \psi_1(2) & \cdots & \psi_1(N) \\ \psi_2(1) & \psi_2(2) & \cdots & \psi_2(N) \\ \vdots & \vdots & \ddots & \vdots \\ \psi_N(1) & \psi_N(2) & \cdots & \psi_N(N) \end{vmatrix}$$
$$= \frac{1}{\sqrt{N!}} \det[\psi_1 \psi_2 \cdots \psi_N] \tag{2.9}$$

Derivation of the Hartree-Fock equation

The expectation value of the Hamiltonian in the HF approximation can be written as

$$<\Phi_A|H|\Phi_A> = \sum_{i=1}^{N} <i|h|i> + \frac{1}{2}\sum_{ij} <ij|\frac{1-\hat{p}_{ij}}{r_{ij}}|ij>$$
$$= \sum_{i=1}^{N} h_i + \frac{1}{2}\sum_{ij}[J_{ij} - K_{ij}] \tag{2.10}$$

where the second term includes both the Coulomb (direct) integral

$$J_{ij} = <ij|\frac{1}{r_{ij}}|ij>$$
$$= \iint d\mathbf{x}_2 d\mathbf{x}_1 |\psi_i(1)|^2 \frac{1}{r_{12}} |\psi_j(2)|^2 \tag{2.11}$$

and the exchange integral

$$K_{ij} = <ij|\frac{1}{r_{ij}}\hat{p}_{ij}|ji>$$
$$= \iint d\mathbf{x}_2 d\mathbf{x}_1 \psi_i^\dagger(1)\psi_j^\dagger(2)\frac{1}{r_{12}}\psi_i(2)\psi_j(1) \tag{2.12}$$

We note here that the restriction $i \neq j$ in the summation $\sum_{ij}$ in Eq. (2.10) is removed because the terms corresponding to $i = j$ in the summation vanish. The Coulomb integral J_{ij} is simply the classical electrostatic energy but the exchange integral K_{ij} has no classical counterpart. It can be proved that $J_{ij} \geq K_{ij}$.

By minimizing the expectation value of the Hamiltonian in Eq. (2.10) with respect to variation of one-electron spin orbitals $\delta\psi_k$, with Lagrange multipliers ϵ_{ij} and the orthogonality condition of Eq. (2.6)

$$\delta[<\Phi_A|H_e|\Phi_A> - \frac{1}{2}\sum_{ij}\epsilon_{ij}(<i|j> - \delta_{ij})] = 0 \tag{2.13}$$

it is not difficult to derive the following HF equation

$$F|\psi_i> = \sum_{j=1}^{N}\epsilon_{ij}|\psi_j> \tag{2.14}$$

Here F is the Fock operator defined as

$$F(1) = h(1) + v^{HF}(1) \tag{2.15}$$

where the Hartree-Fock potential $v^{HF}(i)$ is defined as

$$v^{HF}(j) = h(j) + \sum_{i=1}^{N}[J_i(j) - K_i(j)] \tag{2.16}$$

The Coulomb operator J_i and exchange operator K_i are defined as

$$J_i(1) = <i|\frac{1}{r_{12}}|i> = \int dx_2|\psi_i(2)|^2\frac{1}{r_{12}} \tag{2.17}$$

and

$$\begin{aligned} K_i(1)\psi_j(1) &= <i|\frac{\hat{p}_{12}}{r_{12}}|i>\ \psi_j(1) \\ &= \left[\int dx_2\psi_i^\dagger(2)\frac{1}{r_{12}}\psi_j(2)\right]\psi_i(1) \end{aligned} \tag{2.18}$$

It is easy to see that the Fock operator defined in Eq. (2.15) is hermitian, i.e.

$$<\psi_i|F|\psi_j> = <\psi_j|F|\psi_i> = \epsilon_{ij} \tag{2.19}$$

where ϵ_{ij} is a real symmetric matrix.

Using the Slater determinant in Eq. (2.9), it is straightforward to prove that the antisymmetrized wavefunction Φ_A is *invariant* with respect to any unitary transformation of spin orbitals. Explicitly, if the new spin orbitals are constructed from a unitary transformation of "old" orbitals, i.e.,

$$\psi_i'(1) = \sum_j \psi_j(1) u_{ji} \tag{2.20}$$

where u_{ji} is the matrix of the unitary transformation, then

$$\Phi_A' = \det|\Psi'| = \det|\mathbf{U}^\dagger \Psi \mathbf{U}| = \det|\Psi| = \Phi_A \tag{2.21}$$

This shows that the HF wavefunction is independent of the specific choice of one-electron spin orbitals. It can be further proved that the Fock operator is also invariant with respect to any unitary transformation of orbitals, i.e.,

$$F'(1) = F(1) \tag{2.22}$$

which can be shown easily from Eqs. (2.15), (2.17), and (2.18). In order words, the spin orbitals are not unique while the HF wavefunction is. It is then possible to choose a particular set of spin orbitals in which the HF equation (2.14) is "diagonal" or "decoupled". This can be done by formally diagonalizing the real symmetric matrix ϵ_{ij} in Eq. (2.14) to obtain the *canonical* (diagonal) Hartree-Fock equation

$$\boxed{F(1)\psi_i(1) = \epsilon_i \psi_i(1)} \tag{2.23}$$

where the special orbitals satisfying the above equation are called *canonical spin orbitals.*

Using Eq. (2.23), we can express the orbital energy as

$$\begin{aligned} \epsilon_i &= \langle i|F|i\rangle \\ &= h_i + \sum_{ij}(J_{ij} - K_{ij}) \end{aligned} \tag{2.24}$$

where the Coulomb and exchange integrals J_{ij} and J_{ij} are defined in Eqs. (2.11) and (2.12). Note since $J_{ii} = K_{ii}$, there is no contribution from the term with $i = j$. However, the correct ground state energy in the HF approximation is not simply the sum of HF orbital energies ϵ_i

$$E_g \neq \sum_i^N \epsilon_i \tag{2.25}$$

but is given by Eq. (2.10)

$$\begin{aligned} E_g &= \sum_{i=1}^{N} h_i + \frac{1}{2}\sum_{ij}(J_{ij} - K_{ij}) \\ &= \sum_{i}^{N} \epsilon_i - \frac{1}{2}\sum_{ij}(J_{ij} - K_{ij}) \end{aligned} \tag{2.26}$$

This is because the summation $E_0 = \sum_{i=1}^{N} \epsilon_i$ is actually the zeroth order energy of the reference Hamiltonian

$$H_0 = \sum_i F(i) \tag{2.27}$$

in the HF approximation. It can be shown that the correct ground state energy in Eq. (2.26) includes a first order correction E_1, i.e.,

$$E_g = E_0 + E_1 \tag{2.28}$$

where the first order term is given by

$$\begin{aligned} E_1 &= <\Phi_A|H - H_0|\Phi_A> \\ &= -\frac{1}{2}\sum_{ij}(J_{ij} - K_{ij}) \end{aligned} \tag{2.29}$$

Thus the HF energy includes the first order perturbation correction already.

2.1.2 Restricted Hartree-Fock (RHF)

For a closed shell system in which the spins of the electrons are completely paired (total spin S=0), the electron spin orbital can be written in the form

$$\psi_i(\mathbf{x}) = \begin{cases} \phi_i(\mathbf{r})\alpha(s) \\ \phi_i(\mathbf{r})\beta(s) \end{cases} \tag{2.30}$$

where ϕ_i is the spatial orbital and $\alpha(\beta)$ is the spin up (down) state. Here the spatial orbitals of two spin-paired electrons are chosen to be the same. Thus for a closed shell system, N electrons in the ground state described by the HF wavefunction occupy $N/2$ spatial orbitals. After integrating over all the spin variables, it is not difficult to show that the Fock operator in the HF equation for the spatial orbital $\phi_i(\mathbf{r})$ can be written as

$$F(1) = h(1) + \sum_{i=1}^{N/2}(2J_i - K_i) \tag{2.31}$$

where the Coulomb and exchange operators are defined in terms of spatial orbitals only, viz.,

$$J_i(1) = \int d\mathbf{r}_2 |\phi_i(2)|^2 \frac{1}{r_{12}} \tag{2.32}$$

$$K_i(1)\phi_j(1) = \left[\int d\mathbf{r}_2 \phi_i^*(2) \frac{1}{r_{12}} \phi_j(2)\right] \phi_i(1) \tag{2.33}$$

The HF equation for the spatial orbital is simply

$$F(1)\phi_i = \epsilon\phi_i \tag{2.34}$$

and the HF ground state energy for a closed shell system can be written as

$$\begin{aligned} E_0 &= 2\sum_{i=1}^{N/2} h_i + \sum_{ij}^{N/2}(2J_{ij} - K_{ij}) \\ &= 2\sum_{i=1}^{N/2} \epsilon_i - \sum_{ij}^{N/2}(2J_{ij} - K_{ij}) \end{aligned} \tag{2.35}$$

where $h_i = \langle i|h|i\rangle$. In the RHF approach, the HF wavefunction is also an eigenfunction of the total electron spin.

2.1.3 Unrestricted Hartree-Fock (UHF)

In the unrestricted Hartree Fock (UHF) approach, e.g., for open shell molecules, the spin orbitals are written as

$$\psi_i(\mathbf{x}) = \begin{cases} \phi_i^\alpha(\mathbf{r})\alpha(s) \\ \phi_i^\beta(\mathbf{r})\beta(s) \end{cases} \tag{2.36}$$

where the spatial orbital ϕ_i^α for α spin is different from the orbital ϕ_i^β for β spin. Thus the HF equation for the spatial orbital of α spin is different from that for the spatial orbital of β spin

$$\begin{cases} F^\alpha(1)\phi_i^\alpha(1) = \epsilon_i^\alpha \phi_i^\alpha(1) \\ F^\beta(1)\phi_i^\beta(1) = \epsilon_i^\beta \phi_i^\beta(1) \end{cases} \tag{2.37}$$

Here the Fock operator for α spin can be written as

$$\begin{aligned} F^\alpha(1) &= \langle\alpha|F(1)|\alpha\rangle \\ &= h(1) + \sum_i^{N_\alpha}(J_i^\alpha - K_i^\alpha) + \sum_j^{N_\beta} J_j^\beta \end{aligned} \tag{2.38}$$

where N_α and N_β are, respectively, the total number of spatial orbitals for α and β spins. The superscript in the Coulomb and exchange operators simply denotes the particular spin that the spatial orbital corresponds to. Similarly, the β spin Fock operator $F^\beta(1)$ is given by

$$\begin{aligned} F^\beta(1) &= <\beta|F(1)|\beta> \\ &= h(1) + \sum_i^{N_\beta}[J_i^\beta - K_i^\beta] + \sum_j^{N_\alpha} J_j^\alpha \end{aligned} \tag{2.39}$$

It can be shown that the ground state energy using unrestricted spin orbitals can be written as

$$\begin{aligned} E_0 &= \sum_{i=1}^{N_\alpha} h_i^\alpha + \sum_{j=1}^{N_\beta} h_j^\beta + \frac{1}{2}\sum_{ij}^{N_\alpha}(J_{ij}^{\alpha\alpha} - K_{ij}^{\alpha\alpha}) + \frac{1}{2}\sum_{ij}^{N_\beta}(J_{ij}^{\beta\beta} - K_{ij}^{\beta\beta}) \\ &+ \sum_i^{N_\alpha}\sum_j^{N_\beta} J_{ij}^{\alpha\beta} \end{aligned} \tag{2.40}$$

where the notation $J_{ij}^{\alpha\beta}$ means that the two spatial orbitals used in the definition of J_{ij} are ϕ_i^α and ϕ_i^β. Similar definitions follow for other J_{ij} and K_{ij}.

Because the spatial orbitals of different spins are allowed to be different in UHF, the use of UHF can generally lower the variational energy compared to RHF. Thus for open shell systems, the UHF are frequently used in numerical calculations. One problem of using the UHF approach is, however, that the HF wavefunction is no longer an eigenfunction of total electron spin. This is called spin contamination. In order to obtain eigenfunctions of total electron spin, one needs to form linear combinations of different UHF wavefunctions.

2.1.4 Koopmans' Theorem

For an N electron system, the ground state energy in the Hartree-Fock treatment is given by Eq. (2.26) which is not a simple sum of orbital energies ϵ_i. Does the orbital energy ϵ_i have any direct physical meaning? The answer is given by Koopmans' theorem. If one electron in a particular spin orbital ψ_k is removed from the N-electron system and the state of the remaining $N-1$ electrons remains unchanged, the Hartree-Fock energy for this $N-1$ electron system is given by

$$E_k^{N-1} = \sum_{i \neq k}^{N} h_i + \frac{1}{2} \sum_{i \neq k, j \neq k}^{N} (J_{ij} - K_{ij}) \tag{2.41}$$

The ionization energy from the k orbital is the difference between E_k^{N-1} in Eq. (2.41) and E_g^N in Eq. (2.26)

$$\begin{aligned} IE &= E_k^{N-1} - E_g^N \\ &= -h_k - \frac{1}{2}\sum_j^N (J_{kj} - K_{kj}) - \frac{1}{2}\sum_i^N (J_{ik} - K_{ik}) \\ &= -h_k - \sum_j^N (J_{kj} - K_{kj}) \end{aligned} \tag{2.42}$$

because $J_{ik} = J_{ki}$ and $K_{ik} = K_{ki}$. Using Eq. (2.24) for the orbital energy, we arrive at a simple relation for the ionization energy

$$IE = -\epsilon_k \tag{2.43}$$

Equation (2.43) states that the ionization energy required to eject an electron from the k orbital is equal to the negative of the spin-orbital energy ϵ_k. A similar result can be obtained for the electron affinity, i.e. the energy for adding an extra electron to an unoccupied s orbital

$$EA = E_g^N - E_s^{N+1} = -\epsilon_s \tag{2.44}$$

Equations (2.43) and (2.44) are the results of Koopmans' theorem, which gives physical meaning to orbital energies and thus a means of calculating approximate ionization energies (potentials) and electron affinities. However, the theorem is very approximate. First, the Koopmans theorem assumes that spin orbitals are frozen after losing or adding an electron. In reality, the spin orbitals will relax and the optimized orbitals will be different from the original ones after losing or adding an electron. Secondly, the Koopmans theorem is based on the HF approximation and neglects electron correlations. As a result, while the ionization energies calculated from Koopmans theorem are often reasonable, the electron affinities given by this theorem are often in gross error.

2.1.5 SCF Solution to HF Equation

The HF equations are nonlinear equations and cannot be solved directly because the Fock operator in Eq. (2.31) also depends on unknown orbitals.

Thus the HF equations are solved by iterative methods. One typically starts with an initial guess for the orbitals ψ_0 to construct the Fock operator and then solves the standard eigenvalue problem. The new orbitals are then used to construct new Fock operators and the eigenvalue problem is solved again. This procedure is repeated until the newly obtained orbitals no longer change from the previous ones. This kind of iterative method is called the self-consistent field method (SCF), and the orbitals obtained from solving the HF equations are called *molecular orbitals* (MO).

Once the Fock operator $F(1)$ is constructed from trial orbitals, Eq. (2.31) becomes a standard eigenvalue problem and can be solved by expanding the new MO orbitals in basis functions

$$\phi_i = \sum_n \xi_n C_{ni} \tag{2.45}$$

Substituting the above expansion in the HF equation and integrating out the basis functions, we arrive at a set of linear algebraic equations

$$\sum_n F_{mn} C_{ni} = \epsilon_i \sum_{mn} O_{mn} C_{ni} \tag{2.46}$$

where $O_{mn} = <\xi_m|\xi_n>$ is the overlap matrix of the basis. Equation (2.46) can be solved by standard matrix diagonalization methods. If the basis functions used to expand the MO's are atomic orbitals, which is usually the case, the expansion method is called linear combination of atomic orbitals (LCAO).

In practice, the atomic orbitals are further expanded in Gaussian basis functions

$$\xi_n = \sum_\mu g_\mu D_{\mu n} \tag{2.47}$$

where g_μ are Gaussian functions. Gaussian functions centered at, say atom A, are defined as

$$g_{lmn}(A) = N_{lmn} x_A^l \; y_A^m \; z_A^n \; e^{-\alpha r_A^2} \tag{2.48}$$

Here N_{lmn} is the normalization constant and the electron coordinate $\mathbf{x}_A$ is centered at atom A. The Gaussian basis function $g_{lmn}(A)$ is classified as an $s, p, d, f, \cdots$ orbital for values of $l + m + n = 0, 1, 2, 3, \cdots$, respectively. The construction of the matrix elements of the Fock operator thus involves the following type of integral

$$<\mu\nu|\frac{1}{r_{12}}|\kappa\lambda> = \iint dr_1 dr_2 g_\mu(1) g_\nu(2) \frac{1}{r_{12}} g_\kappa(1) g_\lambda(2) \tag{2.49}$$

which can be efficiently calculated by analytical methods involving Gaussian functions. Thus the use of Gaussian basis functions greatly facilitates the integration in Eq. (2.49). However, the calculated electron energy depends on the quality or completeness of the basis functions. In the limit of infinite basis functions, the energy calculated from the SCF method is called the HF energy.

2.2 Electron Correlation

2.2.1 Multiconfiguration

The HF method is the cornerstone in electron structure calculations due to its transparent physical interpretation and relatively inexpensive computational cost. Although HF calculations often give very useful and even accurate results for quantities like equilibrium geometries of molecules, the HF approach is inherently a very approximate method of solving electron problems. In essence, the HF method is a mean field approximation in which each electron is assumed to move under the influence of a mean field due to all other electrons. As a result, it neglects the ***instantaneous*** or *correlated* motions of electrons. For example, in reality when two electrons approach each other, the Coulomb force will cause strong repulsion and thus the physically correct wavefunction will be vanishingly small as r_{12} approaches zero. This is not reflected in the HF wavefunction, however.

It is useful to define the difference between the exact energy E_{exact} of the electron system and HF energy E_{HF} as electron *correlation energy*

$$E_{co} = E_{ex} - E_{HF} \tag{2.50}$$

Although the relative error of the HF energy is often small, the correlation energy E_{co} is often very substantial in absolute terms. Thus, the HF method is often inadequate to give quantities such as energies. In order to recover the correlation energy, it is necessary to go beyond the HF approximation. The general approach to calculate the electron correlation energy is to include more than one Slater determinant in the expansion of the electron wavefunction

$$\Psi = \sum_k c_k D_k \tag{2.51}$$

where D_0 is the Slater determinant for the ground state HF wavefunction composed of the N lowest MO's and $D_k (k > 1)$ are Slater determinants with one or more electrons in *excited* orbitals. This approach is called configuration interaction or CI. Thus, one usually solves the HF equation using a large (atomic) basis set to expand molecular orbitals. The lowest occupied orbitals calculated are used to construct the ground state determinant $\det|\Psi_0|$ and excited orbitals are used to construct excited determinants $\det|\Psi_k|$. In principle, if the basis set is large enough, one could obtain the exact energy of the system. In practice, however, the CI expansion in Eq. (2.51) is very slowly convergent and quickly becomes computationally prohibitive. Thus, accurate calculation of correlation energy for electrons of molecular systems is a very difficult task.

2.2.2 Perturbation Methods

The simplest approach for including electron correlation is to use perturbation theory. If we use the noninteracting Hamiltonian

$$H_0 = \sum_i F(i) = \sum_i [h(i) + v^{HF}(i)] \tag{2.52}$$

as the zeroth order Hamiltonian, the zeroth order energy is simply the sum of HF orbital energies

$$E_0^{(0)} = <\Psi_0|H_0|\Psi_0> = \sum_i \epsilon_i \tag{2.53}$$

Using the first order correction, we obtain the ground state HF energy

$$\begin{aligned} E_0^{HF} &= E_g^{(0)} + <\Psi_0|H - H_0|\Psi_0> \\ &= \sum_i \epsilon_i - \frac{1}{2}\sum_{ij}(J_{ij} - K_{ij}) \end{aligned} \tag{2.54}$$

as already mentioned on page 18. Thus the lowest order perturbation needed to recover the electron correlation energy is the second order or MP2 (Möller-Plesset perturbation)

$$E_0^{(2)} = {\sum_n}' \frac{| <\Psi_0|H - H_0|\Psi_n> |^2}{E_0^{(0)} - E_n^{(0)}} \tag{2.55}$$

where Ψ_n are HF wavefunctions (or Slater determinants) with excited spin orbitals. It can be shown (Brillouin's theorem) that the Hamiltonian matrix element $<\Psi_0|H|\Psi_n>$ vanishes if Ψ_n includes only single excitations (with only one electron in the excited orbital). The matrix element also vanishes if Ψ_n includes triple or more electron excitations because the Hamiltonian contains only two-electron interactions. Thus the only non-vanishing contributions to $E_0^{(2)}$ are from doubly excited spin orbitals with two electrons in the excited orbitals. If we use i and j to denote two occupied orbitals from which two electrons are excited to two unoccupied orbitals u and v, then Eq. (2.55) can be written more explicitly as

$$E_0^{(2)} = \sum_{i<j,u<v} \frac{|<ij|r_{ij}^{-1}|uv> - <ij|r_{ij}^{-1}|uv>|^2}{\epsilon_i + \epsilon_j - \epsilon_u - \epsilon_v} \tag{2.56}$$

Further reading

References [5–9] are good sources for further reading in quantum chemistry.

Chapter 3

Rovibrational Motions of Molecules

3.1 Vibration in One-Dimension

3.1.1 Harmonic Potential

The ideal molecular vibrational motion is that of a harmonic oscillator, governed by a Hamiltonian with a quadratic potential

$$H = -\frac{\hbar^2}{2m}\frac{d^2}{dx^2} + \frac{1}{2}m\omega^2 x^2 \tag{3.1}$$

where m is the mass of the oscillator and x is the displacement from equilibrium. The eigenenergies of the above Hamiltonian are given by

$$E_n = (n + \frac{1}{2})\hbar\omega \tag{3.2}$$

and are evenly spaced. The normalized eigenfunctions of the harmonic oscillator are given by

$$\psi_n(x) = \sqrt{\frac{\alpha}{\sqrt{\pi}2^n n!}} e^{-\frac{1}{2}\alpha^2 x^2} H_n(\alpha x) \tag{3.3}$$

where α is defined as

$$\alpha = \sqrt{\frac{m\omega}{\hbar}} \tag{3.4}$$

and $H_n(z)$ are the Hermite polynomials discussed in Appendix A.1. The Hermite polynomials satisfy the recursion relation

$$H_{n+1} = 2zH_n - 2nH_{n-1} \tag{3.5}$$

and thus the entire sequence of Hermite polynomials can be generated through the recursion relation with the starting value of $H_0 = 1$ and $H_1 = 2z$. In particular, the ground state wavefunction of the harmonic oscillator is a Gaussian function

$$\psi_0(x) = \sqrt{\frac{\alpha}{\sqrt{\pi}}} e^{-\frac{1}{2}\alpha^2 x^2} \tag{3.6}$$

In some applications, it is often useful to use the creation and annihilation operators defined by

$$a^\dagger = \frac{1}{\sqrt{2}}(X - iP) \tag{3.7}$$

and

$$a = \frac{1}{\sqrt{2}}(X + iP) \tag{3.8}$$

where $X = \alpha x$ and $P = p/(\alpha\hbar)$ where x and p are the position and momentum operators, respectively. In terms of a and a^+, the Hamiltonian can be expressed in a simple form

$$\begin{aligned} H &= \left(a^\dagger a + \frac{1}{2}\right)\hbar\omega \\ &= \left(\hat{N} + \frac{1}{2}\right)\hbar\omega \end{aligned} \tag{3.9}$$

where $\hat{N}$=$a^\dagger a$ is the number operator. Using the commutation relation $[x, p] = i\hbar$, it can be shown that the operators a and $a^\dagger$ satisfy the commutation relation

$$[a, a^\dagger] = aa^\dagger - a^\dagger a = 1 \tag{3.10}$$

If the harmonic oscillator eigenstate is denoted $|n>$, then it is not difficult to obtain the relations

$$\begin{aligned} a^\dagger|n> &= \sqrt{n+1}|n+1> \\ a|n> &= \sqrt{n}|n-1> \\ \hat{N}|n> &= n|n> \end{aligned} \tag{3.11}$$

for $n = 0, 1, 2, \cdots$.

The harmonic potential is only an ideal model and does not really pertain to actual molecules. The energy levels of a harmonic oscillator are equally spaced and the dissociation energy is infinite. In a real molecule, however, the vibrational energy levels are unequally spaced and the dissociation energy is finite. Therefore, the harmonic oscillator model cannot be used to describe the process of bond breaking and bond formation. Nevertheless, the harmonic oscillator is very useful for modeling low-energy vibrations of strongly bound or tight vibrational motions for which normal mode analysis is often a reasonable approximation.

3.1.2 Morse Potential

A more realistic vibrational model is that of the Morse oscillator potential

$$V_M = D_e \left[e^{-2\alpha(r-r_e)} - 2e^{-\alpha(r-r_e)} \right] \tag{3.12}$$

where r_e is the equilibrium distance and D_e is the well depth of the potential. The eigenvalue of the Morse oscillator is given analytically [10]

$$E_n = -D_e + (n + \frac{1}{2})\hbar\omega - (n + \frac{1}{2})^2 \frac{\hbar^2\omega^2}{4D_e} \tag{3.13}$$

where the harmonic frequency is given by

$$\omega = \sqrt{\frac{2D_e\alpha^2}{m}} \tag{3.14}$$

The Morse oscillator has the correct behavior of being anharmonic and having a finite dissociation energy of

$$\begin{aligned} D_0 &= -E_0 \\ &= -D_e - \frac{1}{2}\hbar\omega + \frac{\hbar^2\omega^2}{16D_e} \end{aligned} \tag{3.15}$$

Because of these desired properties, the Morse potential is widely used in chemistry to model realistic vibrational motions of molecules as well as bond breaking and bond formation processes.

3.1.3 General Potential

More realistic potentials are neither harmonic nor Morse type. For a general diatomic potential such as one obtained by numerical fitting of calculated *ab initio* points, one has to resort to numerical methods to solve for eigenvalues and eigenfunctions. The Schrödinger equation for an arbitrary potential $V(r)$

$$\left[-\frac{\hbar^2}{2m}\frac{d^2}{dr} + V(r)\right]\psi_v(r) = \epsilon_v \psi_v(r) \tag{3.16}$$

can be solved by expanding the eigenfunction $\psi_v(r)$ in a predefined basis set $\varphi_n (n = 1, \cdots, N)$

$$\psi_v(r) = \sum_n \varphi_n(r) C_n \tag{3.17}$$

Substituting Eq. (3.17) in Eq. (3.16) and integrating over the basis functions φ_n, one obtains the matrix equation

$$\mathbf{HC} = \lambda \mathbf{SC} \tag{3.18}$$

In the above equation, **H** and **S** are, respectively, the Hamiltonian and overlap matrices with matrix elements defined by

$$\mathbf{H}_{mn} = <\varphi_m|\hat{H}|\varphi_n> \tag{3.19}$$

$$\mathbf{S}_{mn} = <\varphi_m|\varphi_n> \tag{3.20}$$

If the basis functions φ_n are chosen to be orthonormal, the matrix equation in Eq. (3.18) simplifies to the standard form

$$\mathbf{HC} = \lambda \mathbf{C} \tag{3.21}$$

which can be directly diagonalized by standard matrix diagonalization methods to obtain the solution for **C**. For the more general case with a nonorthogonal basis φ_n, one needs to solve the generalized eigenvalue equation of (3.18). This can be done by first diagonalizing the overlap matrix to yield

$$\mathbf{S} = \mathbf{U}^T \mathbf{S}_d \mathbf{U} \tag{3.22}$$

where **U** is a unitary (or orthogonal) matrix and $\mathbf{S}_d$ is a diagonal matrix. Since the overlap matrix **S** is positive definite, the elements of $\mathbf{S}_d$ are all positive. Thus the general eigenvalue Eq. (3.18) can be transformed to the standard eigenvalue equation

$$\mathbf{H}'\mathbf{C}' = \lambda \mathbf{C}' \tag{3.23}$$

where

$$\mathbf{C}' = \mathbf{S}_d^{1/2}\mathbf{U}\mathbf{C} \tag{3.24}$$

and

$$\mathbf{H}' = \mathbf{S}_d^{-1/2}\mathbf{U}^T\mathbf{H}\mathbf{U}\mathbf{S}_d^{-1/2} \tag{3.25}$$

The eigenvalue Eq. (3.23) can be solved by straightforward diagonalization of $\mathbf{H}'$ to yield eigenenergies λ and eigenvectors $\mathbf{C}'$ which can then be transformed back to obtain eigenvectors $\mathbf{C}$ through the inverse relation of Eq. (3.24).

In principle any complete basis set φ_n can be chosen to expand the wavefunction. In practice, however, one would like to choose an efficient basis such that accurate results can be achieved with the use of as few basis functions φ_n as possible. A general method is to use the eigenfunctions of a reference Hamiltonian as φ_n which are either analytically known or easy to solve numerically. For example, if the potential $V(x)$ of the problem in question has a single minimum at r_0, one could use the second order expansion of $V(r)$ around the minimum as a reference potential

$$V_0(r) = V(r_0) + \frac{1}{2}(r - r_0)^2 V''(r_0). \tag{3.26}$$

Thus the reference Hamiltonian would simply be that of a harmonic oscillator, whose solutions are analytically known.

Although the theoretical treatments and discussions in this section use a one-dimensional model as an example, they can be generalized to multi-dimensional eigenvalue problems in a straightforward fashion.

3.1.4 Discrete Variable Representation

The DVR (discrete variable representation) is a very general and powerful method which is widely used in quantum mechanics calculations [11]. It is applied to one-dimensional problems or direct product basis functions in multidimensional problems. To state it simply, DVR is a localized (in coordinate space) *but* discrete representation. For any given finite basis set $\phi_n(x)(n = 1, 2, 3, \cdots N)$, one can define a unique DVR by diagonalizing the matrix

$$x_{mn} = <\phi_m|\hat{x}|\phi_n> \tag{3.27}$$

which generates N eigenvalues x_n and eigenfunctions

$$|X_n> = \sum_m |\phi_m> C_{mn} \tag{3.28}$$

such that

$$\hat{x}|X_n\rangle = x_n|X_n\rangle \tag{3.29}$$

Equation (3.29) implies that in this N-dimensional vector space, the coordinate operator $\hat{x}$ is approximated by

$$\begin{aligned}\hat{x} &= \sum_{m=1}^{N}\sum_{n=1}^{N} |\phi_m\rangle\, x_{mn}\, \langle\phi_n| \\ &= \sum_{n=1}^{N} |X_n\rangle\, x_n\, \langle X_n|\end{aligned} \tag{3.30}$$

With this prescription for the operator $\hat{x}$, $|X_n\rangle$ is also an eigenstate of any operator function $F(\hat{x})$, i.e.,

$$F(\hat{x})|X_n\rangle = F(x_n)|X_n\rangle \tag{3.31}$$

Since the DVR basis set $|X_n\rangle$ is related to the finite basis set $\phi_n(x)$ through a unitary or orthogonal transformation of Eq. (3.28), it is an equivalent basis set to $\phi_n(x)$ in this N-dimensional vector space. The DVR basis functions are highly localized in coordinate space, i.e., $\langle x|X_n\rangle$ is highly peaked near $x = x_n$. Due to this particular local property of the DVR basis, the matrix element of any local operator in the DVR basis is approximately diagonal. For example, the matrix element of the potential energy operator in the DVR basis is approximated by

$$\langle X_m|V(\hat{x})|X_n\rangle = \delta_{mn} V(x_n). \tag{3.32}$$

This result applies to any local operator which is a function of coordinates only, and should be understood in the sense that the coordinate operator is approximated by Eq. (3.30) in the N-dimensional vector space. As the dimension of the vector space increases, the approximation in Eq. (3.32) becomes better and better. Since most potential energy operators are local functions of coordinates, they are diagonal in the DVR representation, and the integration over the coordinates to construct the potential matrix can be eliminated.

DVR and Gaussian quadrature

If the basis functions used to define DVR are polynomials $P_n(x)$ that are orthogonal with the weighting function $W(x)$

$$\langle P_m|P_n\rangle = \int_a^b W(x)P_m(x)P_n(x)dx = \delta_{mn} \tag{3.33}$$

then we can use Gaussian quadrature to evaluate the integral

$$<P_m|P_n>= \sum_{k=1}^{N} P_m(x_k)P_n(x_k)w_k = \delta_{mn} \tag{3.34}$$

where x_k and w_k are Gaussian nodes and weights, respectively. The property of Gaussian quadrature guarantees that Eq. (3.34) is *exact* for $m, n = 0, 1, 2, \cdots, N-1$. Thus for orthogonal polynomials, the associated DVR points are just Gaussian nodes, and the orthogonal transformation between the DVR and the polynomial basis is simply given by the relation

$$<P_n|X_k>= \sqrt{w_k}P_n(x_k) \tag{3.35}$$

If we evaluate the matrix element of a local operator such as a potential $V(x)$ in the basis of orthogonal polynomials by Gaussian quadrature, we have

$$\begin{aligned} <P_m|V|P_n> &= \sum_{k=1}^{N} P_m(x_k)V(x_k)P_n(x_k)w_k \\ &= \sum_{k=1}^{N} <P_m|X_k>\, V(x_k)\, <X_k|P_n> \end{aligned} \tag{3.36}$$

which is equivalent to inserting a complete DVR set $\sum_k |X_k><X_k|$ directly into the integral and using the DVR result

$$V|X_k>= V(x_k)|X_k> \tag{3.37}$$

This means that we can think of Gaussian quadrature as a special case of DVR when the associated basis functions are orthogonal polynomials. Thus we see that evaluating the potential matrix element of a local operator such as the potential operator in an orthogonal polynomial basis by the DVR method is equivalent to using Gaussian quadrature to evaluate the matrix element. This directly relates the approximation used in the DVR evaluation of the potential matrix to that of numerical quadrature.

3.1.5 Gaussian Basis Functions

Gaussian functions are very useful as basis functions for expanding molecular wavefunctions. They are attractive as basis functions primarily due to their local and analytical properties. In molecular applications, one often uses Gaussian functions as flexible bases for radial coordinates. The normalized Gaussian functions are defined as

$$g_i(x) = \left(\frac{2a_i}{\pi}\right)^{1/4} \exp[-a_i(x - x_i)^2] \tag{3.38}$$

where x_i specifies the position of the maximum and a_i the width of the ith Gaussian function.

One of the most important analytical properties of Gaussian functions is that the product of two individual Gaussian functions is also a Gaussian function, viz.,

$$g_i(x)g_j(x) = \left(\frac{4a_ia_j}{\pi^2}\right)^{1/4} \exp\left[-a_{ij}(x_i - x_j)^2\right] \exp\left[-A_{ij}(x - x_c)^2\right] \tag{3.39}$$

which is a new Gaussian function centered at

$$x_c = \frac{a_ix_i + a_jx_j}{a_i + a_j} \tag{3.40}$$

with parameters given by $A_{ij} = a_i + a_j$ and

$$a_{ij} = \frac{a_ia_j}{a_i + a_j} \tag{3.41}$$

It is this property that make Gaussian functions the basis of choice in electronic structure calculations. In molecular dynamics calculations, however, the local property of Gaussian functions is one of the main reasons for their popular use.

In computing matrix elements involving Gaussian functions, one often encounters the following Gaussian integral

$$\int_{-\infty}^{\infty} e^{-\alpha x^2} dx = \sqrt{\pi\alpha} \qquad (\alpha > 0) \tag{3.42}$$

This result is used to derive many analytical integrals involving Gaussian functions. For example, using the result of (3.42), the overlap integral involving two Gaussian bases is simply given by

$$\begin{aligned} O_{ij} &= \langle g_i | g_j \rangle \\ &= \left[\frac{4a_ia_j}{(a_i + a_j)^2}\right]^{1/4} \exp\left[-\frac{a_ia_j}{a_i + a_j}(x_i - x_j)^2\right] \end{aligned} \tag{3.43}$$

It is well known that any integral of a Gaussian function multiplied by a polynomial can be analytically integrated. For example, the matrix element of x in a Gaussian basis can be easily evaluated as

$$\begin{aligned} <g_i|x|g_j> &= <g_i|x - x_c + x_c|g_j> \\ &= <g_i|x_c|g_j> \\ &= x_c O_{ij} \\ &= \frac{a_i x_i + a_j x_j}{a_i + a_j} O_{ij} \end{aligned} \tag{3.44}$$

Similarly, we can evaluate the integral

$$\begin{aligned} <g_i|x^2|g_j> &= <g_i|(x - x_c + x_c)^2|g_j> \\ &= <g_i|[(x - x_c)^2 + 2x_c(x - x_c) + x_c^2]|g_j> \\ &= <g_i|(x - x_c)^2|g_j> + x_c^2 O_{ij} \\ &= \frac{1}{a_i + a_j}\left[\frac{1}{2} + \frac{(a_i x_i + a_j x_j)^2}{a_i + a_j}\right] O_{ij} \end{aligned} \tag{3.45}$$

Since the derivative of a Gaussian function with respect to the argument is given by

$$g_i'(x) = -2a_i(x - x_i)g_i(x) \tag{3.46}$$

one can also evaluate the overlap integral of the derivative of a Gaussian basis function by similar manipulations

$$\begin{aligned} O'_{ij} &= <g_i'|g_j'> \\ &= 4a_i a_j <g_i|(x - x_i)(x - x_j)|g_j> \\ &= 4a_i a_j <g_i|[(x - x_c)^2 + (x_c - x_i)(x_c - x_j)]|g_j> \\ &= 4a_i a_j \left[\frac{1}{2(a_i + a_j)} - \frac{a_i a_j}{(a_i + a_j)^2}(x_i - x_j)^2\right] O_{ij} \\ &= \frac{2a_i a_j}{a_i + a_j}\left[1 - \frac{2a_i a_j}{a_i + a_j}(x_i - x_j)^2\right] O_{ij} \end{aligned} \tag{3.47}$$

Equation (3.47) is useful because the matrix element of the kinetic energy operator in the Gaussian basis is

$$<g_i|T|g_j> = -\frac{\hbar^2}{2\mu} <g_i|g_j''> \tag{3.48}$$

which can be written in a symmetric form by performing a partial integration,

$$<g_i|T|g_j> = \frac{\hbar^2}{2\mu} <g_i'|g_j'> = \frac{\hbar^2}{2\mu} O'_{ij} \tag{3.49}$$

3.2 Vibration-Rotation in Many Dimensions

3.2.1 Diatomic Molecules

In real molecules, the vibrational motions of molecules are generally very complex. In addition, the vibrational motions of different modes are often strongly coupled, especially for highly excited vibrational states. Thus, normal mode analysis is generally inadequate. For example, the normal mode approximation is grossly in error for weakly bound vibrational motions of "floppy" molecules which can exercise large amplitude anharmonic motions. Thus in general, one needs to solve multidimensional vibrational problems for molecules. Since the vibrational motion is also coupled to the rotational motion of the molecule, the rigorous treatment of rotations and vibrations of molecules should include explicitly the coupling of all rovibrational degrees of freedom.

The simplest case of rovibrational motion in realistic molecules is that of a diatomic molecule, AB such as H_2, CO,$\cdots$, etc. and consists of three degrees of freedom (excluding the center-of-mass motion). By using spherical coordinates, the effective Hamiltonian for a diatomic molecule can be written as

$$H = -\frac{\hbar^2}{2\mu}\frac{1}{r}\frac{\partial^2}{\partial r^2}r + \frac{\mathbf{j}^2}{2\mu r^2} + v(r) \tag{3.50}$$

where $v(r)$ is the diatomic (internuclear) potential, μ is the reduced mass of the diatom defined as

$$\mu = \frac{m_A m_B}{m_A + m_B} \tag{3.51}$$

and the angular momentum operator is defined as

$$\mathbf{j}^2 = -\frac{\hbar^2}{\sin^2\theta}\left[\sin\theta\frac{\partial}{\partial\theta}\left(\sin\theta\frac{\partial}{\partial\theta}\right) + \frac{\partial^2}{\partial\varphi^2}\right] \tag{3.52}$$

The rovibrational eigenfunctions of diatomics are direct products of spherical harmonics and a radial function

$$\Phi_{vjm}(\mathbf{r}) = \frac{\phi_{vj}(r)}{r}y_{jm}(\theta,\varphi) \tag{3.53}$$

where the spherical harmonics $y_{jm}(\theta,\varphi)$ are eigenfunctions of the angular momentum operator

$$\mathbf{j}^2 y_{jm}(\theta,\varphi) = j(j+1)\hbar^2 y_{jm}(\theta,\varphi) \tag{3.54}$$

The j-dependent radial vibrational functions $\phi_{vj}(r)$ are found to satisfy the radial Schrödinger equation

$$\left[-\frac{\hbar^2}{2\mu}\frac{d^2}{dr^2} + \frac{j(j+1)^2}{2\mu r^2} + v(r)\right]\phi_{vj}(r) = \epsilon_{vj}\phi_{vj}(r) \tag{3.55}$$

where ϵ_{vj} are the rovibrational energies of diatomics. Thus, the vibrational motion of a diatomic molecule is coupled to its rotational motion and vice versa through the centrifugal potential in the Hamiltonian. The one-dimensional radial Eq. (3.55) can be solved numerically using the basis set method outlined in section (3.1.3).

3.2.2 Triatomic Molecules

The problem for solving for rovibrational eigenvalues and eigenfunctions is relatively trivial for diatomic molecules but it becomes drastically more complicated for triatomic molecules. There are a total of six rovibrational degrees of freedom (excluding the center-of-mass motion) for a triatomic molecule. A nonlinear triatomic molecule has three vibrational modes and three rotational modes. For linear triatomic molecules, there are four vibrational modes and two rotational modes since one of the rotational modes becomes a vibrational mode. Unlike diatomic molecules, there are many ways to choose coordinates to represent the Hamiltonian for triatomic molecules. Thus the first thing to consider in calculating the rovibrational spectrum of a triatomic molecule is to select a coordinate system which should represent the rovibrational motions correctly and is also convenient to use in numerical calculations. Out of many choices, the Jacobi coordinates are the simplest to use in most applications. Although, Jacobi coordinates are the "correct" coordinates to describe asymptotic wavefunctions for scattering problems, they are also very useful for bound state calculations as well. Other coordinates such as Radau or hyperspherical coordinates are also frequently used in bound state calculations for triatomic molecules.

For a triatomic molecule, there are total of three sets of Jacobi coordinates. In terms of any one of the three sets such as the one shown in Fig. 3.1, the kinetic energy operator is diagonal, and the triatomic Hamiltonian is given by the following

$$\begin{aligned} H &= -\frac{\hbar^2}{2\mu_R}\nabla^2_{\mathbf{R}} - \frac{\hbar^2}{2\mu_r}\nabla^2_{\mathbf{r}} + V(\mathbf{R},\mathbf{r}) \\ &= -\frac{\hbar^2}{2\mu_R}\frac{1}{R}\frac{\partial^2}{\partial R^2}R + \frac{\mathbf{L}^2}{2\mu_R R^2} - \frac{\hbar^2}{2\mu_r}\frac{1}{r}\frac{\partial^2}{\partial r^2}r + \frac{\mathbf{j}^2}{2\mu_r r^2} + V(\mathbf{R},\mathbf{r}) \end{aligned} \tag{3.56}$$

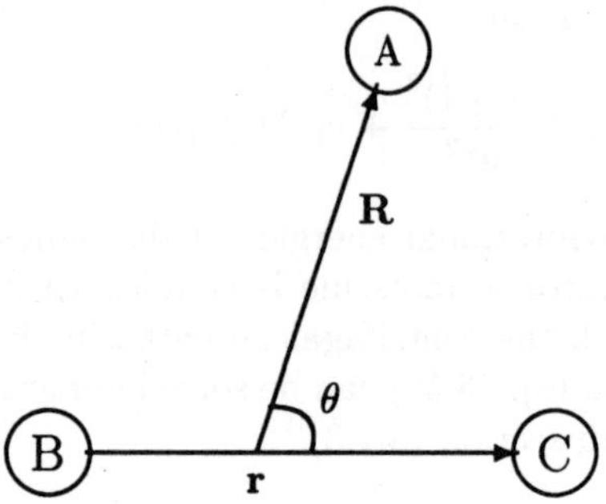

Figure 3.1: Jacobi coordinates for a triatomic system.

where **L** and **j** are, respectively, the orbital and diatomic angular momentum operators. Here the reduced masses are defined by

$$\mu_r = \frac{m_B m_C}{m_B + m_C} \tag{3.57}$$

$$\mu_R = \frac{m_A(m_B + m_C)}{m_A + m_B + m_C} \tag{3.58}$$

Since the total angular momentum $\mathbf{J} = \mathbf{L} + \mathbf{j}$ is conserved, we can use an angular momentum basis in the coupled-angular momentum representations to expand the wavefunction. There are two equivalent and widely used representations; the space-fixed (SFF) or body-fixed (BF) representation. In the SF representation, the basis functions are eigenfunctions of (J^2, J_Z, j^2, L^2) operators where J_Z is the projection of **J** along the space-fixed Z axis. In the BF representation, the basis functions chosen are eigenfunctions of (J^2, J_Z, j^2, J_z) operators where L^2 is replaced by the projection of **J** along the body-fixed z axis. Usually, the body-fixed z axis is chosen to be along the vector **R** as shown in Fig. D.1 in the appendix. In this choice of BF frame, the projection of **J** along the BF z axis is the same as that of **j** because **L** has zero projection along the **R** axis. More detailed discussions on this subject are given in Appendix D.

Since the BF representation has some attractive features, we choose to use the BF representation in this book. A general multidimensional expansion of eigenfunctions of a triatomic molecule $\psi(\mathbf{R}, \mathbf{r})$ can be written as

$$\psi^{JMp}(\mathbf{R},\mathbf{r}) = \sum_{nvjK} C_{nvjK} \frac{u_n(R)}{R} \frac{\phi_v(r)}{r} \mathcal{Y}_{jK}^{JMp}(\hat{\mathbf{R}},\hat{\mathbf{r}}) \tag{3.59}$$

where J and M are the quantum numbers of the total angular momentum and its space-fixed Z component, respectively, and p is the parity of the system defined in the appendix. The functions $u_n(R)$ and $\phi_v(r)$ are basis functions for the two radial coordinates R and r. As shown in Appendix D, the parity-adapted BF angular momentum eigenfunction is defined as

$$\mathcal{Y}_{jK}^{JMp} = \frac{1}{\sqrt{2(1+\delta_{K0})}} \left[\mathcal{Y}_{jK}^{JM} + (-1)^P \mathcal{Y}_{j-K}^{JM}\right] \tag{3.60}$$

where the *total* parity is $P = (-1)^{J+p}$ and $\mathcal{Y}_{jK}^{JMp}$ is the product of a normalized rotation matrix and an associated Legendre polynomial

$$\mathcal{Y}_{jK}^{JM} = \tilde{D}_{MK}^{J} P_{jK} \tag{3.61}$$

Details can be found in Appendix D. The BF angular momentum eigenfunctions defined in Eq. (3.60) are also eigenfunctions of the total parity P in which the $(2J+1)$-manifold of K states is split into a $(K+1)$-manifold of K states with $K = 0, 1, 2, \cdots, J$ for even *total* parity (P=even) and a K-manifold of K states with $K = 1, 2, \cdots, J$ for odd *total* parity (P=odd) (see Appendix D).

Next we need to construct the Hamiltonian matrix $\mathbf{H}$ which is the sum of the kinetic energy and potential matrices.

$$\mathbf{H} = \mathbf{T}_R + \mathbf{T}_r + \mathbf{V} + \mathbf{V}_R^c + \mathbf{V}_r^c \tag{3.62}$$

For simplicity, we assume that all the basis functions are orthogonal. Thus the kinetic energy matrices are given by

$$[\mathbf{T}_R]_{nvjK,n'v'j'K'} = <u_n| - \frac{\hbar^2}{2\mu_R} \frac{d^2}{dR^2} |u_{n'}> \delta_{vv'} \delta_{jj'} \delta_{KK'} \tag{3.63}$$

$$[\mathbf{T}_r]_{nvjK,n'v'j'K'} = <\phi_v| - \frac{\hbar^2}{2\mu_r} \frac{d^2}{dr^2} |\phi_{v'}> \delta_{nn'} \delta_{jj'} \delta_{KK'} \tag{3.64}$$

By utilizing the orthogonality property of the BF angular momentum functions and the fact that the potential V depends only on three internal coordinates (R, r, θ), we can evaluate the potential matrix as

$$[\mathbf{V}]_{nvjK,n'v'j'K'} = <u_n \phi_v P_{jK} |V| P_{j'K} \phi_{v'} u_{n'}> \delta_{KK'} \tag{3.65}$$

We note that the potential matrix is diagonal in the BF (K) representation which is one of the main reasons for the popular use of the BF representation.

The potential matrix $\mathbf{V}_r^c$ is simply given by

$$[\mathbf{V}_r^c]_{nvjK,n'v'j'K'} = <\phi_v|\frac{\hbar^2}{2\mu_r r^2}|\phi_{v'}> j(j+1)\delta_{nn'}\delta_{jj'}\delta_{KK'} \tag{3.66}$$

However, the evaluation of the centrifugal potential matrix $\mathbf{V}_R^c$ in BF representation is a little more tricky. First, the orbital angular momentum is expressed as

$$L^2 = (\mathbf{J}-\mathbf{j})^2 = J^2 + j^2 - 2J_z j_z - J_+ j_- - J_- j_+ \tag{3.67}$$

where the various angular momentum operators are defined in the BF frame. Utilizing the anti-commutation relation given in appendix D.4 for the total angular momentum operator in the body-fixed frame, it can be shown that $J_\pm$ in the BF frame behaves like $J_\mp$ in the SF frame while $j_\pm$ still has normal behavior in the BF frame.

Using the result of Eq. (D.52) derived in Appendix D.4 for matrix elements of L^2 in the BF angular momentum basis, one can readily write out the result for the full matrix element of the centrifugal potential in the body-fixed frame

$$\begin{aligned}[\mathbf{V}_R^c]_{nvjK,n'v'j'K'} &= \frac{\hbar^2}{2\mu_R R^2} <nvjK|(\mathbf{J}-\mathbf{j})^2|n'v'j'K'> \\ &= \delta_{v,v'}\delta_{j,j'} <u_n|\frac{\hbar^2}{2\mu_R R^2}|u_{n'}> W^{Jj}_{KK'}\end{aligned} \tag{3.68}$$

where the matrix element $W^{Jj}_{KK'}$ is defined by

$$\begin{aligned}W^{Jj}_{KK'} = &\left[J(J+1)+j(j+1)-2K^2\right]\delta_{KK'} - \lambda^+_{JK}\lambda^+_{jK}\sqrt{1+\delta_{K0}}\delta_{K+1,K'} \\ &-\lambda^-_{JK}\lambda^-_{jK}\sqrt{1+\delta_{K1}}\delta_{K-1,K'}\end{aligned} \tag{3.69}$$

where the symbols $\lambda^\pm_{JK}$ and $\lambda^\pm_{jK}$ are defined as

$$\lambda^\pm_{JK} = \sqrt{J(J+1)-K(K\pm1)} \tag{3.70}$$

$$\lambda^\pm_{jK} = \sqrt{j(j+1)-K(K\pm1)} \tag{3.71}$$

Thus in the BF frame, the centrifugal coupling is tridiagonal in the K states.

After the Hamiltonian matrix is constructed, the bound state problem is then solved by diagonalizing the Hamiltonian matrix of Eq. (3.62) to obtain eigenvalues and eigenfunctions as discussed in section (3.1.3).

CS approximation

In the BF representation, the interaction potential is diagonal while the centrifugal potential is tridiagonal in the K states. In other words, the coupling of different K states is caused by the centrifugal potential, not the interaction potential. This is the main reason for the popular use of the BF representation because it is often a good approximation to simply neglect the K-coupling in Eq. (3.68) to arrive at a diagonal representation of the Hamiltonian in the quantum number K.

$$[\mathbf{V}_R^c]_{nvjK,n'v'j'K'} \overset{cs}{\approx} \delta_{v,v'}\delta_{j,j'}\delta_{K,K'} <u_n|\frac{\hbar^2}{2\mu_R R^2}|u_{n'}> \times[J(J+1)+j(j+1)-2K^2] \quad (3.72)$$

The neglect of centrifugal coupling is called the CS (centrifugal sudden or coupled states) approximation [12,13]. Since for a given quantum number J and parity ϵ, there are either $J+1$ or J manifolds of K states depending on whether the total parity is even or odd, the CS approximation decouples these K states and thus significantly reduces the computational cost in the diagonalization of the Hamiltonian matrix for total angular momentum $J > 0$.

3.2.3 Tetraatomic Molecules

Hamiltonian and basis set

A tetraatomic molecule has 9 rovibrational degrees of freedom (excluding the three center-of-mass coordinates) and six of them are internal vibrational coordinates (assuming a nonlinear molecule). If we use the Jacobi coordinates shown in Fig. 3.2, the Hamiltonian can be written as

$$\begin{aligned} H &= -\frac{\hbar^2}{2\mu_R}\nabla^2_{\mathbf{R}} - \frac{\hbar^2}{2\mu_1}\nabla^2_{\mathbf{r}_1} - \frac{\hbar^2}{2\mu_2}\nabla^2_{\mathbf{r}_2} + V(\mathbf{R},\mathbf{r}_1,\mathbf{r}_2) \\ &= -\frac{\hbar^2}{2\mu_R}\frac{1}{R}\frac{\partial^2}{\partial R^2}R + \frac{\mathbf{L}^2}{2\mu_R R^2} - \frac{\hbar^2}{2\mu_1}\frac{1}{r_1}\frac{\partial^2}{\partial r_1^2}r_1 + \frac{\mathbf{j}_1{}^2}{2\mu_1 r_1^2} \\ &\quad -\frac{\hbar^2}{2\mu_1}\frac{1}{r_2}\frac{\partial^2}{\partial r_2^2}r_2 + \frac{\mathbf{j}_2{}^2}{2\mu_2 r_2^2} + V(\mathbf{R},\mathbf{r}_1,\mathbf{r}_2) \end{aligned} \quad (3.73)$$

Now we have three angular momentum operators $(\mathbf{L},\mathbf{j}_1,\mathbf{j}_2)$ that are coupled to form the total angular momentum $\mathbf{J}$ which is conserved. Typically, one can first couple two internal rotations $\mathbf{j}_1$ and $\mathbf{j}_2$ to form a new angular

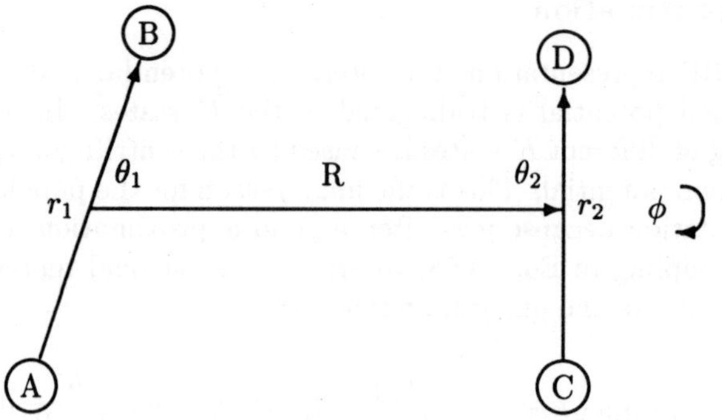

Figure 3.2: Jacobi coordinates for a diatom-diatom molecule.

momentum $\mathbf{j}_{12}$. The new $\mathbf{j}_{12}$ is then coupled with the orbital angular momentum $\mathbf{L}$ to form the total angular momentum $\mathbf{J}$. Of course, one could also couple these angular momentum operators in different ways.

The method of choosing suitable basis sets to expand the wavefunction is essentially similar to that of triatomic systems. For example, similar to Eq. (3.59) for expansion of triatomic molecules, we can expand the wavefunction for a tetraatomic molecule as

$$\psi^{JM\epsilon}(\mathbf{R},\mathbf{r}_1,\mathbf{r}_2,t) = \sum_{nvjK} C_{nvjK}\frac{u_n(R)}{R}\frac{\phi_v(r_1,r_2)}{r_1r_2}\mathcal{Y}^{JMp}_{j_{12}K}(\hat{\mathbf{R}},\hat{\mathbf{r}}_1,\hat{\mathbf{r}}_2) \quad (3.74)$$

Here we have used the composite indexes v and j to denote collections of quantum numbers, i.e., $v = (v_1v_2)$ and $j = (j_1, j_2, j_{12})$ in order to simplify notation. j_{12} is the quantum number of angular momentum $\mathbf{j}_1 + \mathbf{j}_2$. The double vibrational basis $\phi_v(r_1,r_2)$ is defined as

$$\phi_v(r_1,r_2) = \phi_{v_1}(r_1)\phi_{v_2}(r_2) \quad (3.75)$$

and $\mathcal{Y}^{JMp}_{j_{12}K}$ is the parity-adapted BF angular momentum basis function.

It can be shown that the operation of the parity operator on the nonparity-adapted BF angular momentum basis yields the result

$$\hat{p}\mathcal{Y}^{JM}_{j_{12}K} = (-1)^{j_1+j_2+j_{12}+J}\mathcal{Y}^{JM}_{j_{12}-K} \quad (3.76)$$

Thus we can define $\mathcal{Y}^{JMp}_{j_{12}K}$ by

$$\mathcal{Y}^{JMp}_{j_{12}K}(\hat{\mathbf{R}},\hat{\mathbf{r}}_1,\hat{\mathbf{r}}_2) = \frac{1}{\sqrt{2(1+\delta_{K0})}}\left[\mathcal{Y}^{JM}_{j_{12}K} + (-1)^{P+j_1+j_2+j_{12}}\mathcal{Y}^{JM}_{j_{12}-K}\right] (3.77)$$

where P is the total parity. The BF basis $\mathcal{Y}^{JM}_{j_{12}K}$ is defined by a method similar to the one described in Sec. 6.6 and Appendix D for triatomic molecules

$$\mathcal{Y}^{JM}_{j_{12}K}(\hat{\mathbf{R}}, \hat{\mathbf{r}}_1, \hat{\mathbf{r}}_2) = \widetilde{D}^J_{K,M}\mathcal{Y}^{j_{12}K}_{j_1 j_2} \tag{3.78}$$

where the normalization rotation matrix and Euler angles remain the same as for triatomics. The BF internal angular momentum function $\mathcal{Y}^{j_{12}K}_{j_1 j_2}$ is defined as

$$\mathcal{Y}^{j_{12}K}_{j_1 j_2}(\theta_1, \theta_2, \phi) = \sqrt{2\pi}\sum_{m_1} <j_1 m_1 j_2 K - m_1|j_{12}K> \times Y_{j_1 m_1}(\theta_1, 0)Y_{j_2 K-m_1}(\theta_2, \phi) \tag{3.79}$$

where Y_{jm} are spherical harmonics and ϕ_{12} is the out-of-plane torsional angle as shown in Fig. 3.2.

The construction of the Hamiltonian matrix is similar to that in the previous section for triatomics. The kinetic energy matrices are simple to evaluate and the treatment is not repeated here. The potential matrix in the angular basis for fixed radial coordinates (R, r_1, r_2) can be calculated similarly to the case for triatomic molecules

$$\begin{aligned} V^{JK}_{j,j'}(R, r_1, r_2) &= <JMjK|V|JMj'K'> \\ &= \sum_{m_1, m_1'} <j_1 m_1 j_2 K - m_1|j_{12}K><j_1' m_1' j_2' K - m_1'|j_{12}' K> \\ &\quad \times \int_0^\pi \sin\theta_1 d\theta_1 \int_0^\pi \sin\theta_2 d\theta_2 P_{j_1 m_1}(\theta_1)P_{j_2 K-m_1}(\theta_2) \\ &\quad \times V_{m_1, m_1'}(R, r_1, r_2, \theta_1, \theta_2)P_{j_1' m_1'}(\theta_1)P_{j_2' K-m_1'}(\theta_2) \end{aligned} \tag{3.80}$$

where

$$V_{m_1, m_1'} = \frac{1}{\pi}\int_0^\pi d\phi \cos[(m_1 - m_1')\phi]V(R, r_1, r_2, \theta_1, \theta_2, \phi) \tag{3.81}$$

and P_{jK} are normalized associated Legendre polynomials.

The matrix element of the centrifugal potential can be calculated by a formula similar to that for triatomic molecules

$$\begin{aligned} V^{cJj}_{K,K'} &= \frac{\hbar^2}{2\mu R^2} <JMjK|(\mathbf{J} - \mathbf{j}_{12})^2|JMj'K'> \\ &= \frac{\hbar^2}{2\mu R^2}\delta_{jj'}W^{Jj_{12}}_{KK'} \end{aligned} \tag{3.82}$$

where $W^{Jj_{12}}_{KK'}$ is defined in Eq. (3.69) with the replacement of j by j_{12}.

The full Hamiltonian matrix can now be constructed from separate pieces and has the following form

$$\begin{aligned} H^J_{nvjK,n'v'j'K'} &= <JMnvjK|H|JMn'v'j'K'> \\ &= \Bigg[<u_n|T_R|u_{n'}> \delta_{vv'} + <\phi_v|T_{r_1} + T_{r_2} + \frac{\hbar^2}{2\mu_1 r_1^2} j_1(j_1+1) \\ &\quad + \frac{\hbar^2}{2\mu_{r_2} r_2^2} j_2(j_2+1)|\phi_{v'}> \delta_{nn'} \Bigg] \delta_{jj'} \delta_{KK'} \\ &\quad + V^{JK}_{nvj,n'v'j'} \delta_{KK'} + V^{cJj}_{nK,n'K'} \delta_{vv'} \delta_{jj'} \end{aligned} \tag{3.83}$$

where T_R, T_{r_1} and T_{r_2} are the kinetic energy operators corresponding to the R, r_1 and r_2 radial coordinates in the Hamiltonian Eq. (3.73). Here the full potential matrices are calculated from the angular part by

$$V^{JK}_{nvj,n'v'j'} = <u_n \phi_v | V^{JK}_{j,j'} | \phi_{v'} u_{n'}> \tag{3.84}$$

and

$$V^{cJj}_{nK,n'K'} = <u_n | V^{cJj}_{K,K'} | u_{n'}> \tag{3.85}$$

The bound state problem is then solved by diagonalizing the Hamiltonian matrix, provided that the size of the matrix is not prohibitively large.

Symmetry for two identical monomers

If AB and CD are identical monomers as in $(HF)_2$ and $(DF)_2$, the rovibrational eigenfunctions $X^{JMp}_{vjK} = \mathcal{Y}^{JMp}_{jK} \phi_v$ should be properly symmetrized [14–19]. The symmetrized rovibrational basis can be written as

$$X^{\pm JMp}_{vjK} = \left[2\left(1 \pm <X^{JMp}_{vjK}|\hat{p}_{ex}|X^{JMp}_{vjK}>\right)\right]^{-1/2} \left[X^{JMp}_{vjK} \pm \hat{p}_{ex} X^{JMp}_{vjK}\right] \tag{3.86}$$

where $\hat{p}_{ex}$ is the exchange operator that changes the coordinates $(\mathbf{r}_1, \mathbf{r}_2, \mathbf{R}) \rightarrow (\mathbf{r}_2, \mathbf{r}_1, -\mathbf{R})$. It can be shown [19] that

$$\hat{p}_{ex} X^{JMp}_{vjK} = (-1)^{p+j_{12}} X^{JMp}_{\tilde{v}\tilde{j}K} \tag{3.87}$$

where $\tilde{v}$ and $\tilde{j}$ denote transposed indexes (j_2, j_1, j_{12}) and (v_2, v_1). Thus the symmetrized basis is

$$X^{\pm JMp}_{vjK} = \Delta_{vj,\tilde{v}\tilde{j}} \left[\mathcal{Y}^{JMp}_{jK} \Phi_v \pm (-1)^{j_{12}+p} \mathcal{Y}^{JMp}_{\tilde{j}K} \Phi_{\tilde{v}}\right] \tag{3.88}$$

where the normalized constant is given by

$$\Delta_{vj,\bar{v}\bar{j}} = [2(1+\delta_{v\bar{v}}\delta_{j\bar{j}})]^{-1/2} = [2(1+\delta_{v_1 v_2}\delta_{j_1 j_2})]^{-1/2}, \tag{3.89}$$

Here we note that quantum numbers v and j are restricted by the requirement of $v_1 \geq v_2$ and $j_1 \geq j_2$ for $v_1 = v_2$. Note that for $v_1 = v_2$ and $j_1 = j_2$, the allowed j_{12} quantum numbers must satisfy the condition $p_{ex}(-1)^{j_{12}+p} = 1$.

The potential matrix elements in the symmetrized basis representation $X_{vjK}^{\pm JMp}$ can be written explicitly in terms of unsymmetrized functions

$$<X_{vjK}^{\pm JMp}|V|X_{v'j'K'}^{\pm JMp}> = 2\Delta_{vj,\bar{v}\bar{j}}\Delta_{v'j',\bar{v}'\bar{j}'}\delta_{KK'}\left[<\mathcal{Y}_{jK}^{JMp}|V_{vv'}|\mathcal{Y}_{j'K}^{JMp}> \right.$$
$$\left. \pm(-1)^{j_{12}+p} <\mathcal{Y}_{\bar{j}K}^{JM}|V_{\bar{v}v'}|\mathcal{Y}_{j'K}^{JMp}>\right] \tag{3.90}$$

where

$$V_{vv'} = <\phi_v|V|\phi_{v'}> \tag{3.91}$$

and

$$V_{\bar{v}v'} = <\phi_{\bar{v}}|V|\phi_{v'}> \tag{3.92}$$

The centrifugal potential in the symmetrized basis representation is given by

$$<X_{vjK}^{\pm JMp}|V_c|X_{v'j'K'}^{\pm JMp}>=$$
$$\Delta_{vj,\bar{v}\bar{j}}\Delta_{v'j',\bar{v}'\bar{j}'}\frac{\hbar^2}{\mu R^2}\left[\delta_{vv'}\delta_{jj'} \pm (-1)^{j_{12}+p}\delta_{\bar{v}v'}\delta_{\bar{j}j'}\right]W_{K,K'}^{Jj_{12}} \tag{3.93}$$

where the centrifugal coupling tridiagonal matrix $W_{KK'}^{Jj_{12}}$ is defined in Eq. (3.69) with the replacement of j by j_{12}. The rest of the Hamiltonian matrix is relatively simple to calculate and is therefore not given here.

3.2.4 Bound State Calculation of $(HF)_2$

Background

Weakly bound molecular species are often characterized by large amplitude vibrational motions and are therefore very "floppy". In addition to being anharmonic, various vibrational modes of such molecules are often coupled. Therefore the standard normal mode analysis of vibration for valence bond molecules is inadequate for weakly bound systems such as van der Waals or hydrogen-bonded complexes. In order to obtain accurate spectroscopic information in relation to the structure and dynamics of weakly bound complexes, it is imperative to perform accurate multi-dimensional quantum mechanical calculations for these species. Recent years have seen rapid progress in accurate quantum mechanical treatment of weakly bound polyatomic complexes and clusters for triatomic and larger systems.

Equally impressive advances in laser spectroscopy and supersonic molecular beams have enabled highly resolved studies of the spectroscopy and state-to-state dissociation dynamics of weakly bound polyatomic complexes. This has allowed direct comparison between theory and experiment, resulting in deeper insight into the complex behavior of these systems. $(HF)_2$ is perhaps one of the most intensely studied hydrogen-bonded dimers. A great deal of experimental effort has been directed at determining its structure and tunneling dynamics, vibrational predissociation lifetimes for the fundamental and overtone excitations, and product rotational state distributions in photofragmentation. It has long been a prototype system for theoretical and experimental investigation of spectroscopy and dynamics for hydrogen-bonded complexes. Theoretical studies of bound states of the HF dimer have been reported [20–27] including full-dimensional 6D calculations for the ground [25,26] and stretch excited intramolecular states [27].

HF dimer

The theoretical approach outlined in Sec. 3.2.3 has been employed to carry out bound state calculations of the HF dimer. The details of the calculations can be found in [22,26,27]. The calculation for $(HF)_2$ and $(DF)_2$ can be separated into four symmetric blocks corresponding to the four combinations of even and odd character with respect to parity and monomer exchange. We use the symbol $(P)^{\pm}$ to denote the symmetry of the bound states where P is the total parity defined as the system parity ϵ times the factor $(-1)^J$, and $\pm$ is the notation for exchange symmetry. Table (3.1) lists the first 15 $(HF)_2$ bound state energies for each symmetry block for the total angular momentum J=0 together with the spectroscopic assignment of the states.

Many excited intermolecular vibrational states have been assigned by inspecting the plots of the wavefunction. The eigenstates marked by question marks in the table are not cleanly assignable. The calculated ground state tunneling splitting is 0.44 cm^{-1}, as compared to the experimental value of 0.659 cm^{-1} for the HF dimer. By plotting various contour cuts of the wavefunction, it is found that although many bending states can be

Table 3.1: (a) 6D bound state energies of $(HF)_2$ for even parity and total angular momentum J=0 on the SQSBDE PES [28]. The energies are relative to the ground state energy of -1057.33 cm^{-1} for 6D.

	$(P=+1)^+$				$(P=+1)^-$			
n	6D	$(\nu_3\nu_4\nu_5\nu_6)$	<R>	ΔR	6D	$(\nu_3\nu_4\nu_5\nu_6)$	<R>	ΔR
1	0.00	(0000)	5.24	.21	0.44	(0010)	5.24	.21
2	126.37	(0100)	5.35	.37	127.35	(0110)	5.35	.37
3	160.58	(0020)	5.27	.24	168.06	(0030)	5.27	.24
4	244.50	(0200)	5.46	.48	246.16	(0210)	5.48	.49
5	274.97	(0120)	5.33	.39	289.08	(0130)	5.40	.40
6	292.65	(0040)	5.33	.31	339.20	(0050)	5.28	.26
7	355.17	(0300)	5.60	.59	357.33	(0310)	5.62	.59
8	384.78	(0220)	5.46	.52	402.46	(0230)	5.53	.53
9	399.91	(0140)	5.48	.46	440.32	(1010)?	5.32	.27
10	425.30	(1000)?	5.31	.26	455.38	(0150)?	5.56	.57
11	457.85	(0400)	5.76	.68	463.39	(0410)	5.62	.61
12	463.59	(0060)?	5.29	.28	507.95	(0330)	5.68	.63
13	488.37	(0320)	5.64	.64	542.56	(0070)?	5.39	.42
14	502.20	(0240)	5.59	.56	552.86	(0510)	5.89	.76
15	537.23	(1100)	5.46	.43	555.55	(1110)	5.40	.38

Table 3.1: (b) 6D bound state energies of $(HF)_2$ for odd parity.

	$(P=-1)^+$				$(P=-1)^-$			
n	6D	$(\nu_3\nu_4\nu_5\nu_6)$	<R>	ΔR	6D	$(\nu_3\nu_4\nu_5\nu_6)$	<R>	ΔR
1	378.72	(0001)	5.34	.24	380.47	(0011)	5.34	.24
2	491.14	(0101)	5.50	.42	493.83	(0111)	5.50	.42
3	544.53	(0021)	5.37	.26	574.71	(0031)	5.36	.27
4	594.32	(0201)	5.67	.55	598.73	(0211)	5.65	.55
5	641.31	(0121)	5.55	.46	679.65	(0131)	5.55	.48
6	688.36	(0301)	5.79	.65	696.03	(0311)	5.80	.65
7	689.93	(0041)	5.42	.38	773.05	(0231)?	5.70	.65
8	731.65	(0221)	5.72	.59	783.25	?	5.65	.64
9	774.14	(0401)	6.04	.80	788.62	?	5.71	.63
10	788.23	(0141)	5.52	.47	846.09	(1011)	5.46	.30
11	817.03	(0321)	5.88	.71	858.63	(0511)	6.03	1.05
12	827.97	(1001)	5.49	.28	872.54	(0331)?	5.88	.87
13	851.46	(0501)	6.10	1.25	889.54	?	5.59	.47
14	878.97	(0241)	5.68	.61	930.16	?	5.60	.80
15	896.51	(0421)	5.97	.85	932.51	(0611)	5.56	2.04

assigned to the so-called "geared" motions, most "anti-geared" states of the complex actually involve highly mixed vibrational motions except for a few special cases in which the anti-geared states are relatively pure.

The fundamental frequencies for various modes are slightly different in different symmetric blocks. For $(P=+1)^+$ states, the intermolecular stretching frequency (ν_4) is 126.37 cm^{-1}, the lowest of all modes, which is in good agreement with the experimental value of 125 cm^{-1} [24]. The splitting of this stretching state is about 1 cm^{-1}. The trans-bending frequency (ν_5) is 160.57 cm^{-1}, and the cis-bending frequency (ν_3) is 425.27 cm^{-1}, the highest of all intermolecular vibrations. The lowest eigenstate of odd parity in Table 3.1b corresponds to the torsionally excited state of the HF dimer with a tunneling splitting of 1.85 cm^{-1}.

Also shown in Table 3.1 are expectation values of R ($<R>$) and root mean square (rms) amplitudes of R (ΔR). Although we can see the influence of the bending excitation on $<R>$ and especially on ΔR, the excitation of the stretching mode can still be identified quite clearly from $<R>$ and ΔR, which also indicates the relatively weak coupling between stretching and bending modes in the HF dimer.

DF dimer

The DF dimer is analogous to the HF dimer with two symmetric wells. However, due to the smaller rotational constant of DF, both the tunneling splitting and bending frequencies are expected to be smaller than in the HF dimer. Table 3.2 lists the first 15 $(DF)_2$ bound states for all symmetry blocks. The dissociation energy is 1165 cm^{-1} for $(DF)_2$, 100 cm^{-1} larger than that for $(HF)_2$. The splitting of the ground state is only 0.05 cm^{-1}, one tenth of the value in $(HF)_2$. Since DF has a smaller rotational constant than HF, the DF monomers are more localized in bending angles, making the overlap of the wavefunction in bending for two DF monomers smaller than that in $(HF)_2$. Therefore it is very reasonable to see this decrease in ground state energy splitting, as well as in other corresponding states.

HF-DF complex

Unlike the symmetric $(HF)_2$ or $(DF)_2$, there is no symmetry in the HF-DF complex with respect to the exchange of two monomers. Due to the difference in rotational constants for HF and DF, the system has two asymmetric double wells. For low-lying states, the complex is localized in one of the two wells corresponding to the two isomers, HF-DF and DF-HF. The computed dissociation energy is 1142.7 and 1078.5 cm^{-1} for the HF-DF and DF-HF

Table 3.2: (a) 6D bound state energies of $(DF)_2$ for even parity and total angular momentum J=0 on the SQSBDE PES. The energies are relative to the ground state energy of -1166.55 cm^{-1}

	$(P=+1)^+$				$(P=+1)^-$			
n	6D	$(\nu_3\nu_4\nu_5\nu_6)$	<R>	ΔR	6D	$(\nu_3\nu_4\nu_5\nu_6)$	<R>	ΔR
1	0.00	(0000)	5.18	.21	0.03	(0010)	5.18	.21
2	112.84	(0020)	5.22	.29	113.25	(0030)	5.22	.30
3	141.71	(0100)	5.28	.31	142.08	(0110)	5.28	.31
4	214.03	(0040)	5.22	.34	217.95	(0050)	5.24	.36
5	243.56	(0120)?	5.32	.40	247.65	(0130)?	5.32	.39
6	270.73	(0200)?	5.39	.41	272.70	(0210)?	5.38	.40
7	301.69	(0060)?	5.17	.30	316.00	(0070)?	5.24	.30
8	324.13	(1000)	5.26	.32	327.42	(1010)	5.31	.41
9	334.48		5.32	.42	354.12		5.35	.46
10	361.48		5.45	.49	375.36		5.39	.45
11	387.06		5.53	.51	393.03		5.49	.47
12	401.70		5.21	.38	424.55		5.34	.47
13	426.10		5.33	.45	437.70		5.41	.50
14	437.90		5.38	.43	451.73		5.30	.31
15	443.75		5.39	.48	462.45		5.46	.57

Table 3.2: (b) 6D bound state energies of $(DF)_2$ for odd parity.

	$(P=-1)^+$				$(P=-1)^-$			
n	6D	$(\nu_3\nu_4\nu_5\nu_6)$	<R>	ΔR	6D	$(\nu_3\nu_4\nu_5\nu_6)$	<R>	ΔR
1	273.88	(0001)	5.22	.22	273.96	(0011)	5.22	.22
2	388.68	(0101)	5.30	.36	389.13	(0111)	5.30	.36
3	411.70	(0021)	5.27	.27	413.83	(0031)	5.26	.27
4	496.86	(0201)	5.37	.46	498.43	(0211)	5.41	.48
5	512.30		5.29	.38	523.05	(0131)	5.36	.42
6	528.00		5.39	.38	545.33	(0051)	5.29	.29
7	598.08		5.33	.46	601.63	(0311)	5.55	.58
8	602.39		5.43	.50	611.81	(1011)	5.29	.27
9	615.23		5.37	.43	625.83	(0231)	5.44	.48
10	627.63		5.49	.53	630.64	(0151)	5.42	.40
11	651.26		5.42	.44	685.69	(0071)	5.27	.27
12	696.53		5.66	.67	698.05	(0411)	5.72	.69
13	700.39		5.42	.53	715.68	(1111)	5.43	.46
14	712.34		5.36	.35	718.79	(1031)	5.35	.27
15	714.44		5.49	.47	726.36	(0331)	5.62	.61

isomers, respectively. For even parity, about half of the first 30 states are localized in one of the isomerization wells if we define 80% probability as a criterion. The mode mixture between trans-bending and stretching is stronger in the DF-HF isomer than in the HF-DF isomer, making almost all the states in the HF-DF isomers assignable while those in the DF-HF isomers are hard to assign. For odd parity, about two thirds of the first 30 states are localized in one of the isomerization wells. The mode mixture in DF-HF isomers is relatively weaker than in the case of even parity, making almost all the localized states in odd parity assignable.

3.3 Time-Dependent Spectrum Method

Motivation

The standard matrix diagonalization method becomes impractical when the size of the Hamiltonian matrix to be diagonalized is prohibitively large. Firstly, the computer memory required to store the Hamiltonian matrix increases quadratically as the dimension of the matrix. For large matrices, the computer can quickly run out of memory. Secondly, the CPU time to diagonalize a matrix increases as N^3, the dimension of the matrix cubed. Thus the limitations of computer memory and CPU time will severely limit the size of the matrix to be diagonalized. Therefore for large molecular systems, we need to use alternative methods that scale more slowly than N^3 in order to solve large scale eigenvalue problems. One approach is to use iterative methods based primarily on the Lanczos method of tri-diagonalization. The Lanczos method allows one to converge large eigenvalues quickly with a relatively short iteration. In addition, sophisticated basis optimization techniques are often needed in order to drastically reduce the total number of basis functions for polyatomic molecules [29]. In the following, however, we discuss the time-dependent approach which has recently become very popular in solving large scale eigenvalue problems.

3.3.1 Autocorrelation Function

Time-dependent wavepacket propagation can be employed to obtain bound state energies and bound state wavefunctions without the need to diagonalize the Hamiltonian matrix. The application of the method to bound state calculation is quite straightforward. If the Hamiltonian H supports bound states Φ_n with eigenenergies E_n, one can expand any given initial wavepacket in this eigen basis set,

$$\Psi(0) = \sum_n C_n \Phi_n \tag{3.94}$$

where the expansion coefficient C_n is given by

$$C_n = <\Phi_n|\Psi(0)> \tag{3.95}$$

The time-dependent solution to the Schrödinger equation

$$i\hbar \frac{\partial}{\partial t}|\Psi(t)> = H|\Psi(t)> \tag{3.96}$$

can be formally written as

$$\begin{aligned} \Psi(t) &= e^{-\frac{i}{\hbar}Ht}\Psi(0) \\ &= \sum_n C_n e^{-\frac{i}{\hbar}E_n t}\Phi_n. \end{aligned} \tag{3.97}$$

We can define a autocorrelation function $A(t)$ by

$$A(t) = <\Psi(0)|\Psi(t)> \tag{3.98}$$

Using Eqs. (3.95) and (3.97), the autocorrelation function $A(t)$ can be written as

$$\begin{aligned} A(t) &= \sum_n C_n e^{-\frac{i}{\hbar}E_n t} <\Psi(0)|\Phi_n> \\ &= \sum_n e^{-\frac{i}{\hbar}E_n t}|C_n|^2 \end{aligned} \tag{3.99}$$

3.3.2 Energy Spectrum

Once the autocorrelation function has been calculated using the time-dependent method, it is then straightforward to calculate the Fourier transform of the autocorrelation function to obtain the energy spectrum $S(E)$

$$\begin{aligned} S(E) &= \mathrm{Re}\left[\frac{1}{T}\int_0^T e^{\frac{i}{\hbar}Et} <\Psi(0)|\Psi(t)> dt\right] \\ &= \sum_n |C_n|^2 \mathrm{Re}\left[\frac{1}{T}\int_0^T e^{\frac{i}{\hbar}(E-E_n)t} dt\right] \\ &= \sum_n |C_n|^2 \frac{\hbar}{(E-E_n)T} \sin\left[\frac{(E-E_n)T}{\hbar}\right] \end{aligned} \tag{3.100}$$

In the limit T→ ∞, the above expression gives rise to a stick spectrum and the height of the stick is equal to $|C_n|^2$ which is just the square of the overlap between the particular eigenfunction Φ_n and the initial wavefunction, i.e.,

$$|C_n|^2 = | <\Psi(0)|\Phi_n> |^2 \tag{3.101}$$

For finite time T, Eq. (3.100) gives the Sinc spectrum which is oscillatory as a function of energy. Due to this oscillation, there is spectral "noise" or extra peaks that do not correspond to actual eigenvalues. A common method to reduce the noise is to use a window function, either in the time domain or in the energy domain, to help eliminate these unwanted peaks from the spectrum. The use of a Gaussian or Lorentzian spectrum was shown to be numerically superior for eigenenergy calculations. Most importantly, the Gaussian or Lorentzian spectrum is "noise-free" and every single spectral peak represents an eigenvalue [30]. For example, by including a window function $\sim \exp(-\alpha^2 t^2)$ in Eq. (3.100), one obtains the Gaussian spectrum

$$\begin{aligned} S_G(E) &= \frac{1}{2\pi}\int_{-\infty}^{\infty} e^{iEt} <\Psi(0)|\Psi(t)> e^{-\alpha^2 t^2} dt \\ &= \frac{1}{2\alpha\sqrt{\pi}} \sum_n e^{-\frac{(E-E_n)^2}{4\alpha^2}} |C_n(E)|^2 \end{aligned} \tag{3.102}$$

which is positive definite and decays exponentially from the spectral peak.

To illustrate the usefulness of the window function used in obtaining the energy spectrum, we show a sample calculation of the energy spectrum for a Morse oscillator. The bottom figure of Fig. 3.3 shows a typical Sinc spectrum where there are many oscillatory peaks that do not correspond to eigenenergies. The corresponding Gaussian spectrum in Fig. 3.3 is positive definite and every peak corresponds to an eigenenergy. The Gaussian spectrum depends on the value of the parameter α. If different values of α in the Gaussian window function are used, the spectrum will look different. If one increases the value of α from zero, the spectrum changes continuously from the Sinc spectrum to the Gaussian spectrum in Fig. 3.3. If α is further increased, the spectrum will be broadened and sufficient resolution will eventually be lost. In actual time-dependent calculations, one first computes the autocorrelation function and then Fourier transforms the autocorrelation function to obtain the energy spectrum. By using the Gaussian window function, one can adjust the exponential parameter α to obtain an optimal spectrum with the least noise. This is similar to the process of focusing in optics.

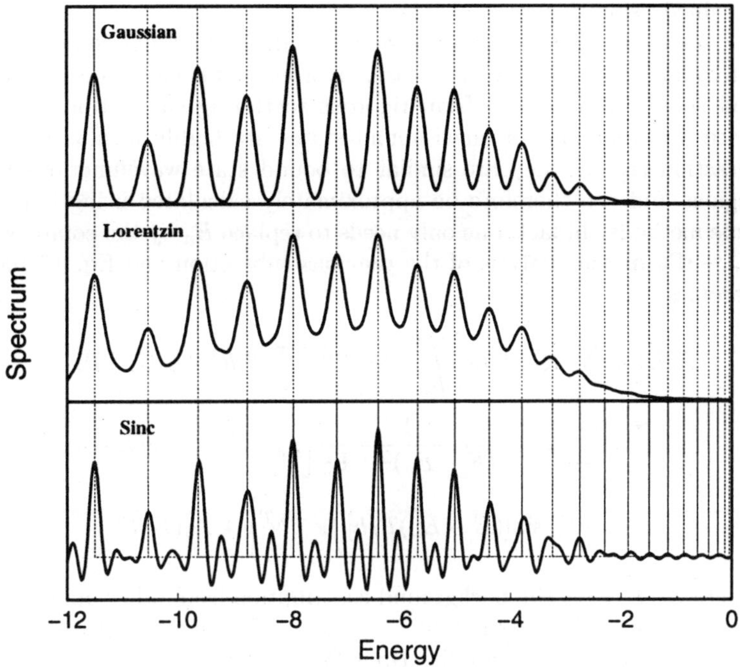

Figure 3.3: Gaussian, Lorentzian, and Sinc spectra for a Morse oscillator obtained from a finite time Fourier transform of the autocorrelation function. The dotted lines denote positions of exact eigenvalues.

Once the eigenenergy is obtained, it is straightforward to obtain the corresponding eigenfunction by performing a Fourier transform

$$|\Phi_n\rangle \propto \frac{1}{2T}\int_{-T}^{T} e^{\frac{i}{\hbar}E_n t}|\Psi(t)\rangle\, dt. \tag{3.103}$$

However, in order to separate the neighboring states E_n and $E_{n'}$, the propagation time T has to be long enough such that $T \gg 1/(E_n - E_{n'})$. If the bound state is degenerate, a single TD propagation from a given initial wavepacket can only give one degenerate state. One needs to perform n separate propagations with n different initial wavepackets in order to obtain all n degenerate states. One approach is to use a mixed TD/TI method to deal with this problem in which one carries out explicit matrix diagonalization based on approximate eigenstates that are obtained from relatively short propagation of the wavepacket. Some encouraging results have been obtained using this method [31].

A similar approach can be applied to quasi-bound or resonance states. Of course, quasi-bound states are, rigorously speaking, continuum states and should be obtained from rigorous scattering calculations. However, since wavefunctions of quasi-bound states are highly localized in the interaction region, they are similar to bound state wavefunctions in these regions and can therefore be approximately calculated using bound state type methods. In fact, one only needs to replace E_n by the complex energy $E_n - i\Gamma_n$ in the analysis of the previous subsection and Eq. (3.100) then becomes

$$\begin{aligned} S(E) &= \sum_n |C_n|^2 Re\left[\int_0^T e^{\frac{i}{\hbar}(E-E_n+i\Gamma_n)t}dt\right] + A_b(E,T) \\ &= \sum_n |C_n|^2 \frac{1}{(E-E_n)^2+\Gamma_n^2}\left[\Gamma_n - \Big(\Gamma_n \cos[(E-E_n)T/\hbar]\right. \\ &\quad \left. - E_n \sin[(E-E_n)T/\hbar]\Big)e^{-\Gamma_n T}\right] + A_b(E,T) \end{aligned} \tag{3.104}$$

where $A_b(E,T)$ is the background contribution from other bound and continuum states. In the limit of $T \to \infty$ and for an isolated resonance, Eq. (3.104) reduces to the familiar Breit-Wigner formula,

$$S(E) = |C_n|^2 \frac{\Gamma_n}{(E-E_n)^2+\Gamma_n^2} \tag{3.105}$$

The calculation of a complex eigenenergy is greatly facilitated by applying complex absorbing potentials at boundaries. Generally, in order to resolve both the resonance energy E_n and resonance width Γ_n using the Breit-Wigner formula (3.105), the time propagation has to be carried out to a sufficiently long time so that the uncertainty relation

$$T \gg \frac{1}{\Gamma_n} \tag{3.106}$$

is satisfied. Thus the direct application of the TD method to obtain accurate resonance widths, especially narrow ones, is quite difficult. However, there are ways to overcome this problem. One simple approach is just to calculate the decay rate of the resonance wavefunction (survival probability) and extract the width from the time-dependence of the survival probability [32,33]. Another possible approach is to directly fit Eq. (3.104) at a relatively short propagation time T to extract complex eigenenergies. Both approaches require that the background contribution to Eq. (3.104) be small enough such that the resonance contribution dominates the spectrum in Eq. (3.104). Thus it is possible to extract a narrow resonance width from only short time propagation if the initial wavepacket is chosen to have a large overlap with the resonance wavefunction.

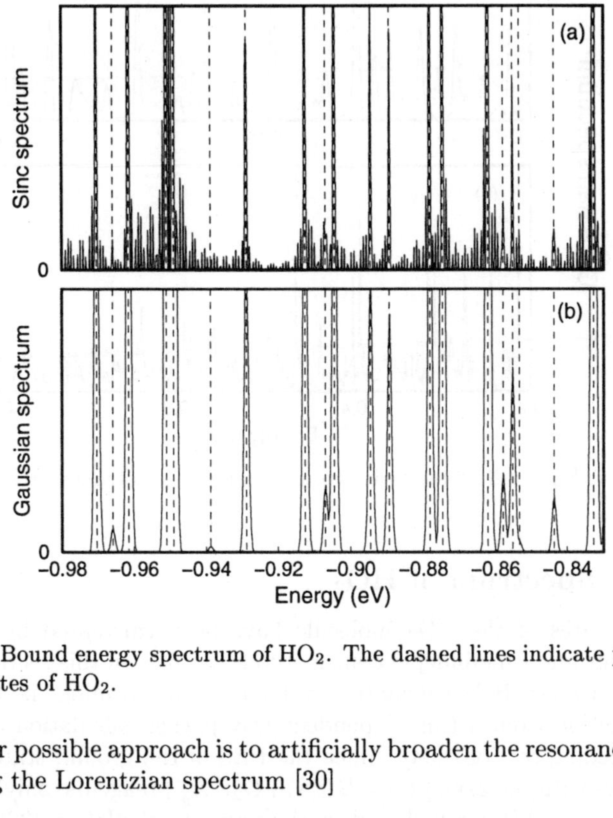

Figure 3.4: Bound energy spectrum of HO_2. The dashed lines indicate positions of bound states of HO_2.

Another possible approach is to artificially broaden the resonance width Γ_n by using the Lorentzian spectrum [30]

$$\begin{aligned} S_L(E) &= \frac{1}{2\pi}\int_{-\infty}^{\infty} e^{iEt} <\Psi(0)|\Psi(t)> e^{-\alpha|t|}dt \\ &= \frac{1}{\pi}\frac{(\alpha+\Gamma_n)}{(\alpha+\Gamma_n)^2+(E-E_n)^2}|C_n(E)|^2 + A_b(E). \end{aligned} \tag{3.107}$$

where α is chosen to be much greater than Γ_n so that condition (3.106) is satisfied at a much shorter time for the total decay width $\Gamma = \alpha + \Gamma_n$. Direct fitting of the spectrum to Eq. (3.107) will yield the total width, and the true width Γ_n is then simply obtained by subtracting out α from the total width Γ.

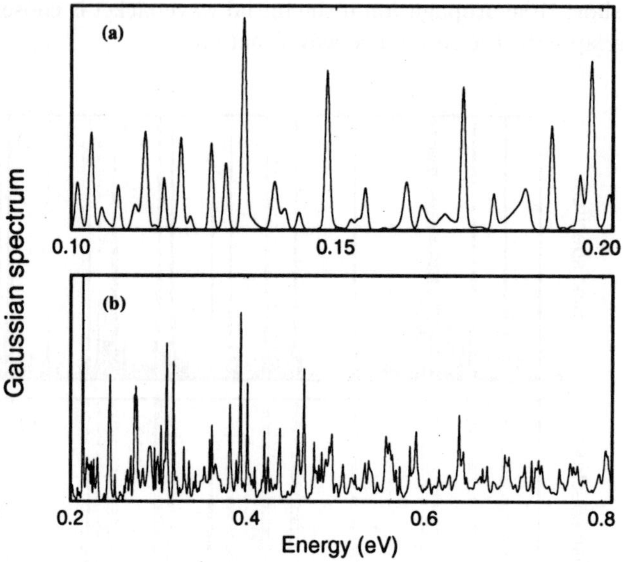

Figure 3.5: Inelastic resonance energy spectrum of HO_2.

3.3.3 Spectrum of HO_2

Bound states of the HO_2 molecule have been calculated by both time-dependent and time-independent methods [34,35]. In this section, we show the spectrum of HO_2 calculated by Fourier transforming the autocorrelation function from a time-dependent wavepacket calculation described in the preceding sections [34]. Since the PES of the ground state $HO_2(^2A'')$ correlates with the asymptotic $H(^2S)+O_2(^3\Sigma_g^-)$ reagents, only odd nuclear rotations (j=odd) are included in the present calculation [36]. The total angular momentum J is set to zero. After propagating the wavefunction to about 200,000 atomic units, a large number of eigenvalues show up as spectral peaks. Figure 3.4a shows a fragment of the Sinc spectrum and the corresponding Gaussian spectrum. As is clear from the Figure, all the peaks in the Gaussian spectrum match exactly the eigenvalues in the energy range. These eigenvalues are in good agreement with TI calculations [37,38].

To calculate resonances, including both inelastic and reactive resonance energies, absorbing potentials are used to eliminate pseudo eigenstates due to boundary reflections. Fig. 3.5 shows a Gaussian spectrum for inelastic scattering covering the whole range of energies starting from the threshold energy of H + O_2 up to the threshold energy of the O + HO channel. The TD calculation revealed a total of 159 inelastic resonance peaks contained in the energy range of Fig. 3.5. Most of the calculated inelastic resonance energies are in good agreement with the time-independent calculation of Ref. [39].

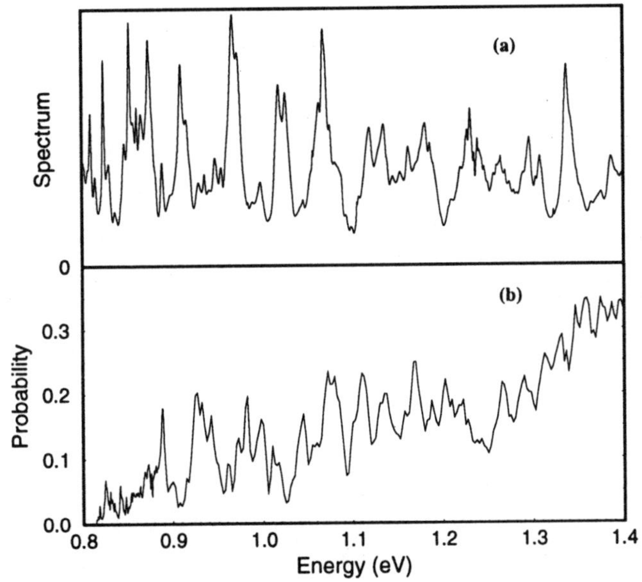

Figure 3.6: Reactive resonance energy spectrum of HO_2 (a) and reaction probabilities for H + O_2 → HO + O reaction (b).

In order to calculate resonance energies, we scanned energies above the reaction threshold (0.81147 eV) up to 1.4 eV. In this energy range 67 resonance peaks have been found, mostly overlapping ones. Figure 3.6a plots the resonance spectrum in the energy range from 0.81147 eV to 1.4 eV. In order to compare the resonance structure, reaction probabilities in the same energy range are also plotted in Fig. 3.6b. Most spectral peaks in the figure match quite well the corresponding peaks in the reaction probability.

Chapter 4

Fundamental Theory of Quantum Scattering

4.1 Time-Dependent Scattering Theory

4.1.1 Mϕller Operator

In order to develop a consistent theory of quantum scattering, we choose to start from the basic definition of scattering using the time-dependent formalism. Later the more familiar time-independent formalism of scattering will be derived from the time-dependent formalism. Let H denote the Hamiltonian of the scattering problem

$$H = H_0 + V \tag{4.1}$$

where H_0 is the "free" or asymptotic Hamiltonian and V is the interaction potential that goes to zero asymptotically. Let Ψ be a scattering solution (no bound state component) of the time-dependent Schrödinger equation at t=0; the time-dependent solution at time t is then given by the formal solution (provided that H is independent of time).

$$\Psi_S(t) = e^{-\frac{i}{\hbar}Ht}\Psi \tag{4.2}$$

We can now define the wavefunction in the *interaction picture* by the unitary transformation

$$\begin{aligned}\Psi_I(t) &= e^{\frac{i}{\hbar}H_0 t}\Psi_S(t)\\ &= e^{\frac{i}{\hbar}H_0 t}e^{-\frac{i}{\hbar}Ht}\Psi\end{aligned} \tag{4.3}$$

It is relatively straightforward to derive the equation for $\Psi_I(t)$,

$$i\hbar\frac{\partial}{\partial t}|\Psi_I>= H_I(t)|\Psi_I(t)>, \tag{4.4}$$

where $H_I(t)$ is the Hamiltonian in the interaction picture

$$H_I(t)>= e^{\frac{i}{\hbar}H_0 t}Ve^{-\frac{i}{\hbar}H_0 t}. \tag{4.5}$$

The time evolution of Ψ_I can be represented by a unitary evolution operator $U_I(t, t_0)$

$$|\Psi_I(t)>= U_I(t, t_0)|\Psi_I(t_0)> \tag{4.6}$$

The operator $U_I(t, t_0)$ can obviously be partitioned as

$$U_I(t, t_0) = U_I(t, 0)U_I(0, t_0) \tag{4.7}$$

where

$$U_I(t, 0) = e^{\frac{i}{\hbar}H_0 t}e^{-\frac{i}{\hbar}Ht} \tag{4.8}$$

and

$$\begin{aligned} U_I(0, t) &= U_I(t, 0)^\dagger \\ &= e^{\frac{i}{\hbar}Ht}e^{-\frac{i}{\hbar}H_0 t} \end{aligned} \tag{4.9}$$

We are now in a position to define an *incoming* asymptote (wavepacket) by

$$\begin{aligned} \Phi_{in} &= \lim_{t\to-\infty} \Psi_I(t) \\ &= \lim_{t\to-\infty} U_I(t, 0)\Psi \end{aligned} \tag{4.10}$$

Since the scattering state Ψ has no overlap with bound states of H (if any exists), the *incoming* asymptote Φ_{in} is a stationary wavepacket. This can be seen easily from the following consideration. Assuming that the Schrödinger wavefunction $\Psi_s(t)$ passes through the interaction region after $t = T$, then Φ_{in} is given by the finite time limit

$$\Phi_{in} = e^{\frac{i}{\hbar}H_0 T}e^{-\frac{i}{\hbar}HT}\Psi \tag{4.11}$$

because for $t > T$, $e^{-\frac{i}{\hbar}Ht}\Psi = e^{-\frac{i}{\hbar}H_0 t}\Psi$ and the forward and backward propagations cancel out exactly. Similarly, we can define an *outgoing* asymptote as the $t = \infty$ limit of $\Psi_I(t)$,

$$\begin{aligned} \Phi_{out} &= \lim_{t\to\infty} \Psi_I(t) \\ &= \lim_{t\to\infty} U_I(t, 0)\Psi \end{aligned} \tag{4.12}$$

Equation (4.10) can be inverted to express the scattering state in terms of the incoming asymptote

$$\begin{aligned}\Psi &= \lim_{t\to-\infty} U_I^+(t,0)\Phi_{in} \\ &= \lim_{t\to-\infty} U_I(0,t)\Phi_{in}\end{aligned} \tag{4.13}$$

Similarly we can invert Eq. (4.12) to yield the equation

$$\Psi = \lim_{t\to\infty} U_I(0,t)\Phi_{out} \tag{4.14}$$

In connection with the definitions for Φ_{in} and Φ_{out}, we can define the so-called Mϕller operator as the limit

$$\begin{aligned}\Omega_\pm &= \lim_{t\to\mp\infty} U_I(0,t) \\ &= U_I(0,\mp\infty)\end{aligned} \tag{4.15}$$

where the limits are assumed to exist. Therefore the scattering state Ψ can be expressed in terms of the Mϕller operators as

$$\boxed{\Psi = \Omega_+\Phi_{in} = \Omega_-\Phi_{out}} \tag{4.16}$$

The Mϕller operators satisfy the following important relation

$$\boxed{H\Omega_\pm = \Omega_\pm H_0} \tag{4.17}$$

which is called the *interwining* relation. This relation can be proved by noting that for any finite time τ, the following relation holds

$$\begin{aligned}e^{\frac{i}{\hbar}H\tau}\Omega_\pm &= \lim_{t\to\mp\infty} e^{\frac{i}{\hbar}H(t+\tau)}e^{\frac{i}{\hbar}H_0(t+\tau)}e^{\frac{i}{\hbar}H_0\tau} \\ &= \Omega_\pm e^{\frac{i}{\hbar}H_0\tau}\end{aligned} \tag{4.18}$$

By differentiating with respect to τ on both sides of the above equation and setting τ=0, we obtain Eq. (4.17). In fact, the inter-wining relation (4.17) holds for any power of H, and thus it also holds for any arbitrary function of H

$$f(H)\Omega_\pm = \Omega_\pm f(H_0) \tag{4.19}$$

At this point, it is useful to discuss the quasi unitary property of the Mϕller operator. The Mϕller operator $\Omega_\pm$ with the relation

$$|\Psi> = \Omega_\pm|\Phi> \tag{4.20}$$

maps each vector Φ in the Hilbert space of H_0 onto a unique vector Ψ in the space of H. Since the asymptotic Hamiltonian H_0 does not support bound states while the scattering Hamiltonian H might support bound states, the vectors in the Hilbert space of H_0 are mapped only to a subspace of H that is composed of only continuous vectors. Thus the unitarity of $\Omega_\pm$ is unidirectional, i.e.,

$$\Omega_\pm^\dagger \Omega_\pm = I \tag{4.21}$$

Equation (4.21) guarantees the conservation of normalization of the scattering wavefunction

$$\begin{aligned} <\Psi|\Psi> &= <\Phi_{out}|\Omega_+^\dagger \Omega_+|\Phi_{out}> \\ &= <\Phi_{out}|\Phi_{out}> \end{aligned} \tag{4.22}$$

or

$$\begin{aligned} <\Psi|\Psi> &= <\Phi_{in}|\Omega_-^\dagger \Omega_-|\Phi_{in}> \\ &= <\Phi_{in}|\Phi_{in}> \end{aligned} \tag{4.23}$$

In general, however,

$$\Omega_\pm \Omega_\pm^\dagger \neq 1 \tag{4.24}$$

This is because the bound states in the Hilbert space of H have no counter part in the Hilbert space of H_0. In order words, any state vector in the space of H_0 can uniquely map onto a continuous state vector in the space of H and vice versa. But any bound state in the space of H could not be mapped onto a state vector in the space of H_0. For example, if Ψ_n is a bound state of H, it can be shown easily that

$$<\Psi_n|\Omega_\pm|\Phi> = \lim_{t\to\mp\infty} e^{\frac{i}{\hbar}Et} <\Psi_n|e^{-\frac{i}{\hbar}H_0 t}|\Phi> \tag{4.25}$$

Since the propagation $e^{-\frac{i}{\hbar}H_0 t}|\Phi>$ is a traveling wave, it will have no overlap with Ψ_n in the asymptotic limit and the above integral is zero for any continuum state Φ in the space of H_0.

However, if the scattering Hamiltonian H does not support any bound states, or if we are restricted to only continuous state vectors in the Hilbert space of H, then we can treat the operators $\Omega_\pm$ as if they were unitary, i.e.,

$$\boxed{\Omega\Omega^\dagger = \Omega^\dagger\Omega = I} \tag{4.26}$$

4.1.2 Scattering Operator

From the definition of the Mϕller operator, we can now define the scattering or S matrix operator by

$$\boxed{S = \Omega_-^{\dagger}\Omega_+} \tag{4.27}$$

which can also be expressed in terms of the asymptotic limit of the evolution operators

$$\begin{aligned} S &= U_I(\infty, 0)U_I(0, -\infty) \\ &= U_I(\infty, -\infty) \end{aligned} \tag{4.28}$$

Using the inter-wining relation of Eq. (4.17), it is straightforward to show that the S matrix operator defined by Eq. (4.27) commutes with the asymptotic Hamiltonian H_0, i.e.,

$$H_0 S = S H_0 \tag{4.29}$$

which guarantees the energy conservation of the scattering process.

The S matrix operator is explicitly unitary, i.e.,

$$\boxed{SS^{\dagger} = S^{\dagger}S = I} \tag{4.30}$$

The proof of the unitarity of the S matrix operator is based on the quasi unitary property of the Mϕller operator Ω, for example,

$$\begin{aligned} SS^{\dagger} &= \Omega_-^{\dagger}\Omega_+\Omega_+^{\dagger}\Omega_- \\ &= \Omega_-^{\dagger}\Omega_- \\ &= I \end{aligned} \tag{4.31}$$

where the condition for the validity of Eq. (4.26) is met because only continuous state vectors of H are present in this case. Similarly, it can be shown that $S^{\dagger}S$=I.

The scattering equation

$$\Psi = \Omega_+\Phi_{in} = \Omega_-\Phi_{out} \tag{4.32}$$

dictates that a given ***incoming*** asymptote Φ_{in} determines a unique ***outgoing*** asymptote Φ_{out} and vice versa. We can therefore express the ***outgoing*** asymptote Φ_{out} in terms of the ***incoming*** asymptote Φ_{in}

$$\begin{aligned} \Phi_{out} &= \Omega_-^{\dagger}\Omega_+\Phi_{in} \\ &= S\Phi_{in} \end{aligned} \tag{4.33}$$

Eq. (4.33) gives a clear physical meaning to the S operator: the S operator transforms an *incoming* asymptote Φ_{in} in the infinite past to an *outgoing* asymptote Φ_{out} in the infinite future. We can also rewrite Eq. (4.33) in terms of the interaction picture wavefunction,

$$\boxed{\Psi_I(\infty) = S\Psi_I(-\infty)} \tag{4.34}$$

in which the S matrix operator transforms the interaction picture wavefunction from the infinite past to the infinite future.

4.2 Time-Independent Scattering Theory

4.2.1 Green's Function

So far the time-dependent definition of the scattering theory involves only $\mathcal{L}^2$ integrable wavefunctions Ψ and Φ that are not energy eigenfunctions of the Hamiltonian H. We now define the ***stationary*** wavefunctions $\psi^\pm$ that are energy eigenfunctions of H. Since the energy of the scattering system is continuous, the stationary wavefunction $\psi^+(E)$ is not $\mathcal{L}^2$ integrable and can be obtained by the following operation

$$|\psi^\pm(E)\rangle = \delta(E - H)|\Psi\rangle \tag{4.35}$$

where the + or − sign denotes whether the state ψ is evolved from a given state in the infinite past or in the infinite future. It is easy to see that the stationary wavefunction $\psi^+(E)$ so defined is an eigenfunction of the Hamiltonian H and satisfies the time-independent Schrödinger equation

$$(E - H)|\psi^\pm\rangle = 0 \tag{4.36}$$

and has the normalization

$$\langle\psi^\pm(E)|\psi^\pm(E')\rangle = \delta(E - E')\ \langle\Psi|\delta(E - H)|\Psi\rangle \tag{4.37}$$

From Eqs. (4.16) and (4.19), we can obtain a formal equation for the *incoming* stationary wavefunction $\psi^+(E)$ for a given free *incoming* state ϕ_{in} by applying a Dirac delta function operator on both sides of Eq. (4.16)

$$|\psi^+(E)\rangle = \Omega_+|\phi_{in}(E)\rangle \tag{4.38}$$

where $\phi_{in}(E)$ is an eigenfunction of H_0 defined by

$$|\phi_{in}(E)\rangle = \delta(E - H_0)|\Phi_{in}\rangle \tag{4.39}$$

Similarly for a given *outgoing* free state ϕ_{out}, we can define an *outgoing* stationary wavefunction $\psi^-(E)$ by the equation

$$|\psi^-(E)\rangle = \Omega_-|\phi_{out}(E)\rangle \tag{4.40}$$

where $\phi_{out}(E)$ is given by

$$|\phi_{out}(E)\rangle = \delta(E - H_0)|\Phi_{out}\rangle \tag{4.41}$$

It is also apparent from the definitions of (4.35), (4.39) and (4.41) that $\psi^\pm(E)$ satisfies the stationary Schrödinger equation

$$(E - H)|\psi^\pm(E)\rangle = 0 \tag{4.42}$$

as does

$$(E - H_0)|\phi(E)\rangle = 0 \tag{4.43}$$

for both *in-* and *outgoing* free states where the subscript on ϕ has been dropped. Thus, Eq. (4.38) can be formally written in the form

$$|\psi^+(E)\rangle = \Omega_+|\phi\rangle = \lim_{t\to-\infty} e^{-\frac{i}{\hbar}(E-H)t}|\phi(E)\rangle \tag{4.44}$$

A similar equation holds for the *outgoing* stationary wavefunction $|\psi^-(E)\rangle$.

It should be noted here that the use of both ψ^+ and ψ^- can sometimes cause confusion. The reason we use both of them is that the scattering wavefunction is continuous and ψ^+ and ψ^- are simply two independent solutions to the Schrödinger equation, very much like $\exp(ikx)$ and $\exp(-ikx)$ in one dimension. Thus we can simply view ψ^+ and ψ^- as some sort of generalized complex conjugate of each other.

In order to derive the familiar time-independent scattering equations from time-dependent definitions, we need to use a very useful mathematical relation. For any function $F(t)$ which is differentiable with respect to t, we can always express it as the integral over its own derivative

$$F(\infty) = F(0) + \int_0^\infty \frac{dF}{dt} dt \tag{4.45}$$

Utilizing this general mathematical relation, We can express the Mϕller operator as

$$\begin{aligned} \Omega_+ &= U_I(0, t = -\infty) \\ &= 1 + \frac{i}{\hbar} \lim_{\epsilon\to 0} \int_0^{-\infty} dt e^{\epsilon t} e^{\frac{i}{\hbar}Ht} V e^{-\frac{i}{\hbar}H_0 t} \end{aligned} \tag{4.46}$$

where a quantity $\lim_{\epsilon\to 0}\exp(\epsilon t)$ is inserted in the integrand in order to damp out the integral at $t = -\infty$ based on physical considerations. Thus we can rewrite the scattering equation as

$$\begin{aligned}\Omega_+|\phi(E)\rangle &= 1 + \frac{i}{\hbar}\lim_{\epsilon\to 0}\int_0^{-\infty} dt e^{\epsilon t} e^{\frac{i}{\hbar}Ht} V e^{\epsilon t} e^{-\frac{i}{\hbar}Et}|\phi(E)\rangle \\ &= [1 + G^+(E)V]|\phi(E)\rangle \end{aligned} \tag{4.47}$$

The operator

$$\boxed{G^{\pm}(E) = \lim_{\epsilon\to 0}(E - H \pm i\epsilon)^{-1}} \tag{4.48}$$

is the definition of the *full* Green's function (operator) which satisfies the equation

$$(E - H)G^{\pm}(E) = I \mp \lim_{\epsilon\to 0} i(E - H \pm i\epsilon)^{-1} \tag{4.49}$$

Since when operating on any L^2 quantity, the second term on the right side vanishes, we can simply drop the second term in Eq. (4.49) to obtain the standard equation for the Green's function

$$(E - H)G^{\pm}(E) = I \tag{4.50}$$

with the implicit assumption of operating on an L^2 quantity.

A similar definition of the Green's function can be made for the asymptotic Hamiltonian H_0

$$G_0^{\pm}(E) = \lim_{\epsilon\to 0}(E - H_0 + \pm i\epsilon)^{-1} \tag{4.51}$$

The limit $\lim_{\epsilon\to 0}$ is often assumed implicitly in the definition of the Green's function rather than given explicitly in the equation for purpose of simplifying the notation.

4.2.2 Lippmann-Schwinger Equation

By substituting Eq. (4.47) in Eq. (4.44), we obtain the Lippmann-Schwinger (LS) equation

$$\boxed{|\psi^+(E)\rangle = |\phi(E)\rangle + G^+(E)V|\phi(E)\rangle} \tag{4.52}$$

If we follow the same procedure but start from Eq. (4.38), we can obtain the LS equation for $\psi^-(E)$

$$|\psi^-(E)> = |\phi(E)> + G^-(E)V|\phi(E)> \tag{4.53}$$

The + or − notation on the scattering wavefunction $\psi^\pm$ is used to distinguish different asymptotic conditions of the scattering. Using the time-dependent definition in section (4.1), we state that ψ^+ determines the output of the scattering process for a given specific input wavefunction while the ψ^- determines the input for a given specific output wavefunction. In scattering applications, ψ^+ is more commonly used than ψ^- because one often wants to know the outcome from a given input state in a scattering process.

If we multiply the Green's function in eq. (4.48) by $(E - H_0 \pm i\epsilon)$ from the left, we obtain the LS equation for the Green's function

$$\boxed{G^\pm = G_0^\pm + G_0^\pm V G^\pm} \tag{4.54}$$

If the multiplication is done from the right side of eq. (4.48) we obtain

$$G^\pm = G_0^\pm + G^\pm V G_0^\pm \tag{4.55}$$

where the energy dependence of the Green's function is implicit. Using the LS equation for the Green's function, the LS Eq. (4.52) for the wavefunction can be written in another familiar form. By multiplying $G_0^+ V$ to the left side of Eq. (4.52) and utilizing Eq. (4.54), we obtain

$$\begin{aligned} G_0^+ V \psi^+ &= G_0^+ V\phi + G_0^+ V G^+ V \phi \\ &= G_0^+ V\phi + (G^+ - G_0^+) V\phi \\ &= G^+ V\phi \\ &= \psi^+ - \phi \end{aligned} \tag{4.56}$$

and therefore we obtain another form of the LS equation

$$\boxed{\psi^+ = \phi + G_0^+ V \psi^+} \tag{4.57}$$

4.2.3 The S Matrix

Using the definition (4.27) for the S matrix operator, the matrix element of the S operator can be written as

$$\begin{aligned} S_{fi}(E', E) &= <\phi_f(E')|S|\phi_i(E)> \\ &= <\Omega_-\phi_f|\Omega_+\phi_i> \\ &= <\psi_f^-(E')|\psi_i^+(E)> \end{aligned} \tag{4.58}$$

The LS equation for $\psi_f^-(E')$ can then be employed in Eq. (4.58) to yield the S matrix element

$$\begin{aligned} S_{fi}(E',E) &= <\phi_f(E')|1+VG^+(E')|\psi_i^+(E)> \\ &= <\phi_f(E')|\psi_i^+(E)> + <\phi_f(E')|V|\psi_i^+(E)> \\ &\quad \times (E'-E+i\epsilon)^{-1} \end{aligned} \tag{4.59}$$

With the application of the LS equation again for $\psi_i^+(E)$, we obtain

$$\begin{aligned} S_{fi}(E',E) &= <\phi_f(E')|\phi_i(E)> + <\phi_f(E')|G_0^+(E)V|\psi_i^+(E)> \\ &\quad + <\phi_f(E')|V|\psi_i^+(E)>\ (E'-E+i\epsilon)^{-1} \\ &= <\phi_f(E')|\phi_i(E)> + \left[(E-E'+i\epsilon)^{-1} + (E'-E+i\epsilon)^{-1}\right] \\ &\quad \times <\phi_f(E')|V|\psi_i^+(E)> \\ &= \delta(E-E')\delta_{fi} - 2\pi i\delta(E-E')\ <\phi_f(E)|V|\psi_i^+(E)> \end{aligned} \tag{4.60}$$

So we finally arrive at an expression for the S matrix element

$$\boxed{S_{fi}(E',E) = \delta(E-E')\left[\delta_{fi} - 2\pi i <\phi_f(E)|V|\psi_i^+(E)>\right]} \tag{4.61}$$

Using similar derivations leading to Eq. (4.61), we can also prove the normalization relation

$$\begin{aligned} <\psi_f^+(E')|\psi_i^+(E)> &= <\psi_f^-(E')|\psi_i^-(E)> \\ &= <\phi_f(E')|\phi_i(E)> \\ &= \delta_{fi}\delta(E-E') \end{aligned} \tag{4.62}$$

assuming the free functions $\phi_i(E)$ are so normalized. Of course, Eq. (4.62) can also be trivially proved based on Eq. (4.38) by utilizing the quasi-unitary property of the Mϕller operator.

The T and K operators

We now define a T operator by

$$T|\phi_i(E)>= V|\psi_i^+(E)> \tag{4.63}$$

With the help of the LS equation for $\psi_i^+(E)$, it is easy to show that the T operator obeys the LS equation

$$\boxed{T = V + VG^+V} \tag{4.64}$$

which can also be written as

$$T = V + VG_0^+ T \tag{4.65}$$

Using the definition of the T operator, the S matrix can be cast into the form

$$\boxed{S = 1 - 2\pi i \delta(E - H_0) T} \tag{4.66}$$

From Eq. (4.64), we can derive the following equation

$$\begin{aligned} T - T^\dagger &= V(G^+ - G_0^+)V \\ &= -i2\pi V \delta(E - H) V \end{aligned} \tag{4.67}$$

Taking the matrix element of the above equation, we obtain

$$T_{ii} - T_{ii}^* = -i2\pi < \phi_i | V | \sum_f |\psi_f^+><\psi_f^+ | \phi_i> \tag{4.68}$$

or

$$\text{Im}[T_{ii}] = -\pi \sum_f |T_{fi}|^2 \tag{4.69}$$

which is the so called *optical theorem.*

Another useful operator is the K operator which is hermitian and defined as

$$\begin{aligned} K &= V + VG^p V \\ &= V + VG_0^p K \end{aligned} \tag{4.70}$$

where G^p is the principal value Green's function defined as

$$G^p = \frac{P}{E - H} = G^+ + i\pi\delta(E - H) \tag{4.71}$$

From the definitions of the T and K operators, we can derive the relation between them

$$\begin{aligned} T - K &= VG_0^+(T - K) - i\pi V \delta(E - H_0) \\ &= -(1 - VG_0^+)^{-1} i\pi V \delta(E - H_0) K \\ &= -i\pi T \delta(E - H_0) K \end{aligned} \tag{4.72}$$

Eq. (4.72) can be rewritten to express the T operator in terms of the K operator

$$T = K\left[1 + i\pi\delta(E - H_0)K\right]^{-1} \tag{4.73}$$

Using the above equation for the T operator, the S matrix operator can be expressed in terms of the K operator by

$$\begin{aligned} S &= 1 - 2\pi i\delta(E - H_0)T \\ &= 1 - 2\pi i\delta(E - H_0)K\left[1 + i\pi\delta(E - H_0)K\right]^{-1} \\ &= \left[1 - i\pi\delta(E - H_0)K\right]\left[1 + i\pi\delta(E - H_0)K\right]^{-1} \end{aligned} \tag{4.74}$$

Since K is a hermitian operator, the above equation shows explicitly the unitarity of the S matrix operator.

On-shell S matrix

It is useful to define an *on-shell* S matrix $S_{fi}(E)$ by

$$<\phi_f(E')|S|\phi_i(E)> = \delta(E - E')S_{fi}(E) \tag{4.75}$$

It is easy to show that the on-shell S matrix $S_{fi}(E)$ is a unitary matrix. Starting from the unitarity of the S operator

$$\begin{aligned} <\phi_f(E_f)|S^\dagger S|\phi_i(E_i)> &= <\phi_f(E_f)|\phi_i(E)> \\ &= \delta_{fi}\delta(E_f - E_i) \end{aligned} \tag{4.76}$$

the left side of the above equation can also be written as

$$<\phi_f(E_f)|S^\dagger S|\phi_i(E_i)> = \delta(E_f - E_i)\sum_k S^\dagger_{fk}(E_i)S_{ki}(E_i) \tag{4.77}$$

where the completeness relation $\sum\!\!\!\!\!\!\int dE_k|\phi_k(E_k)><\phi_k(E_k)| = I$ is used in Eq. (4.77). Comparing Eq. (4.76) with Eq. (4.77), one obtains the unitarity condition for the *on-shell* S matrix

$$\boxed{\mathbf{S}^\dagger(E)\mathbf{S} = \mathbf{I}} \tag{4.78}$$

Assuming the asymptotic states are normalized as

$$<\phi_f(E')|\phi_i(E)> = \delta_{fi}\delta(E - E') \tag{4.79}$$

then from Eq. (4.66), the on-shell S matrix can be written as

$$\boxed{S_{fi}(E) = \delta_{fi} - 2\pi i T_{fi}} \tag{4.80}$$

where

$$T_{fi} = <\phi_f|T|\phi_i> \tag{4.81}$$

We can also express the on-shell S matrix in terms of the K matrix

$$\boxed{S(E) = (1 - i\pi K)(1 + i\pi K)^{-1}} \tag{4.82}$$

where the K matrix is given by

$$T_{fi} = <\phi_f|K|\phi_i> \tag{4.83}$$

4.2.4 Distorted Wave

The LS equation (4.57) can often be used iteratively to obtain an approximate solution to ψ^+, provided that the iteration converges. A straightforward approximation is to simply replace ψ^+ on the right side of eq. (4.57) by ϕ to obtain the Born approximation

$$\psi^+ = \phi + G_0^+ V\phi \tag{4.84}$$

The Born approximation is often a good approximation for high energy scattering. This is because at high collision energy, the kinetic energy is much larger than the potential energy. Therefore the change of scattering wavefunction due to interaction potential should be relatively small and the replacement of ψ^+ by ϕ on the right side of eq. (4.57) is reasonable.

In practice, however, it is often desirable to include some potential interaction or distortion in the reference Hamiltonian H_0. For this purpose, it is often desirable to partition the full Hamiltonian as

$$\begin{aligned} H &= H_0 + V \\ &= H_0 + V_0 + V - V_0 \\ &= H_0' + V' \end{aligned} \tag{4.85}$$

where H_0' is the free Hamiltonian H_0 plus a distortion V_0 and V' is the residual interaction $V - V_0$. From the definition of the Mϕller operator, we can rewrite Eq. (4.16) as

$$\begin{aligned}\Psi^+ &= \lim_{t\to-\infty} e^{\frac{i}{\hbar}Ht} e^{-\frac{i}{\hbar}H_0 t}\Phi_{in} \\ &= \lim_{t\to-\infty} e^{\frac{i}{\hbar}Ht} e^{-\frac{i}{\hbar}H_0' t} e^{\frac{i}{\hbar}H_0' t} e^{-\frac{i}{\hbar}H_0 t}\Phi_{in} \\ &= \Omega'_+ \chi^+ \end{aligned} \tag{4.86}$$

where the distorted state χ^+ is defined as

$$\chi^+ = \Omega'_{0+}\Phi_{in} \tag{4.87}$$

Here the Mϕller operators are defined as

$$\Omega'_+ = \lim_{t\to-\infty} e^{\frac{i}{\hbar}Ht} e^{-\frac{i}{\hbar}H_0' t} \tag{4.88}$$

and

$$\Omega'_{0+} = \lim_{t\to-\infty} e^{\frac{i}{\hbar}H_0' t} e^{-\frac{i}{\hbar}H_0 t} \tag{4.89}$$

From Eq. (4.87) and the derivation of the Lippmann-Schwinger equation for the stationary state in section (3.2.2), we can straightforwardly obtain the corresponding LS equation for the distorted state $\chi^+ = \delta(E - H_0')\chi^+$

$$\chi^+(E) = \phi(E) + G_0^+(E)V_0\chi^+(E) \tag{4.90}$$

where the free Green's function is defined by

$$G_0^+(E) = (E - H_0 + i\epsilon)^{-1} \tag{4.91}$$

and $\phi(E) = \delta(E - H_0)\Phi_{in}$. Similarly from Eq. (4.86), we can obtain the LS equation for the full stationary state $\psi^+(E) = \delta(E - H)\Psi^+$

$$\begin{aligned}\psi^+(E) &= \chi^+(E) + G_0'^+(E)V'\psi^+(E) \\ &= \chi^+(E) + G^+(E)V'\chi^+(E)\end{aligned} \tag{4.92}$$

where the distorted wave Green's function is defined as

$$G_0'^+(E) = (E - H_0' + i\epsilon)^{-1} \tag{4.93}$$

From the normalization conservation in Eq. (4.62), we have the general relation for conservation of normalization

$$\begin{aligned}<\psi_f^+(E')|\psi_i^+(E)> &= <\chi_f^+(E')|\chi_i^+(E)> \\ &= <\phi_f(E')|\phi_i(E)>\end{aligned} \tag{4.94}$$

Using the distorted wave, we can express the full T matrix as

$$\begin{aligned} T_{fi} &= <\phi_f|V|\psi_i^+> \\ &= <\phi_f|V_0|\psi_i^+> + <\phi_f|V'|\psi_i^+> \end{aligned} \tag{4.95}$$

By utilizing the LS equation for ψ_i^+ and χ_f^-, the above equation can be rewritten as

$$\begin{aligned} T_{fi} &= <\phi_f|V_0|\chi_i^+> + <\phi_f|V_0 G_0'^+ V'|\psi_i^+> + <\phi_f|V'|\psi_i^+> \\ &= <\phi_f|V_0|\chi_i^+> + <\phi_f|(1 + V_0 G_0'^+)V'|\psi_i^+> \\ &= <\phi_f|V_0|\chi_i^+> + <\chi_f^-|V'|\psi_i^+> \end{aligned} \tag{4.96}$$

We now identify the first term as the T^0 matrix due to scattering by the potential V_0 and the second term the T' matrix due to scattering by the potential V'. Thus the full T matrix is simply the sum of two T matrices

$$\boxed{T_{fi} = T_{fi}^0 + T_{fi}'} \tag{4.97}$$

where

$$T_{fi}^0 = <\phi_f|V_0|\chi_i^+> \tag{4.98}$$

and

$$T_{fi}' = <\chi_f^-|V'|\psi_i^+> \tag{4.99}$$

Equation (4.97) is very useful because it states that the full scattering T matrix is a simple sum of separate T matrices resulting from scattering by V_0 and V', respectively. For example, if V_0 is chosen to include only elastic scattering, then the inelastic T matrix can be obtained from the calculation of scattering caused by V' alone.

Distorted wave Born approximation

If one can partition the Hamiltonian such that the distorted scattering state χ^+ is much easier to solve than the full scattering state ϕ^+ and the residual interaction V' is small, it is then possible to treat V' as a small perturbation. In such a case, the LS equation of (4.92) can be approximated to yield the distorted wave Born approximation

$$\psi^+(E) \approx \chi^+(E) + G_0'^+(E)V'\chi^+(E) \tag{4.100}$$

or in terms of the T matrix

$$T_{fi} \approx T^0_{fi} + <\chi^-_f|V'|\chi^+_i> + <\chi^-_f|V'G'^+_0(E)V'|\chi^+_i> \tag{4.101}$$

which is often further simplified to yield

$$\boxed{T_{fi} \approx T^0_{fi} + <\chi^-_f|V'|\chi^+_i>} \tag{4.102}$$

The distorted wave Born approximation is more general and useful than the standard Born approximation which simply approximates the full wavefunction by the free (noninteracting) wavefunction. In many situations, the full potential interaction V is quite large and the standard Born approximation cannot be applied. However, it may be possible to partition the Hamiltonian such that the residual interaction V' is small. The distorted wavefunction is also very useful in practical calculations to help speed up the convergence. For example, if a scattering process is dominated by strong elastic scattering but relatively weak inelastic scattering, one could solve the elastic scattering state $\chi^+(E)$ exactly and use Eq. (4.102) to solve the inelastic scattering problem approximately.

4.3 Elastic Scattering

4.3.1 Radial Schrödinger Equation

Elastic scattering is generally represented by the scattering of a structureless particle from a central potential $V(R)$. The Hamiltonian for a particle in a central potential is given by

$$H = -\frac{\hbar^2}{2m}\frac{1}{r}\frac{\partial^2}{\partial r^2}r + \frac{\mathbf{L}^2}{2mr^2} + V(r) \tag{4.103}$$

where m is the mass of the particle and $\mathbf{L}$ is the orbital angular momentum operator. The stationary wavefunction for elastic scattering is usually represented by a partial wave expansion

$$\Psi^+ = \sum_{l=0}^{\infty} P_l(\cos\theta)\frac{\psi_l^+(r)}{r} \tag{4.104}$$

where $P_l(\cos\theta)$ are Legendre polynomials. The radial function ψ_l^+ for a given energy E can be shown to satisfy the radial Schrödinger equation

$$\frac{d^2\psi_l^+}{dr^2} + \left[k^2 - \frac{l(l+1)}{r^2} - \frac{2m}{\hbar^2}V(r)\right]\psi_l^+ = 0 \tag{4.105}$$

where the momentum k is defined by

$$k^2 = \frac{2m}{\hbar^2} E \tag{4.106}$$

One can directly solve the differential equation (4.105) by numerical methods and match the solution to the asymptotic boundary condition to extract the scattering information (phase shift). The detailed properties of the radial Eq. (4.105) are discussed in the following sections.

4.3.2 Free Radial Functions

Equation (4.105) can be written in the standard form of

$$(E - H_{0l})\psi_l^+ = V\psi_l^+ \tag{4.107}$$

where the radial Hamiltonian is defined as

$$H_{0l} = -\frac{\hbar^2}{2m}\frac{d^2}{dr^2} + \frac{\hbar^2}{2m}\frac{l(l+1)}{r^2} \tag{4.108}$$

Thus the LS equation for Eq. (4.107) is

$$\psi_l^+ = j_l + g_{0l}^+ V \psi_l^+ \tag{4.109}$$

where $j_l(r)$ is the Ricatti-Bessel function satisfying the free radial equation

$$\left[\frac{d^2}{dr^2} - \frac{l(l+1)}{r^2} + k^2\right] j_l(kr) = 0 \tag{4.110}$$

There are two linearly independent solutions to the second order differential equation (4.110). The Ricatti-Bessel function $j_l(z)$ is the regular solution which vanishes as $r \to 0$ and has the asymptotic expansion

$$j_l(kr) \overset{r\to\infty}{\longrightarrow} \sin(kr - \frac{l\pi}{2}) \tag{4.111}$$

The presence of a phase shift $\pi/2$ in Eq. (4.111) is the result of elastic scattering of a free particle by the centrifugal potential. In fact, we could treat the centrifugal potential as a scattering potential and solve the LS equation for the scattering state $j_l(kr)$. Thus, we could think of $j_l(kr)$ as a distorted state χ_l which solves the problem of centrifugal scattering exactly as is discussed in Sec. 4.2.4. Since in any practical three-dimensional scattering problem, the centrifugal potential is almost always present, it is thus desirable to solve the LS equation using the distorted function $j_l(kr)$ rather than the free sine function.

Normalization of radial functions

Using the normalization of the Ricatti-Bessel function from the appendix

$$\begin{aligned} <j_l(k)|j_l(k')> &= \int_0^\infty j_l(kr) j_l(k'r) dr \\ &= \frac{\pi}{2}\delta(k-k') \end{aligned} \tag{4.112}$$

we can define the energy-normalized free radial function

$$\boxed{\bar{\jmath}_l(E) = \sqrt{\frac{2m}{\pi\hbar^2 k}}\, j_l(kr)} \tag{4.113}$$

which normalizes to a δ-function in energy

$$<\bar{\jmath}_l(E)|\bar{\jmath}_l(E')> = \delta(E-E') \tag{4.114}$$

For the special case of $l = 0$, the Ricatti-Bessel function reduces to a sine function

$$\bar{\jmath}_0(E) = \sqrt{\frac{2m}{\pi\hbar^2 k^2}} \sin(kr) \tag{4.115}$$

Due to the quasi-unitary property of the Mϕller operator, the radial wavefunction given in Eq. (4.109) has the same normalization as j_l, i.e.,

$$\begin{aligned} <\psi_l^+(k)|\psi_l^+(k')> &= <j_l(k)|\Omega_+^\dagger \Omega_+|j_l(k')> \\ &= <j_l(k)|j_l(k')> \end{aligned} \tag{4.116}$$

4.3.3 Radial Green's Function

The radial Green's function satisfies the equation

$$(E - H_{0l}) g_{0l}^+ = I \tag{4.117}$$

whose coordinate representation is given by

$$\left[\frac{d^2}{dr^2} - \frac{l(l+1)}{r^2} + k^2\right] g_{0l}^+(r|r') = \frac{2m}{\hbar^2}\delta(r-r') \tag{4.118}$$

It is straightforward to see that the coordinate representation of the Green's function $g_{0l}^+(r|r') = <r|g_{0l}^+|r'>$ is symmetric, i.e., $g_{0l}^+(r|r') = g_{0l}^+(r'|r)$. Equation (4.118) can be solved by separating the solution into regions of $r < r'$

and $r > r'$. With the regular boundary condition at the origin and the asymptotic outgoing boundary condition, we can write the solution of Eq. (4.118) as

$$g_{0l}^{+}(r|r') = \begin{cases} j_l(kr)A(r') & (r < r') \\ h_l^{+}(kr)B(r') & (r > r') \end{cases} \tag{4.119}$$

where h_l^+ is the Ricatti-Hankel function with the outgoing spherical wave. The $h_l^\pm$ is defined as

$$h_l^{\pm} = n_l \pm i j_l \tag{4.120}$$

where n_l is the Ricatti-Neumann function with the cosine asymptotic condition. Since the Ricatti-Neumann function behaves asymptotically as

$$n_l(kr) \overset{r\to\infty}{\longrightarrow} \cos(kr - \frac{l\pi}{2}), \tag{4.121}$$

the Ricatti-Hankel function has the asymptotic expansion

$$h_l^{\pm}(kr) \overset{r\to\infty}{\longrightarrow} \exp\left(kr - \frac{l\pi}{2}\right) \tag{4.122}$$

At the boundary of $r = r'$, the two forms of the Green's function in Eq. (4.119) are matched by the continuity conditions

$$\begin{cases} j_l(kr')A(r') = h_l^{+}(kr')B(r') \\ h_l^{+'}(kr')B(r') - j_l'(kr')A(r') = \dfrac{2m}{\hbar^2} \end{cases} \tag{4.123}$$

We then find

$$\begin{cases} A(r') = -\dfrac{2m}{\hbar^2}W^{-1}h_l^{+}(kr') = -\dfrac{2m}{\hbar^2 k}h_l^{+}(kr') \\ B(r') = -\dfrac{2m}{\hbar^2}W^{-1}j_l(kr') = -\dfrac{2m}{\hbar^2 k}j_l(kr') \end{cases} \tag{4.124}$$

where the Wronskian can be evaluated to be

$$W = j_l'(kr)h_l^{+}(kr) - j_l(kr)h_l^{+'}(kr) = k \tag{4.125}$$

Thus the radial Green's function can be written as

$$\boxed{g_{0l}^{+}(r|r') = -\frac{2m}{\hbar^2 k}j_l(kr_<)h_l^{+}(kr_>)} \tag{4.126}$$

where $r_<$ and $r_>$ denote, respectively, the smaller and larger of r and r'.

It is often useful to define the principal value Green's function g_{0l}^p by the relation

$$g_{0l}^+ = g_{0l}^p - i\pi\delta(E - H_{0l}) \tag{4.127}$$

By rewriting Eq. (4.126) as

$$g_{0l}^+(r|r') = -\frac{2m}{\hbar^2 k} j_l(kr_<) n_l(kr_>) - i\frac{2m}{\hbar^2 k} j_l(kr) j_l(kr') \tag{4.128}$$

and comparing Eq. (4.127) with Eq. (4.128), we obtain the coordinate representation for the principal value Green's function

$$\boxed{g_{0l}^p(r|r') = -\frac{2m}{\hbar^2 k} j_l(kr_<) n_l(kr_>)} \tag{4.129}$$

and also the relation

$$\boxed{<r|\delta(E - H_{0l})|r'> = \frac{2m}{\pi\hbar^2 k} j_l(kr) n_l(kr')} \tag{4.130}$$

It is often advantageous to use the principal value Green's function in numerical calculations for scattering problems because it is a real function.

4.3.4 Scattering Phase Shift

The radial LS equation (4.109) can now be written in coordinate representation as

$$\begin{aligned} \psi_l^+(r) &= j_l(kr) + \int_0^\infty g_{0l}^+(r|r') V(r') \psi_l^+(r') dr' \\ &= j_l(kr) - \frac{2m}{\hbar^2 k} \int_0^\infty j_l(kr_<) h_l^+(kr_>) V(r') \psi_l^+(r') dr' \end{aligned} \tag{4.131}$$

In the limit $r \to \infty$, Eq. (4.131) becomes

$$\begin{aligned} \psi_l^+(r \to \infty) &= j_l(kr) - \frac{2m}{\hbar^2 k} h_l^+(kr) \int_0^\infty j_l(kr) V(r') \psi_l^+(r') dr' \\ &= j_l(kr) - T_l h_l^+(kr) \end{aligned} \tag{4.132}$$

where the T matrix is defined by

$$\boxed{T_l = \frac{2m}{\hbar^2 k} \int_0^\infty j_l(kr) V(r') \psi_l^+(r') dr'} \tag{4.133}$$

Equation (4.132) can also be written in terms of the incoming and outgoing waves $h_l^{\pm}$

$$\begin{aligned}\psi_l^+(r \to \infty) &= j_l(kr) - h_l^+(kr)T_l \\ &= (2i)^{-1}\left[-h_l^-(kr) + h_l^{(+)}(kr)S_l\right]\end{aligned} \tag{4.134}$$

where the S matrix S_l is given by

$$\boxed{S_l = 1 - 2iT_l} \tag{4.135}$$

We note that a factor π in Eq. (4.135) is missing from the formal relation (4.80) between the S and T matrices. This is simply due to the particular normalization of the wavefunction used to calculate T_l in Eq. (4.133). If we use the energy normalized free function of Eq. (4.113), Eq. (4.133) can be rewritten as

$$T_l = \pi \int_0^\infty \bar{\jmath}_l(kr)V(r')\sqrt{\frac{2m}{\pi\hbar^2 k}}\psi_l^+(r')dr' = \pi <\bar{\jmath}_l|V|\bar{\psi}_l^+> \tag{4.136}$$

where both $\bar{\jmath}_l$ and $\bar{\psi}_l^+$ are energy normalized, viz.,

$$\begin{aligned}<\bar{\psi}_l^+(E)|\bar{\psi}_l^+(E')> &= <\bar{\jmath}_l(E)|\bar{\jmath}_l(E')> \\ &= \delta(E - E')\end{aligned} \tag{4.137}$$

Thus, Eq. (4.135) becomes

$$S_l = 1 - 2i\pi <\bar{\jmath}_l|V|\bar{\psi}_l^+> \tag{4.138}$$

which is exactly the standard definition of (4.80).

Since ψ_l^+ is a regular solution of the radial equation, it must be a real function apart from a complex phase factor. Asymptotically we can express ψ_l^+ as a linear combination of two independent free radial functions

$$\begin{aligned}\psi_l^+(r \to \infty) &= a_l j_l(kr) + b_l n_l(kr) \\ &= A_l \sin\left(kr - \frac{l\pi}{2} + \delta_l\right)\end{aligned} \tag{4.139}$$

where δ_l is called the ***phase shift*** resulting from the potential interaction. Obviously, if $V(r)$=0, the phase shift will be zero as well. Comparing Eq. (4.132) with (4.139), we obtain the expression for the T and S matrices

$$T_l = -\sin\delta_l e^{i\delta_l} \tag{4.140}$$

which in turn gives the S matrix via Eq. (4.135)

$$\boxed{S_l = e^{2i\delta_l}} \tag{4.141}$$

The asymptotic condition (4.134) for the radial wavefunction corresponds to the so called S matrix boundary condition. We can rewrite Eq. (4.132) or Eq. (4.134) in terms of the K matrix boundary condition

$$\psi_l^+(r \to \infty) = [j_l(kr) + n_l(kr)K_l]\,(1 - iT_l) \tag{4.142}$$

where the K matrix is related to the T or S matrices through the relations

$$\boxed{K_l = -\frac{T_l}{1 - iT_l} = i\frac{1 - S_l}{1 + S_l} = \tan\delta_l} \tag{4.143}$$

In fact we can define a real radial wavefunction

$$\psi_l(r) = \psi_l^+(r)(1 - iT_l)^{-1} \tag{4.144}$$

which, according to Eq. (4.142), will have the real asymptotic boundary condition

$$\psi_l(r \to \infty) = j_l(kr) + n_l(kr)K_l \tag{4.145}$$

Thus we can directly express the K matrix or phase shift in terms of the real integral

$$\begin{aligned} K_l &= -\frac{T_l}{1 - iT_l} \\ &= -\frac{2m}{\hbar^2 k}\frac{<j_l|V|\psi_l^+>}{1 - iT_l} \\ &= -\frac{2m}{\hbar^2 k}<j_l|V|\psi_l> \end{aligned} \tag{4.146}$$

This result can also be obtained directly if we use the principal value Green's function G_0^p instead of the outgoing wave Green's function G_0^+ in the Lippman-Schwinger equation for the real radial wavefunction

$$\psi_l = j_l + g_{0l}^p V \psi_l \tag{4.147}$$

which directly gives the K matrix in terms of the real function by definition.

The variable phase method

The variable phase method provides an attractive way to understand the change of phase shift in relation to the scattering potential directly. It also provides a means for directly calculating the phase shift without having to solve the radial Schrödinger equation [40]. For any given scattering potential $V(r)$, we can define a new potential $V_\rho(r)$, which is just the original potential V but is cut off after a fixed radial distance $r = \rho$, or explicitly,

$$V_\rho(r) = \begin{cases} V(r) & (r \leq \rho) \\ 0 & (r > \rho) \end{cases} \tag{4.148}$$

Denoting $\phi_\rho(r)$ as the s-wave radial wavefunction and $\delta(\rho)$ as the corresponding phase shift for the potential defined in Eq. (4.148), it is obvious that $\delta(\rho)$ approaches the correct phase shift of the original potential scattering δ as $\rho \to \infty$

$$\delta(\rho) \longrightarrow \delta \tag{4.149}$$

Also if ρ=0, then $V_\rho(r)$ is zero everywhere and $\delta(0)$=0.

It is possible to derive the equation for the function $\delta(\rho)$ and directly integrate $\delta(\rho)$ outward from origin to $\rho = \infty$ to yield the desired phase shift δ for the original scattering problem. We first note that in the inner region $r \leq \rho$, the radial wavefunction $\psi(r)$ for the original potential is the same as $\phi_\rho(r)$

$$\psi(r) = \phi_\rho(r) \qquad (r \leq \rho) \tag{4.150}$$

In the outer region $r > \rho$, $\phi_\rho(r)$ is a free particle function given by

$$\phi_\rho(r) = A(\rho)\sin[kr + \delta(\rho)] \qquad (r > \rho) \tag{4.151}$$

and its derivative is

$$\phi'_\rho(r) = kA(\rho)\cos[kr + \delta(\rho)] \qquad (r > \rho) \tag{4.152}$$

Since the wavefunction is continuous at $r = \rho$, we can construct the log derivative of the wavefunction at $r = \rho$

$$\begin{aligned} F(\rho) &= \frac{\psi'(\rho)}{\psi(\rho)} = \left.\frac{\phi'_\rho(r)}{\phi_\rho(r)}\right|_{r=\rho} \\ &= \frac{k\cos[kr + \delta(\rho)]}{\sin[kr + \delta(\rho)]} \end{aligned} \tag{4.153}$$

for any value of ρ. Since $\psi(\rho)$ satisfies the Schrödinger equation, it is not difficult to obtain the following nonlinear equation for the log derivative function $F(\rho)$

$$F'(\rho) + F^2(\rho) + K^2(\rho) = 0 \tag{4.154}$$

where

$$K^2(\rho) = \frac{2m}{\hbar^2}\left[E - V(\rho)\right] \tag{4.155}$$

From Eq. (4.154), one can derive a nonlinear equation for the phase shift $\delta(\rho)$

$$\boxed{\delta'(\rho) = -\frac{1}{k}V(\rho)\sin^2[k\rho + \delta(\rho)]} \tag{4.156}$$

This equation, together with the boundary condition $\delta(0) = 0$ can be integrated outward in ρ to obtain the phase shift for the original scattering potential.

Equation (4.156) shows that the phase change is negative for a repulsive potential and positive for an attractive potential. It can be shown that, if $V_1(r) \geq V_2(r)$ in the entire space, then the corresponding phase shift satisfies

$$\delta_1 \leq \delta_2 \tag{4.157}$$

Equation (4.156) also shows that in the high energy limit $k \to \infty$, the phase shift is zero.

4.3.5 Scattering Cross Section

According to Eqs. (4.104) and (4.139), the asymptotic form of the three dimensional wavefunction can now be written as

$$\Psi \overset{r\to\infty}{\longrightarrow} \sum_{l=0}^{\infty} \frac{A_l}{r} P_l(\cos\theta)\sin(kr - \frac{l\pi}{2} + \delta_l) \tag{4.158}$$

On the other hand, the physically relevant wavefunction can be expressed asymptotically as the superposition of a plane wave along the z axis plus a scattered spherical wave, i.e.,

$$\Psi \to e^{ikz} + f(\theta)\frac{e^{ikr}}{r} \tag{4.159}$$

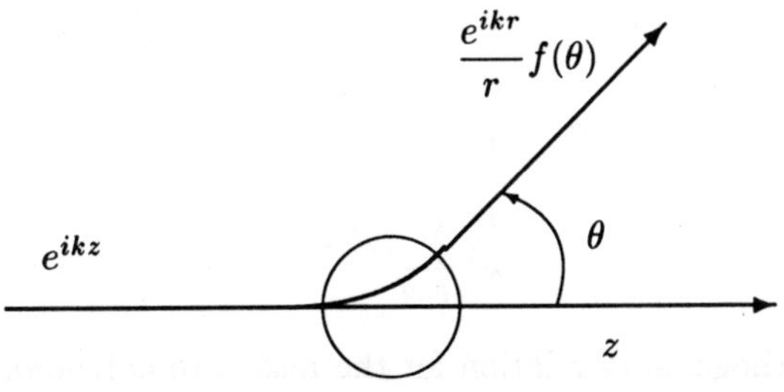

Figure 4.1: Physical scattering process.

which corresponds to the physical scattering process shown in Fig. 4.1. In Fig. 4.1, the plane wave carries an incoming flux $I_i = \hbar k/m$ and the spherical wave carries an outgoing flux $I_f = v|f(\theta)|^2$. From Eq. (A.55) in the appendix, the plane wave e^{ikz} can be expanded in terms of spherical waves

$$e^{ikz} \overset{r\to\infty}{\longrightarrow} \frac{1}{kr}\sum_{l=0}^{\infty} i^l(2l+1)P_l(\cos\theta)\sin(kr - \frac{l\pi}{2}) \tag{4.160}$$

By equating the two expressions in (4.158) and (4.159), we find

$$A_l = \frac{1}{k}(2l+1)i^l e^{i\delta_l} \tag{4.161}$$

and

$$\boxed{f(\theta) = \frac{1}{k}\sum_{l=0}^{\infty}(2l+1)T_l P_l(\cos\theta)} \tag{4.162}$$

The experimentally measurable scattering differential cross section is defined as the outgoing flux from the spherical wave of the second term in Eq. (4.159) divided by the incoming flux from the plane wave of the first term, viz.

$$d\sigma = \frac{I_f}{I_i}d\Omega = |f(\theta)|^2 d\Omega \tag{4.163}$$

where $d\Omega$ is the solid angle. Since the wavefunction is axially symmetric, we can integrate over the azimuthal angle ϕ to obtain

$$d\sigma = 2\pi\sin\theta|f(\theta)|^2 d\theta \tag{4.164}$$

for scattering angles in the range of θ to $\theta + d\theta$. The integral cross section is obtained by integrating over the angle θ

$$\begin{aligned}\sigma &= \int_0^{\pi} 2\pi \sin\theta |f(\theta)|^2 d\theta \\ &= \frac{4\pi}{k^2}\sum_{l=0}^{\infty}(2l+1)\sin^2\delta_l\end{aligned} \tag{4.165}$$

where the orthogonality relation for the Legendre polynomial

$$\begin{aligned}<P_l|P_{l'}> &= \int_0^{\pi} \sin\theta d\theta P_l(\cos\theta)P_{l'}(\cos\theta) \\ &= \frac{2}{2l+1}\delta_{ll'}\end{aligned} \tag{4.166}$$

has been used.

In the low energy limit $k \to 0$, only the $l = 0$ term (s wave scattering) contributes to the cross section,

$$\sigma \simeq \frac{4\pi}{k^2}\sin^2\delta_0 \simeq \frac{4\pi}{k^2}\delta_0^2 \tag{4.167}$$

and we can express σ in the form

$$\lim_{k\to 0}\sigma = 4\pi a^2 \tag{4.168}$$

where the quantity

$$a = -\lim_{k\to 0}\frac{\delta_0(k)}{k} \tag{4.169}$$

is called the *scattering length*.

The general approach to an elastic scattering problem is to solve the radial equation (4.105) to determine the corresponding phase shift δ_l for each partial wave l. The maximum partial wave l_{max} that contributes to the cross section is determined by the range of the interaction potential. Classically, the quantum number l is related to the impact parameter b by $l \approx kb$ where k is the wave number. Thus one can estimate l_{max} by the relation $l_{max} \approx kb_{max}$ where b_{max} is the maximum range of the potential. Thus in general, long range potentials have larger scattering cross sections than short range potentials because there are more partial waves contributing to the cross section.

Further reading

For formal discussions on quantum scattering theory, refs. [48–51] are good sources for reading on the topic.

4.4 Inelastic Scattering

4.4.1 Coupled Channel Equations

If the scattering particle is a molecule with internal structure such as rotational and vibrational states which are changed during or after the collision, we have an inelastic scattering. In that case, the interaction potential depends also on the internal coordinates as well and there can be energy transfer among different coordinates. In general we can classify the coordinates of the scattering object as internal coordinates q and the scattering or radial coordinate r. The Hamiltonian of the scattering system can be generically expressed as

$$\begin{aligned} H &= K(r) + H_{int}(q) + V(r,q) \\ &= H_0 + V \end{aligned} \tag{4.170}$$

where $K(r)$ is the kinetic energy associated with the motion of the scattering coordinate, $H_{int}(q)$ is the Hamiltonian describing the motions of all internal degrees of freedom, and $V(r,q)$ is the interaction potential that couples the motions between r and q coordinates. In particular, $V(r,q) \to 0$ as $r \to \infty$. For generality, we assume that the reference Hamiltonian H_0 also includes a distortion potential V_0.

Following similar steps in the previous section for elastic scattering, we can expand the scattering wavefunction Ψ in terms of a complete basis set $\varphi_n(q)$ for the internal degrees of freedom,

$$\Psi^+ = \sum_n \varphi_n(q)\psi_n^+(r)/r \tag{4.171}$$

The basis functions $\varphi_n(q)$ are often chosen to be eigenfunctions of H_{int} with eigenvalues ϵ_n, viz,

$$H_{int}\varphi_n = \epsilon_n\varphi_n \tag{4.172}$$

By using the expansion of (4.171) for Ψ^+ to solve the Schrödinger equation with the Hamiltonian of (4.170), we can obtain the coupled equation for the radial wavefunction

$$\frac{d^2\psi_m^+}{r^2} + \sum_n \left[k_m^2\delta_{mn} - \frac{2m}{\hbar^2}V_{mn}(r)\right]\psi_n^+ = 0 \tag{4.173}$$

or in matrix form

$$\left[\mathbf{I}\frac{d^2}{dr^2} + \mathbf{k}^2 - \frac{2m}{\hbar^2}\mathbf{V}\right]\mathbf{\Psi}^+(r) = \mathbf{0} \tag{4.174}$$

where

$$k_m^2 = \frac{2m}{\hbar^2}(E - \epsilon_m) \tag{4.175}$$

and the potential matrix element is defined as

$$V_{mn}(r) = <\varphi_m|V|\varphi_n> \tag{4.176}$$

Since one normally solves the equation for all possible initial states, we usually treat the wavefunction $\mathbf{\Psi}^+$ as a matrix in which the ith column represents the scattering wavefunction from the initial ith state. The inelastic scattering problem is typically solved by propagating the coupled radial Eq. (4.173) and matching the solution at a large radial distance r_L (where the potential V vanishes) to its asymptotic form.

4.4.2 Multichannel Green's Function

In solving the LS equation

$$\Psi^+ = \Phi + G_0^+ V \Psi^+ \tag{4.177}$$

one first needs to calculate the Green's function G_0^+ by solving the equation

$$(E - H_0)G_0 = I \tag{4.178}$$

where the superscript on the Green's function is dropped to represent a general Green's function with the desired boundary conditions. The Green's function can be expanded in channel basis functions φ_n

$$G_0 = \sum_{mn} |\varphi_m> \mathbf{G}_{0mn} <\varphi_n| \tag{4.179}$$

By substituting Eq. (4.179) in Eq. (4.178) and integrating over the channel functions φ_n, we arrive at a matrix equation for the radial Green's function $\mathbf{G}_{0mn}$

$$(\mathbf{E} - \mathbf{H}_0)\mathbf{G}_0 = \mathbf{I} \tag{4.180}$$

which is written in coordinate representation as

$$\left[\mathbf{I}\frac{d^2}{dr^2} + \mathbf{k}^2 - \frac{2m}{\hbar^2}\mathbf{V}_0\right]\mathbf{G}_0(r|r') = \frac{2m}{\hbar^2}\delta(r-r')\mathbf{I} \tag{4.181}$$

where the diagonal matrix $\mathbf{k}^2$ is given by Eq. (4.175)

The procedure used to solve the multi-channel Green's function $\mathbf{G}_0$ is essentially similar to that described in solving the single-channel Green's function g_{0l} in Sec. 4.3.3 except now the wavefunctions are given in matrix form. Thus, we can generalize the single-channel result (4.119) to

$$\mathbf{G}_0(r|r') = \begin{cases} \mathbf{J}(kr)\mathbf{A}(r') & (r < r') \\ \mathbf{N}(kr)\mathbf{B}(r') & (r > r') \end{cases} \tag{4.182}$$

where $\mathbf{J}$ and $\mathbf{N}$ are, respectively, the regular and irregular solutions of the equation

$$\left[\mathbf{I}\frac{d^2}{dr^2} + \mathbf{k}^2 - \frac{2m}{\hbar^2}\mathbf{V}_0\right]\left\{\begin{array}{c}\mathbf{J}(r)\\ \mathbf{N}(r)\end{array}\right\} = 0 \tag{4.183}$$

Since the irregular solution $\mathbf{N}(r)$ is not unique as discussed previously for the single channel case, we need to choose a particular solution whose asymptotic boundary condition matches the type of Green's function we need. For example, if we set $\mathbf{V}_0$ to be diagonal and it contains only the centrifugal potential in its diagonal elements

$$\mathbf{V}_0 = \frac{l_i(l_i+1)\hbar^2}{2mr^2}\delta_{ij}, \tag{4.184}$$

then $\mathbf{J}(r)$ is diagonal and is simply the Ricatti-Bessel function

$$\begin{aligned}\mathbf{J}_{ii}(r) &= j_{l_i}(k_i r) \\ &\xrightarrow{r\to\infty} \frac{\sin(k_i r - l_i\pi/2)}{\sqrt{k_i}}\end{aligned} \tag{4.185}$$

and $\mathbf{N}(r)$ is the corresponding Ricatti-Neumann function

$$\begin{aligned}\mathbf{N}_{ii}(r) &= n_{l_i}(k_i r) \\ &\xrightarrow{r\to\infty} \frac{\cos(k_i r - l_i\pi/2)}{\sqrt{k_i}}\end{aligned} \tag{4.186}$$

The coefficient matrices $\mathbf{A}$ and $\mathbf{B}$ in Eq. (4.182) are determined by the continuity condition at $r = r'$

$$\begin{cases} \mathbf{J}(r')\mathbf{A}(r') = \mathbf{N}(r')\mathbf{B}(r') \\ \mathbf{N}'(r')\mathbf{B}(r') - \mathbf{J}'(r')\mathbf{A}(r') = \dfrac{2m}{\hbar^2}\mathbf{I} \end{cases} \tag{4.187}$$

which gives solutions for the coefficient matrices

$$\begin{cases} \mathbf{A}(r') = -\frac{2m}{\hbar^2}\mathbf{W}^{-1}\mathbf{N}^T(r') \\ \mathbf{B}(r') = -\frac{2m}{\hbar^2}\mathbf{W}^{-1}\mathbf{J}^T(r') \end{cases} \tag{4.188}$$

where the Wronskian is defined as

$$\mathbf{W} = \mathbf{N}^T(r)\mathbf{J}'(r) - \mathbf{N}'^T(r)\mathbf{J}(r). \tag{4.189}$$

We can thus express the multi-channel Green's function as

$$\mathbf{G}_0(r|r') = \begin{cases} -\frac{2m}{\hbar^2}\mathbf{J}(kr)\mathbf{W}^{-1}\mathbf{N}^T(kr') & (r < r') \\ -\frac{2m}{\hbar^2}\mathbf{N}(kr)\mathbf{W}^{-1}\mathbf{J}^T(kr') & (r > r') \end{cases} \tag{4.190}$$

Using Eq. (4.190), it is straightforward to show that the Green's function is complex symmetric, viz.,

$$[\mathbf{G}_0]_{mn}(r|r') = [\mathbf{G}_0]_{nm}(r'|r) \tag{4.191}$$

This is a general property of a *complex symmetric* operator as discussed more generally in Sec. 8.2.1.

4.4.3 S, T and K Matrices

The multi-channel LS equation for the radial wavefunction is given in matrix form

$$\mathbf{\Psi}^+ = \mathbf{J} + \mathbf{G}_0^+\mathbf{V}\mathbf{\Psi}^+ \tag{4.192}$$

or in coordinate representation

$$\mathbf{\Psi}^+(r) = \mathbf{J}(r) + \int_0^\infty \mathbf{G}_0^+(r|r')\mathbf{V}(r')\mathbf{\Psi}^+(r')dr' \tag{4.193}$$

In the limit $r \to \infty$, we can utilize Eq. (4.190) for the Green's function to obtain the asymptotic form of the inelastic radial wavefunction

$$\begin{aligned} \mathbf{\Psi}^+(r \to \infty) &= \mathbf{J}(r) - \frac{2m}{\hbar^2}\int_0^\infty \mathbf{N}^+(r)\mathbf{J}^T(r')V(r')\mathbf{\Psi}^+(r')dr' \\ &= \mathbf{J}(r) - \mathbf{N}^+(r)\mathbf{T} \\ &= \frac{1}{2i}[-\mathbf{N}^-(r) + \mathbf{N}^+(r)\mathbf{S}] \end{aligned} \tag{4.194}$$

where the irregular wavefunction $\mathbf{N}^{\pm}$ is defined as

$$\mathbf{N}^{\pm}(r) = \mathbf{N}(r) \pm i\mathbf{J}(r) \tag{4.195}$$

The S matrix is given by the relation

$$\mathbf{S} = \mathbf{I} - 2i\mathbf{T} \tag{4.196}$$

and the T matrix is defined as

$$\begin{aligned}\mathbf{T} &= \frac{2m}{\hbar^2}\int_0^{\infty} \mathbf{J}^T(r)\mathbf{V}(r)\mathbf{\Psi}^+(r)dr \\ &= \frac{2m}{\hbar^2} <\mathbf{J}|\mathbf{V}|\mathbf{\Psi}^+>\end{aligned} \tag{4.197}$$

Equation (4.194) can also be written in terms of the real wavefunction $\mathbf{\Psi}(r)$

$$\mathbf{\Psi}^+(r) = \mathbf{\Psi}(r)(\mathbf{I} - i\mathbf{T}) \tag{4.198}$$

where the real scattering wavefunction $\mathbf{\Psi}$ has the real boundary condition

$$\mathbf{\Psi}(r \to \infty) = \mathbf{J}(r) + \mathbf{N}(r)\mathbf{K} \tag{4.199}$$

Here the K matrix is given by

$$\begin{aligned}\mathbf{K} &= -\mathbf{T}(\mathbf{I} - i\mathbf{T}) \\ &= -\frac{2m}{\hbar^2}\int_0^{\infty} \mathbf{J}^{\dagger}(r)V(r)\mathbf{\Psi}^+(r)dr(\mathbf{I} - i\mathbf{T})^{-1} \\ &= -\frac{2m}{\hbar^2} <\mathbf{J}|V|\mathbf{\Psi}>\end{aligned} \tag{4.200}$$

which is related to the S matrix by the relation

$$\mathbf{K} = -\frac{\mathbf{T}}{\mathbf{I} - i\mathbf{T}} = i\frac{\mathbf{I} - \mathbf{S}}{\mathbf{I} + \mathbf{S}} \tag{4.201}$$

These relations are simply generalizations of elastic scattering results discussed in Sec. 4.3.

4.4.4 Scattering Cross Section

Scattering of a rigid rotor

We discuss a simple but rather useful example of inelastic scattering for a rigid rotor. Because it involves angular momentum coupling, most results obtained for rigid rotor scattering can be easily generalized to more complicated cases. The Hamiltonian for a rigid rotor is given by

$$\begin{aligned} H &= -\frac{\hbar^2}{2m}\frac{1}{R}\frac{\partial^2}{\partial R^2}R + \frac{\mathbf{L}^2}{2mR^2} + B\mathbf{j}^2 + V(R,\theta) \\ &= H_0 + V \end{aligned} \tag{4.202}$$

where $B=\hbar^2/(2mr_0^2)$ and $\theta=\cos^{-1}(\hat{\mathbf{R}}\cdot\hat{\mathbf{r}})$. If we use the coupled angular momentum representation in the space fixed (SF) frame discussed in the appendix, we can expand the scattering wavefunction for a fixed total angular momentum state JM

$$\Psi_{jL}^{JM} = \sum_{j'L'} Y_{j'L'}^{JM}(\hat{\mathbf{R}},\hat{\mathbf{r}})\psi_{j'L',jL}^{JM}(R)/R \tag{4.203}$$

With the choice of H_0 in Eq. (4.202), the free functions are diagonal

$$\begin{cases} \mathbf{J}_{j'L',jL}(R) = \dfrac{j_L(k_jR)}{\sqrt{k_j}}\delta_{j'j}\delta_{L'L} \\ \\ \mathbf{N}_{j'L',jL}^{+}(R) = \dfrac{h_L^+(k_jR)}{\sqrt{k_j}}\delta_{j'j}\delta_{L'L} \end{cases} \tag{4.204}$$

where $j_L(k_jR)$ and $h_L^+(k_jR)$ are Ricatti-Bessel and Ricatti-Hankel functions defined in the appendix and

$$k_j = \frac{1}{\hbar}\sqrt{2m(E - Bj(j+1))} \tag{4.205}$$

From the asymptotic condition in Eq. (4.194), we can explicitly write out the asymptotic expression for the radial wavefunction

$$\begin{aligned} \psi_{j'L',jL}^{JM} &\overset{R\to\infty}{\longrightarrow} \frac{j_L(k_jR)}{\sqrt{k_j}}\delta_{j'j}\delta_{L'L} - \frac{h_L^+(k_{j'}R)}{\sqrt{k_{j'}}}T_{j'L',jL}^{JM} \\ &\longrightarrow \frac{\sin(k_jR - L\pi/2)}{\sqrt{k_j}}\delta_{j'j}\delta_{L'L} - \frac{e^{(ik_{j'}R - L'\pi/2)}}{\sqrt{k_{j'}}}T_{j'L',jL}^{JM} \end{aligned} \tag{4.206}$$

which is given in the T matrix boundary condition. The T matrix is related to the S matrix via Eq. (4.196)

$$S_{j'L',jL} = \delta_{j'j}\delta_{L'L} - 2iT_{j'L',jL} \tag{4.207}$$

Substituting the asymptotic expression of the radial function (4.206) in the full wavefunction expansion in Eq. (4.203), we obtain the asymptotic form for the full wavefunction

$$\Psi_{jL}^{JM} = \sum_{j'L'} Y_{j'L'}^{JM}(\hat{\mathbf{R}}, \hat{\mathbf{r}})\psi_{j'L',jL}^{JM}$$

$$\xrightarrow{R\to\infty} Y_{jL}^{JM}(\hat{\mathbf{R}}, \hat{\mathbf{r}})\frac{\sin(k_j R - L\pi/2)}{\sqrt{k_j}R} - \sum_{j'L'} Y_{j'L'}^{JM}(\hat{\mathbf{R}}, \hat{\mathbf{r}})$$

$$\times \frac{e^{(ik_{j'}R - L'\pi/2)}}{\sqrt{k_{j'}}R} T_{j'L',jL}^{JM} \tag{4.208}$$

Open and closed channels

The summation in Eq. (4.208) includes only open states or energetically accessible states whose eigenenergies ϵ_n are smaller than the total scattering energy E. This is because the closed channel components of the wavefunction must vanish asymptotically from simple physical considerations. For open channels, the kinetic energies are positive, or

$$k_n^2 = \frac{2m}{\hbar^2}(E - \epsilon_n) > 0 \tag{4.209}$$

and for closed channels, the above kinetic energy is negative. Physically, the closed channel components of the wavefunction must vanish asymptotically from simple physical considerations and only open channels are energetically accessible asymptotically. However, the expansion of the wavefunction in Eq. (4.203) is based on the requirement for completeness of the basis functions. Thus it needs to include both open and closed channels.

Scattering amplitude

To obtain general expressions for multichannel scattering cross sections, we give here specific derivations for the scattering of a rigid rotor. However, the final expressions for scattering cross sections can be generalized to general cases of scattering including reactive scattering in a straightforward fashion. The simplest method to derive the expression for the scattering amplitude is to start from asymptotic expressions of the scattering wavefunction. The basic procedure is essentially the same as that leading to the expression of the cross section for elastic scattering in Sec. 4.3.5, but the derivation for inelastic scattering is a little more tedious.

The standard inelastic scattering process for a rigid rotor is described by the asymptotic wavefunction

$$\Psi_{j_0 m_0}^{+}(\mathbf{R}, \hat{\mathbf{r}}) \longrightarrow e^{i\mathbf{k}_{j_0}\cdot\mathbf{R}} Y_{j_0 m_0}(\hat{\mathbf{r}}) + \sum_{jm} f_{jm,j_0 m_0}(\vec{\mathbf{R}})\frac{e^{ik_j R}}{R} Y_{jm}(\hat{\mathbf{r}}) \tag{4.210}$$

By using the plane wave expansion of Eq. (A.55) in the appendix, the first term in Eq. (4.210) becomes

$$e^{i\mathbf{k}_{j_0}\cdot\mathbf{R}}Y_{j_0m_0}(\hat{\mathbf{r}}) = \sum_{L_0m}\frac{4\pi i^{L_0}}{k_{j_0}R}j_{L_0}(k_{j_0}R)Y^*_{L_0m}(\hat{\mathbf{k}}_{j_0})Y_{L_0m}(\hat{\mathbf{R}})Y_{j_0m_0}(\hat{\mathbf{r}})$$
$$\to \sum_{L_0m}4\pi i^{L_0}\frac{\sin(k_{j_0}R - L_0\pi/2)}{k_{j_0}R}Y^*_{L_0m}(\hat{\mathbf{k}}_{j_0})$$
$$\times Y_{L_0m}(\hat{\mathbf{R}})Y_{j_0m_0}(\hat{\mathbf{r}}) \tag{4.211}$$

Using the coupled angular momentum representation discussed in Sec. D.3 in the appendix, we can rewrite it as

$$e^{i\mathbf{k}_{j_0}\cdot\mathbf{R}}Y_{j_0m_0}(\hat{\mathbf{r}}) \to 4\pi\sum_{JML_0}i^{L_0}\frac{\sin(k_{j_0}R - \frac{L_0}{2}\pi)}{k_{j_0}R} <L_0M - m_0j_0m_0|JM>$$
$$\times Y^*_{L_0M-m_0}(\hat{\mathbf{k}}_{j_0})Y^{JM}_{j_0L_0}(\hat{\mathbf{R}},\hat{\mathbf{r}}) \tag{4.212}$$

where $Y^{JM}_{j_0L_0}(\hat{\mathbf{R}},\hat{\mathbf{r}})$ is the SF coupled angular momentum eigenfunction defined in the appendix.

On the other hand, the asymptotic form of the scattering wavefunction $\Psi^{JM}_{j_0L_0}$ for a given initial quantum state j_0L_0 in the space-fixed coordinate system can be expressed in terms of the T matrix boundary condition of Eq. (4.208)

$$\Psi^{JM}_{j_0L_0} \xrightarrow{R\to\infty} Y^{JM}_{j_0L_0}(\hat{\mathbf{R}},\hat{\mathbf{r}})\frac{\sin(k_{j_0}R - L_0\pi/2)}{\sqrt{k_{j_0}}R} - \sum_{jL}Y^{JM}_{jL}(\hat{\mathbf{R}},\hat{\mathbf{r}})$$
$$\times\frac{e^{(ik_jR-L\pi/2)}}{\sqrt{k_j}R}T^{JM}_{jL,jL} \tag{4.213}$$

We can thus express $\Psi^+_{j_0m_0}(\mathbf{R},\hat{\mathbf{r}})$ in Eq. (4.210) as a linear combination of $\Psi^{JM}_{j_0L_0}$

$$\Psi^+_{j_0m_0} = \sum_{JML_0}A^{JML_0}_{j_0m_0}\Psi^{JM}_{j_0L_0} \tag{4.214}$$

By equating the first terms in Eqs. (4.210) and (4.214), we obtain the expansion coefficient

$$A^{JML_0}_{j_0m_0} = \frac{4\pi i^{L_0}}{\sqrt{k_{j_0}}}Y^*_{L_0M-m_0}(\hat{\mathbf{k}}_{j_0}) <L_0M - m_0j_0m_0|JM> \tag{4.215}$$

Further equating the second terms in Eqs. (4.210) and (4.214) gives rise to the expression for the scattering amplitude

$$\begin{aligned}
f_{jm,j_0m_0}(\hat{\mathbf{R}}) &= -\sum_{JMLL_0} \frac{A^{JML_0}_{j_0m_0}}{\sqrt{k_j}} i^{-L} <LM-mjm|JM> Y_{LM-m}(\hat{\mathbf{R}}) T^{JM}_{jL,j_0L_0} \\
&= -\sum_{JMLL_0} \frac{4\pi i^{L_0-L}}{\sqrt{k_j k_{j0}}} <LM-mjm|JM> \\
&\quad \times <JM|L_0M-m_0j_0m_0> Y^*_{L_0M-m_0}(\hat{\mathbf{k_{j_0}}}) \\
&\quad \times Y_{LM-m}(\hat{\mathbf{R}}) T^{JM}_{jL,j_0L_0}
\end{aligned} \tag{4.216}$$

If we choose the relative incident motion to be in the space-fixed Z direction, then $\hat{\mathbf{k}}_{j_0} = (0,0)$, and

$$Y^*_{LM-m_0}(0,0) = \sqrt{\frac{2L+1}{4\pi}}\delta_{Mm_0} \tag{4.217}$$

and we then obtain the standard expression for the scattering amplitude originally given by Arthur-Dalgarno [41]

$$\begin{aligned}
f_{jm,j_0m_0}(\theta,\phi) = -\sum_{JLL_0} & \frac{\sqrt{4\pi} i^{L_0-L}\sqrt{2L_0+1}}{\sqrt{k_j k_{j0}}} <Lm_0-mjm|Jm_0> \\
& \times <Jm_0|L_00j_0m_0> Y_{Lm_0-m}(\hat{\mathbf{R}}) T^{JM}_{jL,j_0L_0}
\end{aligned} \tag{4.218}$$

In molecular scattering, one often uses the *helicity* representation in which the BF z axis points along the $\mathbf{R}$ direction asymptotically. From Eqs. (4.210) and (4.218) we can write

$$\begin{aligned}
\sum_m f_{jm,j_0m_0} Y_{jm}(\hat{\mathbf{r}}) = \sum_{JLL_0} & \frac{\sqrt{4\pi} i^{L_0-L}\sqrt{2L_0+1}}{\sqrt{k_j k_{j0}}} <Jm_0|L_00j_0m_0> \\
& \times Y^{Jm_0}_{jL}(\hat{\mathbf{R}},\hat{\mathbf{r}}) Y_{Lm_0-m}(\hat{\mathbf{R}}) T^{JM}_{jL,j_0L_0}
\end{aligned} \tag{4.219}$$

Using the following relation from the appendix

$$\begin{aligned}
Y^{Jm_0}_{jL}(\hat{\mathbf{R}},\hat{\mathbf{r}}) = \sum_m & <Lm_0-mjm|Jm_0> \sqrt{\frac{2L+1}{4\pi}} \\
& \times D^J_{mm_0}(\phi,\theta,0) Y_{jm}(\hat{\mathbf{r}}|\hat{\mathbf{R}})
\end{aligned} \tag{4.220}$$

and defining the body-fixed T matrix through a unitary transformation from the SF T

$$
\begin{aligned}
T^{JM}_{jm,j_0m_0} &= \sum_{LL_0} i^{L_0-L} \frac{\sqrt{(2L_0+1)(2L+1)}}{2J+1} <L0jm|Jm> \\
&= <Jm_0|L0j_0m_0>\, T^{JM}_{jL,j_0L_0} \qquad (4.221)
\end{aligned}
$$

Eq. (4.219) can be rewritten in the helicity representation

$$\sum_m f_{jm,j_0m_0} Y_{jm}(\hat{\mathbf{r}}) = \sum_m \bar{f}_{jm,j_0m_0} Y_{jm}(\mathbf{r}|\hat{\mathbf{R}}) \qquad (4.222)$$

Here the scattering amplitude in the helicity representation is given by

$$\bar{f}_{jm,j_0m_0}(\theta,\phi) = -\frac{e^{im\phi}}{\sqrt{k_j k_{j_0}}} \sum_J (2J+1) d^J_{mm_0}(\theta) T^{JM}_{jm,j_0m_0} \qquad (4.223)$$

where $d^J_{mm_0}(\theta)$ is the reduced rotation matrix defined in the appendix. Thus the asymptotic form of the full scattering wavefunction for the rigid rotor can be written in the helicity representation as

$$\Psi^+_{j_0m_0}(\mathbf{R},\hat{\mathbf{r}}) \longrightarrow e^{i\mathbf{k}_{j_0}\cdot\mathbf{R}} Y_{j_0m_0}(\hat{\mathbf{r}}) + \sum_{jm} \bar{f}_{jm,j_0m_0}(\theta) \frac{e^{ik_jR}}{R} Y_{jm}(\hat{\mathbf{r}}|\hat{\mathbf{R}}) \quad (4.224)$$

where $Y_{jm}(\hat{\mathbf{r}}|\hat{\mathbf{R}})$ is the angular momentum eigenfunction of the rotator in the BF frame and is given by

$$Y_{jm}(\hat{\mathbf{r}}|\hat{\mathbf{R}}) = Y_{jm}(\cos^{-1}(\hat{\mathbf{r}}\cdot\hat{\mathbf{R}}), 0) \qquad (4.225)$$

Scattering cross section

The differential cross section is given by the flux formula

$$
\begin{aligned}
\frac{d\sigma_{jm,j_0m_0}}{d\Omega} &= \frac{v_j}{v_{j_0}} \left| \bar{f}_{jm,j_0m_0}(\theta,\phi) \right|^2 \\
&= \frac{1}{k^2_{j_0}} \left| \sum_J (2J+1) d^J_{mm_0}(\theta_R) T^J_{jm,j_0m_0} \right|^2 \qquad (4.226)
\end{aligned}
$$

where the superscript M has been dropped because the T matrix is independent of M. The integral cross section is obtained by integrating over the solid angle

$$\begin{aligned}\sigma_{jm,j_0m_0} &= \int_\Omega d\sigma_{jm,j_0m_0} \\ &= \frac{4\pi}{k_{j_0}^2}\sum_J (2J+1)|T^J_{jm,j_0m_0}|^2\end{aligned} \quad (4.227)$$

where the orthogonality condition for the rotation matrix

$$\int \sin(\theta)d\theta\ d^J_{mm_0}(\theta)d^{J'}_{mm_0}(\theta) = \frac{2}{2J+1} \quad (4.228)$$

has been used.

These results can be easily extended to scattering systems more complicated than the rigid rotor by proper augmentation of the channel index. For example, for rotating-vibrating atom-diatom scattering, the scattering amplitude can be obtained by a straightforward generalization of Eq. (4.224)

$$\bar{f}_{vjm,v_0j_0m_0}(\theta,\phi) = -\frac{e^{im\phi}}{\sqrt{k_{vj}k_{v_0j_0}}}\sum_J (2J+1)d^J_{mm_0}(\theta_R)T^J_{vjm,v_0j_0m_0} \quad (4.229)$$

and the integral cross section is given by

$$\boxed{\sigma_{vjm,v_0j_0m_0} = \frac{4\pi}{k_{v_0j_0}^2}\sum_J (2J+1)|T^J_{vjm,v_0j_0m_0}|^2} \quad (4.230)$$

4.4.5 R Matrix Method

One of the earliest computational methods for calculating the scattering matrix in physics was the so called R matrix method [42]. The R matrix method is very general but the numerical result converges rather slowly. Since it is an important computational technique, it is worthwhile to present it here. We use a one-dimensional radial Hamiltonian

$$H = -\frac{\hbar^2}{2m}\frac{\partial^2}{\partial R^2} + V(r) \quad (4.231)$$

to illustrate the methodology, although the result can be generalized to multidimensional scattering problems in a straightforward fashion. In the R matrix method, the scattering wavefunction ψ is expanded in an orthogonal basis set u_λ which is defined as eigenfunctions of the Hamiltonian H in a finite range of the scattering (radial) coordinate $r \in [0, L]$ with *Dirichlet* boundary condition

$$u'_\lambda(L) = 0 \tag{4.232}$$

(or with the more general boundary condition of $u'_\lambda(L) + au_\lambda(L) = 0$). Thus within the finite range $r \in [0, L]$, we can solve the eigenvalue problem

$$H|u_\lambda\rangle = E_\lambda |u_\lambda\rangle \tag{4.233}$$

Within this coordinate range, the scattering wavefunction ψ can be expanded in terms of the complete set u_λ

$$|\psi\rangle = \sum_\lambda |u_\lambda\rangle\, c_\lambda \tag{4.234}$$

which gives the expansion coefficient as the integral

$$c_\lambda = \langle u_\lambda|\psi\rangle = \int_0^L dr u^*_\lambda(r)\psi(r) \tag{4.235}$$

In order to calculate the expansion coefficient c_λ we consider the following integral

$$\langle u_\lambda|H|\psi\rangle = -\frac{\hbar^2}{2m} \langle u_\lambda|\psi''\rangle + \langle u_\lambda|V|\psi\rangle \tag{4.236}$$

which can be integrated by parts to yield

$$\begin{aligned} \langle u_\lambda|H|\psi\rangle &= \langle Hu_\lambda|\psi\rangle - \frac{\hbar^2}{2m}[u_\lambda(L)\psi'(L) - u'_\lambda(L)\psi(L)] \\ &= E_\lambda \langle u_\lambda|\psi\rangle - \frac{\hbar^2}{2m} u_\lambda(L)\psi'(L) \\ &= E \langle u_\lambda|\psi\rangle \end{aligned} \tag{4.237}$$

We note here that for integration in the finite range $[0, L]$

$$\langle u_\lambda|H|\psi\rangle \neq \langle u_\lambda|H^\dagger|\psi\rangle = E_\lambda \langle u_\lambda|\psi\rangle \tag{4.238}$$

because the scattering wavefunction ψ does not satisfy the standard boundary condition at $r = L$.

Using Eq. (4.237), we obtain

$$(E - E_\lambda) \langle u_\lambda|\psi\rangle = -\frac{\hbar^2}{2m} u_\lambda(L)\psi'(L) \tag{4.239}$$

or

$$\begin{aligned} \langle u_\lambda|\psi\rangle &= \frac{1}{E_\lambda - E}\frac{\hbar^2}{2m}u_\lambda(L)\psi'(L) \\ &= \frac{1}{E_\lambda - E}\langle u_\lambda|\mathcal{L}|\psi\rangle \end{aligned} \tag{4.240}$$

where the Bloch operator is defined as [43]

$$\mathcal{L} = \frac{\hbar^2}{2m}\delta(r-L)\frac{d}{dr} \tag{4.241}$$

Thus, the scattering wavefunction in the range $r \in [0, L]$ can be written as

$$\begin{aligned} |\psi\rangle &= \sum_\lambda |u_\lambda\rangle\langle u_\lambda|\psi\rangle \\ &= \frac{\hbar^2}{2m}\sum_\lambda \frac{|u_\lambda\rangle}{E_\lambda - E}u_\lambda(L)\psi'(L) \end{aligned} \tag{4.242}$$

Finally, we obtain the so called R matrix

$$\begin{aligned} R &= \psi(L)/\psi'(L) \\ &= \frac{\hbar^2}{2m}\sum_\lambda \frac{|u_\lambda(L)|^2}{E_\lambda - E} \end{aligned} \tag{4.243}$$

Once the R matrix is obtained for L sufficiently large such that the asymptotic (K matrix) boundary condition can be applied at L for the wavefunction

$$\psi(L) = \sin(kr) + K\cos(kr) \tag{4.244}$$

we then obtain the equation

$$R(L) = \frac{\psi(L)}{\psi'(L)} = \frac{\sin(kr) + K\cos(kr)}{k\cos(kr) - kK\cos(kr)} \tag{4.245}$$

which can be inverted easily to yield the K matrix

$$K = \frac{R(L)k\cos(kr) - \sin(kr)}{\cos(kr) - R(L)k\sin(kr)} \tag{4.246}$$

Thus the calculation of the scattering matrix becomes the problem of calculating eigenvalues and eigenfunctions of the Hamiltonian E_λ and u_λ. However, special attention must be paid to the boundary condition of u_λ.

Equation (4.242) can also be formally obtained from the Schrödinger equation by adding the Bloch operator $\mathcal{L}$ to the Hamiltonian

$$[E - H - \mathcal{L}]\psi = -\mathcal{L}\psi \tag{4.247}$$

to make the operator $\bar{H} = H + \mathcal{L}$ hermitian in the finite range $[0, L]$, and thus directly obtain the result

$$\begin{aligned} |\psi\rangle &= \frac{1}{H + \mathcal{L} - E}\mathcal{L}|\psi\rangle \\ &= \sum_{\lambda} \frac{|u_\lambda\rangle}{E_\lambda - E} \langle u_\lambda|\mathcal{L}|\psi\rangle \\ &= \frac{\hbar^2}{2m}\sum_{\lambda} \frac{|u_\lambda\rangle\, u_\lambda(L)}{E_\lambda - E}\psi'(L) \end{aligned} \tag{4.248}$$

The results of the R matrix method can be easily generalized to the multi-dimensional case. If $|n\rangle$ denotes internal channel states, then Eq. (4.241) is generalized to

$$\mathcal{L} = \sum_{n} \frac{\hbar^2}{2m}|n\rangle\, \delta(r - L)\frac{d}{dr} \langle n| \tag{4.249}$$

and the u_λ are replaced by multidimensional eigenfunctions of the Hamiltonian H.

Buttle correction

The convergence of the R matrix method with the size of the basis set is very slow due to finite basis truncation error. Buttle has proposed a method to improve the numerical accuracy by adding a correction term to account for the truncation error of finite basis [44]. Considering a scattering problem with an unperturbed Hamiltonian (V=0), the exact R matrix is simply given by

$$R_0 = \frac{\psi_0}{\psi_0'} \tag{4.250}$$

where ψ^0 is the exact eigenfunction of the unperturbed Hamiltonian. However, the same quantity can also be calculated by using the R matrix method with N bases

$$R_0^N = \sum_{\lambda=1}^{N} \frac{|u_\lambda^0(L)|^2}{E - E_\lambda^0} \tag{4.251}$$

The difference between R_0 and R_0^N is purely the truncation error with finite N bases. We expect a similar truncation error exists in the calculated R matrix for the perturbed Hamiltonian equation and thus can subtract it out by adding a correction term to the original expression for the R matrix

$$R_B = R^N + R_0 - R_0^N \tag{4.252}$$

The Buttle corrected expression for the R matrix gives much better numerical results than the uncorrected one.

4.4.6 Other Methods

In molecular collisions, two efficient methods are often used for numerical calculations of multichannel inelastic scattering. These two methods are called the R matrix propagation method [45] and the log derivative method [46, 47]. Both methods solve the coupled channel inelastic scattering problem by propagating either the local R matrix or the log derivative of the scattering wavefunction along the radial coordinate from the origin to the asymptotic region. Because the wavefunction itself is not directly propagated, these methods can avoid the numerical instability problem associated with closed channels. Both methods are efficient solvers of the coupled channel equation and are widely used to solve inelastic scattering problems. These methods are highly technical and specific details can be found in the corresponding references.

4.5 Reactive Scattering

4.5.1 Partitioning of the Hamiltonian

In elastic or inelastic scattering, the Hamiltonian is partitioned as the sum of an unperturbed Hamiltonian H_0 and an interaction potential V which vanishes as the scattering or radial coordinate becomes large. Since there is only one asymptotic arrangement, this is the only partitioning possible. In reactive or rearrangement scattering, however, the *full* Hamiltonian H can be partitioned in different ways corresponding to different asymptotic arrangements. If we use γ as the arrangement channel label for a reactive collision with N possible arrangements, the Hamiltonian can be partitioned as

$$H = H_\gamma + V_\gamma \qquad (\gamma = 1, 2, 3, ..,) \tag{4.253}$$

where H_γ is the asymptotic Hamiltonian of the fragments written as the sum of the relative kinetic energy operator $\hat{K}_\gamma$ plus an internal Hamiltonian H_γ^{int}

$$H_\gamma = \hat{K}_\gamma + H_\gamma^{int} \tag{4.254}$$

and V_γ is the interaction potential between the fragments. If we use R_γ to denote the radial distance between the centers of the two fragments in the γ arrangement, then

$$\lim_{R_\gamma \to \infty} V_\gamma = 0 \qquad (\gamma = 1, 2, 3, ..,) \tag{4.255}$$

Thus the scattering wavefunction takes different asymptotic forms in different arrangements, or in other words, the wavefunction bifurcates. Because of the bifurcation of the scattering wavefunction, we need to be very careful in deriving the appropriate Lippmann Schwinger equation and in solving the reactive scattering problem.

If we consider a reactive scattering process initiated in arrangement α, we can use the partition $H = H_\alpha + V_\alpha$ to obtain the Lippman-Schwinger equation

$$\boxed{\psi_\alpha^+ = \phi_\alpha + G^+ V_\alpha \phi_\alpha} \tag{4.256}$$

Here we use the LS equation with the full Green's function rather than the LS equation involving the free Green's function

$$\psi_\alpha^+ = \phi_\alpha + G_\alpha^+ V_\alpha \psi_\alpha^+ \tag{4.257}$$

The reason for this is that although Eq. (4.257) is formally correct, it is essentially useless in practical calculations. This is because the standard expansion of G_α^+ in terms of channel basis functions in Eq. (4.179) includes discrete states only. Thus G_α^+ contains the outgoing boundary condition in the α arrangement only and will give rise to only inelastic scattering in the α arrangement. To obtain reactive scattering, it is necessary to include continuous states in the channel expansion of Eq. (4.179) in order to describe the bond breaking process in reactive scattering. This is, however, not a viable approach in practical applications due to numerical difficulties in dealing with the double continuum.

4.5.2 Scattering Matrix and Cross Section

Scattering matrices

The main mathematical difficulty in treating reactive scattering problems is the bifurcation or splitting of the full scattering wavefunction ψ into different arrangement channels (rearrangement process). As a result, the wavefunction ψ has *multiple* asymptotic boundary conditions as does the full Green's function G^+. To examine this in more detail, we examine the Lippman-Schwinger equation for the full Green's function

$$G^+ = G_\gamma^+ + G_\gamma^+ V_\gamma G^+ \qquad (\gamma = 1, 2, 3, ..,) \tag{4.258}$$

If we choose $\gamma = \alpha$ in Eq. (4.258), we can use it to rewrite the LS equation for the scattering wavefunction as

$$\psi_\alpha^+ = \phi_\alpha + G_\alpha^+ V_\alpha \psi_\alpha^+$$

Similarly, if we choose $\gamma = \beta$ in Eq. (4.258), we obtain

$$\psi_\alpha^+ = G_\beta^+ V_\beta \psi_\alpha^+ \qquad (\alpha \neq \beta) \tag{4.259}$$

Although Eqs. (4.257) and (4.259) are just formal solutions and are not very useful for actual numerical calculations for reasons discussed above, they do give the correct outgoing asymptotic boundary conditions in the α and β arrangements, respectively. This is because in the asymptotic region, say, $R_\gamma \to \infty$, the continuum state contribution to G_γ^+ vanishes and only discrete (open) states survive. Thus asymptotically, G_γ^+ is simply the inelastic Green's function discussed in Sec. 4.4.2 and has the asymptotic form

$$\mathbf{G}_\gamma(R_\gamma|R'_\gamma) \overset{R_\gamma\to\infty}{\longrightarrow} -\frac{2}{\hbar^2}\mathbf{N}_\gamma^+(R_\gamma)\mathbf{J}^T\gamma(R'_\gamma) \qquad (\gamma = 1, 2, 3, ..) \tag{4.260}$$

We note here that unlike in inelastic scattering, the Wronskian is not taken to be unit but rather

$$W[\mathbf{J}, \mathbf{N}_\gamma^+] = \mu_\gamma \tag{4.261}$$

and asymptotically

$$\mathbf{J}_{mn} \overset{R\to\infty}{\longrightarrow} \frac{\sin(k_n R - L_n\pi/2)}{\sqrt{v_n}}\delta_{mn} \tag{4.262}$$

Here the denominator is the speed $v_n = \hbar k_n/m_\gamma$ instead of k_n because the mass for the reactant and product translational motion is different. This definition of the asymptotic functions facilitates the definition of a unitary S matrix

From the discussion on inelastic scattering leading to Eq. (4.194) in Sec. 4.4.3, we can combine Eqs. (4.257) and (4.259) for the asymptotic scattering wavefunction to be

$$\psi_\alpha^+(R_\beta \to \infty) = \mathbf{J}_\alpha(R_\alpha)\delta_{\alpha\beta} - \mathbf{N}_\beta^+(R_\beta)\mathbf{T}_{\beta\alpha} \tag{4.263}$$

where the reactive T matrix is defined as

$$\begin{aligned}\mathbf{T}_{\beta\alpha} &= \frac{2}{\hbar^2}\int_0^{\infty} \mathbf{J}_\beta^T(R'_\beta)\mathbf{V}_\beta(R'_\beta)\mathbf{\Psi}_\alpha^+(R'_\beta)dR'_\beta \\ &= \frac{2}{\hbar^2} <\mathbf{J}_\beta|\mathbf{V}_\beta|\mathbf{\Psi}_\alpha^+> \end{aligned} \tag{4.264}$$

We can then define an $\alpha \to \beta$ reactive T matrix operator by

$$T_{\beta\alpha}|\phi_\alpha>= V_\beta|\psi_\alpha^+> \tag{4.265}$$

By using Eq. (4.257), we obtain

$$\boxed{T_{\beta\alpha} = V_\beta + V_\beta G^+ V_\alpha} \tag{4.266}$$

which is the general expression for the T matrix operator. Since the potential satisfies the relation

$$\begin{aligned}<\phi_\beta|V_\beta|\phi_\alpha> &= <\phi_\beta|H - E|\phi_\alpha> \\ &= <\phi_\beta|V_\alpha|\phi_\alpha> \end{aligned} \tag{4.267}$$

the reactive T matrix operator can also be defined as

$$T_{\beta\alpha} = V_\alpha + V_\beta G^+ V_\alpha \tag{4.268}$$

which gives the same on-shell T matrix element as that defined in Eq. (4.266). Similarly, we can define the reactive K operator by

$$K_{\beta\alpha} = V_\beta + V_\beta G^{p+} V_\alpha = V_\alpha + V_\beta G^p V_\alpha \tag{4.269}$$

Note that the general relations between the S, T, and K matrices have remained the same as in inelastic scattering. For example, the S matrix operator is given by

$$S_{\beta\alpha} = \delta_{\beta\alpha} - 2\pi i T_{\beta\alpha} \tag{4.270}$$

and is unitary.

Reaction cross sections

From Eq. (4.263), the reactive scattering wavefunction has the asymptotic expansion

$$\begin{aligned}\psi_{\alpha i}^+ \overset{R_\gamma\to\infty}{\longrightarrow} & \frac{\sin(k_{\alpha i}R_\alpha - L_{\alpha i}\pi/2)}{\sqrt{v_{\alpha i}}}\varphi_{\alpha i}\delta_{\alpha\gamma} \\ & - \sum_n \frac{e^{i(k_{\gamma n}R_\gamma - L_{\gamma n}\pi/2)}}{\sqrt{v_{\gamma n}}}\varphi_{\gamma n}T_{\gamma n,\alpha i}\end{aligned} \tag{4.271}$$

where $\varphi_{\gamma m}$ are the arrangement channel eigenfunctions satisfying the eigenvalue equation

$$H_\gamma^{int}(q_\gamma)\varphi_{\gamma m}(q_\gamma) = \epsilon_\gamma \varphi_{\gamma m}(q_\gamma) \tag{4.272}$$

From the Lippman-Schwinger Eq. (4.258)), the full Green's function G^+ also has the outgoing asymptotic boundary condition in all arrangements, i.e.,

$$G^+(R_\gamma|R'_\gamma) \overset{R_\gamma\to\infty}{\longrightarrow} \mathbf{N}_\gamma^+(R_\gamma)\mathbf{A}_\gamma(R'_\gamma) \qquad (\gamma = 1, 2, 3, ..,) \tag{4.273}$$

The mathematical procedure to obtain cross sections in terms of the scattering matrix is essentially the same as in inelastic scattering. For example, the scattering amplitude in the helicity representation for atom-diatom reactive scattering now becomes

$$\begin{aligned} \bar{f}_{\beta vjm,\alpha v_0 j_0 m_0}(\theta_\beta, \phi_\beta) = & \frac{\sqrt{\mu_\beta}}{\sqrt{\mu_\alpha k_{\beta vj} k_{\alpha v_0 j_0}}} \sum_J e^{im\phi_\beta}(2J+1) \\ & \times d_{mm_0}^J(\theta_\beta) T_{\beta vjm,\alpha v_0 j_0 m_0}^J \end{aligned} \tag{4.274}$$

and the differential cross section is therefore given by

$$\begin{aligned} \frac{d\sigma_{\beta vjm,\alpha v_0 j_0 m_0}}{d\Omega} &= \frac{v_{\beta vj}}{v_{\alpha v_0 j_0}} \left| \bar{f}_{\beta vjm,\alpha v_0 j_0 m_0}(\theta_\beta, \phi_\alpha) \right|^2 \\ &= \frac{1}{k_{\alpha v_0 j_0}^2} \left| \sum_J (2J+1) d_{mm_0}^J(\theta_\beta) T_{\beta vjm,\alpha v_0 j_0 m_0}^J \right|^2 \end{aligned} \tag{4.275}$$

In reactive scattering, one often measures the molecular product and its angular distribution instead of the atomic projectile. In the center-of-mass frame, the molecular product moves in the opposite direction from the atomic projectile, i.e., $\theta^M = 180 - \theta$. Thus the differential cross section is often expressed in terms of the angle of the molecular product θ^M, or

$$\begin{aligned} \frac{d\sigma_{\beta vjm,\alpha v_0 j_0 m_0}(\theta_\beta^M)}{d\Omega} &= \frac{v_{\beta vj}}{v_{\alpha v_0 j_0}} \left| \bar{f}_{\beta vjm,\alpha v_0 j_0 m_0}(\theta_\beta, \phi_\alpha) \right|^2 \\ &= \frac{1}{k_{\alpha v_0 j_0}^2} \left| \sum_J (2J+1) d_{mm_0}^J(\pi - \theta_\beta^M) \right. \\ & \left. \times T_{\beta vjm,\alpha v_0 j_0 m_0}^J \right|^2 \end{aligned} \tag{4.276}$$

The integral cross section is given by

$$\boxed{\sigma_{\beta vjm,\alpha v_0 j_0 m_0} = \frac{4\pi}{k^2_{\alpha v_0 j_0}} \sum_J (2J+1) \left| T^J_{\beta vjm,\alpha v_0 j_0 m_0} \right|^2} \tag{4.277}$$

4.5.3 Jacobi Coordinates

In reactive scattering (without breakup), the different rearrangements or fragmentations of the collision system are naturally described by the Jacobi coordinates corresponding to the particular arrangements. In terms of the Jacobi coordinates, the kinetic energy operators of different fragments are decoupled (or diagonal). There is at least one set of Jacobi coordinates corresponding to each rearrangement for polyatomic reactions. We have already encountered Jacobi coordinates in Chapter 3 for solving bound state problems. Here, we give transformation relations between different sets of Jacobi coordinates for triatomic systems. The transformation relations for tetraatomic systems are very similar to those for triatomic systems and are thus not presented here.

Triatomic systems

For a triatomic system with atomic masses m_1, m_2, m_2, there are three sets of Jacobi coordinates as shown in Fig. 4.2. To discuss the transformation properties between different sets of Jacobi coordinates, it is often convenient to use mass-scaled Jacobi coordinates. The mass-scaled Jacobi coordinates $(\bar{\mathbf{R}}_\alpha, \bar{\mathbf{r}}_\alpha)$ are simply scaled Jacobi coordinates $(\mathbf{R}_\alpha, \mathbf{r}_\alpha)$ defined by

$$\begin{cases} \bar{\mathbf{R}}_\alpha = \lambda_\alpha \mathbf{R}_\alpha \\ \bar{\mathbf{r}}_\alpha = \lambda_\alpha{}^{-1} \mathbf{r}_\alpha \end{cases} \tag{4.278}$$

for $(\alpha = 1, 2, 3)$. The mass scaling factor is defined as

$$\lambda_\alpha = \sqrt{\frac{\mu_{R_\alpha}}{\mu}} = \sqrt{\frac{\mu}{\mu_{r_\alpha}}} \tag{4.279}$$

where the reduced masses are defined as

$$\mu_{r_\alpha} = \frac{m_\beta m_\gamma}{m_\beta + m_\gamma}$$

$$\mu_{R_\alpha} = \frac{m_\alpha(m_\beta + m_\gamma}{M}$$

$$\mu = \sqrt{\frac{m_\alpha m_\beta m_\gamma}{M}}$$
$$M = m_\alpha + m_\beta + m_\gamma \tag{4.280}$$

and the cyclic relation (α, β, γ) is assumed.

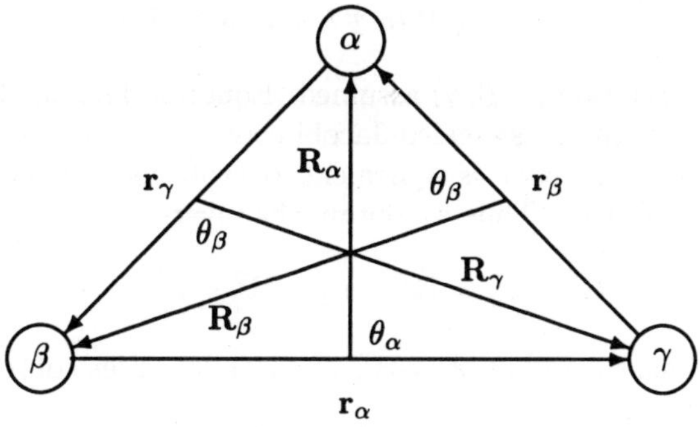

Figure 4.2: Three sets of Jacobi coordinates for triatomic systems.

It is easy to see from the definitions of the mass-scaled Jacobi coordinates in Eq. (4.279) that

$$\mu \bar{\mathbf{R}}_\alpha^2 = \mu_{R_\alpha} \mathbf{R}_\alpha^2$$
$$\mu \bar{\mathbf{r}}_\alpha^2 = \mu_{r_\alpha} \mathbf{r}_\alpha^2 \tag{4.281}$$

and thus the kinetic energy operator can be written in terms of the mass-scaled Jacobi coordinates as

$$T = -\frac{\hbar^2}{2\mu_{R_\alpha}} \nabla^2_{\mathbf{R}_\alpha} - \frac{\hbar^2}{2\mu_{r_\alpha}} \nabla^2_{\mathbf{r}_\alpha}$$
$$= -\frac{\hbar^2}{2\mu} \left[\nabla^2_{\bar{\mathbf{R}}_\alpha} + \nabla^2_{\bar{\mathbf{R}}_\alpha} \right] \tag{4.282}$$

which involves only the system reduced mass μ.

By using mass-scaled Jacobi coordinates, it is not difficult to work out the coordinate transformation relations between different arrangements

$$\begin{pmatrix} \bar{\mathbf{R}}_\beta \\ \bar{\mathbf{r}}_\beta \end{pmatrix} = \begin{pmatrix} \cos\chi_{\beta\alpha} & \sin\chi_{\beta\alpha} \\ -\sin\chi_{\beta\alpha} & \cos\chi_{\beta\alpha} \end{pmatrix} = \begin{pmatrix} \bar{\mathbf{R}}_\alpha \\ \bar{\mathbf{r}}_\alpha \end{pmatrix} \tag{4.283}$$

where the skewing angle $\chi_{\beta\alpha}$ is defined as

$$\cos\chi_{\beta\alpha} = -\sqrt{\frac{m_\beta m_\alpha}{(m_\beta + m_\gamma)(m_\alpha + m_\gamma)}} \tag{4.284}$$

$$\sin\chi_{\beta\alpha} = -\sqrt{\frac{m_\gamma M}{(m_\beta + m_\gamma)(m_\alpha + m_\gamma)}} \tag{4.285}$$

with the cyclic relation (α, β, γ) assumed. Equation (4.283) shows that the transformation of the mass-scaled Jacobi coordinates between different arrangements is orthogonal; it is equivalent to a rotation in a plane containing the vectors $\bar{\mathbf{R}}_\alpha$ and $\bar{\mathbf{r}}_\alpha$. Thus we obtain the invariant relations

$$\bar{\mathbf{R}}_\alpha^2 + \bar{\mathbf{r}}_\alpha^2 = \bar{\mathbf{R}}_\beta^2 + \bar{\mathbf{r}}_\beta^2 = \bar{\mathbf{R}}_\gamma^2 + \bar{\mathbf{r}}_\gamma^2 \tag{4.286}$$

or in other words, the radius $\rho^2 = \bar{\mathbf{R}}_\alpha^2 + \bar{\mathbf{r}}_\alpha^2$ is independent of rearrangement.

4.5.4 A Note on Reactive Scattering

Although reactive scattering may appear as a simple generalization of inelastic scattering by simply augmenting the channel index n in inelastic scattering to include the arrangement index γ, the practical theoretical treatment and calculation of reactive scattering problems is much more involved than that for inelastic problems. This is because the natural coordinates describing the asymptotic scattering wavefunction in arrangement α are the Jacobi coordinates corresponding to the α arrangement. Thus the presence of multiple arrangements generally requires different sets of Jacobi coordinates to describe the wavefunction in different arrangements. It is thus problematic to choose just a single set of coordinates to properly describe the wavefunction in all the arrangement channels. This is the source of the notorious coordinate problem in reactive scattering and the standard theoretical methods for inelastic scattering simply do not work in reactive scattering. We will describe a number of practical approaches to solving reactive scattering problems in the following two chapters.

Chapter 5

Time-Independent Approach to Reactive Scattering

5.1 Introduction

The traditional approach to solving the stationary Schrödinger equation for nonreactive (inelastic) scattering problems is to propagate the wavefunction along the scattering (radial) coordinate from the origin outward to the asymptotic region where the interaction vanishes. By matching the numerical wavefunction to its asymptotic form at any large radial distance, one can extract all the scattering information such as the S, T, and K matrices. Naturally, one would like to extend this kind of approach to reactive scattering as well. However, due to the coordinate problem discussed in the previous chapter, it is not practical to use a single set of Jacobi coordinates to propagate the wavefunction in the entire coordinate space because the bifurcation of the wavefunction into different arrangements would require one to use basis functions with a double continuum. An early approach to circumvent the coordinate problem was to propagate the wavefunction separately in each arrangement by using the coordinates appropriate for each arrangement. One could then match the wavefunctions obtained by propagations in different arrangements on a dividing surface. This approach met with some initial success, and in fact produced the first accurate result for the benchmark H + H_2 reaction in three dimensions [52]. However, this method of wavefunction matching was found to give large errors and, it became increasingly difficult to obtain accurate results at high energies.

In the following we describe two general computational approaches for treating reactive scattering: the hyperspherical coordinate approach and the algebraic variational approach with detailed presentation for the latter. Both approaches have been successfully developed during the past decade to solve reactive scattering problems and both have been widely applied to the calculation for many atom-diatom reactive scattering systems with great success. In this chapter, we will focus on the theory and application of the variational approach to atom-diatom reactive scattering.

5.2 Distorted-Wave Born Approximation

A simple approach to treat the reactive scattering problem is to employ the distorted wave Born approximation (DWBA). From Eq. (4.102) in Sec. 4.2.4, we can obtain an approximate reactive T matrix $T_{\beta\alpha}(\alpha \neq \beta)$ by replacing the full wavefunction by the distorted wave

$$\begin{aligned} T_{\beta f,\alpha i} &\approx <\chi_{\beta f}^{-}|V_\beta|\chi_{\alpha i}^{+}> \\ &= <\chi_{\beta f}^{-}|H - E|\chi_{\alpha i}^{+}> \end{aligned} \tag{5.1}$$

where $\chi_{\alpha i}^{+}$ and $\chi_{\beta f}^{-}$ are, respectively, the full *inelastic* scattering wavefunctions in the α and β arrangement. For example, the inelastic wavefunction $\chi_{\alpha i}^{+}$ is obtained by solving the complete inelastic scattering problem in the α arrangement. In this approach, one only needs to solve inelastic scattering problems for each arrangement channel and then carries out numerical integration in Eq. (5.1) to obtain the approximate reactive T matrix. The DWBA approach is reasonably accurate if the collision is dominated by inelastic scattering and the reaction probability is relatively small, such as in collisions at low energies that are below or near the reaction threshold [53]. When the reaction probability is large, the reactive wavefunction becomes substantially different from its inelastic approximation and the DWBA approach fails. Since it is a perturbation method, one cannot expect the DWBA method to give very accurate results but only qualitative or semi-quantitative results. In order to obtain numerically exact reactive scattering results, it is necessary to solve the scattering problem exactly, as is discussed in the following sections.

5.3 Hyperspherical Coordinate Approach

Definition of hyperspherical coordinates

The basic idea in the hyperspherical coordinate approach is to transform the reactive scattering problem into a nonreactive-like scattering problem and then use nonreactive scattering methods to solve the reactive problem. This is achieved by defining a new set of coordinates called hyperspherical coordinates which consist of a hyperradius ρ and several hyperspherical angles. The hyperradius ρ serves as a scattering coordinate that is arrangement-independent while the hyperspherical angles make up the rest of the coordinates. In the limit $\rho \to \infty$, the scattering wavefunction can reach all the asymptotic arrangements as the hyperspherical angles sweep through the coordinate space. Thus, one only has to deal with a single continuous scattering coordinate ρ and the reactive scattering problem formally becomes an inelastic scattering problem in terms of the hyperspherical coordinates. In the two dimensional case of collinear atom-diatom scattering, the hyperspherical coordinates are equivalent to the familiar polar coordinates.

In the following, we discuss the hyperspherical coordinates for a specific atom-diatom system. The hyperspherical coordinates are conveniently defined in terms of the mass-scaled Jacobi coordinates defined in Sec. 4.5.3. If we let $\bar{R}_\alpha$ and $\bar{r}_\alpha$ denote mass scaled radial Jacobi coordinates for arrangement α, the hyperradius is then defined as

$$\begin{aligned}\rho &= \sqrt{\bar{R}_\alpha^2 + \bar{r}_\alpha^2} \\ &= \sqrt{\bar{R}_\beta^2 + \bar{r}_\beta^2}\end{aligned} \tag{5.2}$$

which is *independent* of the arrangement as shown in Sec. 4.5.3. In the asymptotic region of arrangement α, $\bar{R}_\alpha \to \infty$ but $\bar{r}_\alpha$ is finite; thus the hyperradius $\rho \simeq \bar{R}_\alpha$. This property holds for other arrangements as well. Thus one can use ρ as a *unique* scattering coordinate to describe the scattering process for all arrangements. This is the main feature of the hyperspherical coordinate method that allows one to formally transform a reactive scattering problem into an inelastic one with a well defined single scattering coordinate ρ. The rest of the hyperspherical coordinates describe the bound motions of the molecular system and are called hyperspherical angles. The definitions of the hyperspherical angles are, however, non-unique. The following is one of the several definitions for the hyperspherical angles [54]

$$\omega_\alpha = 2\tan^{-1}\left(\frac{\bar{r}_\alpha}{\bar{R}_\alpha}\right) \qquad (0 \le \omega_\alpha \le \pi) \tag{5.3}$$

$$\theta_\alpha = 2\cos^{-1}\left(\frac{\mathbf{R}_\alpha \cdot \mathbf{r}_\alpha}{R_\alpha r_\alpha}\right) \qquad (0 \le \theta_\alpha \le \pi) \tag{5.4}$$

where α can be chosen to be any one of the arrangements. In this definition, the hyperspherical angles are dependent on the particular arrangement used to define them. There is another commonly used definition of hyper-angles called APH (adjusted principal axis hyperspherical coordinates) in which the two hyper-angles are independent of arrangement labels [55].

Using $(\rho, \omega_\alpha, \theta_\alpha)$ as three internal coordinates plus three Euler angles for the overall orientation of the triatomic system, the Hamiltonian of an atom-diatom system can be written as

$$H = -\frac{\hbar^2}{2\mu}\rho^{-5/2}\frac{\partial^2}{\rho^2}\rho^{5/2} + \frac{15\hbar^2}{8\mu\rho^2} + H_S \tag{5.5}$$

where H_S is called the surface Hamiltonian. For atom-diatom systems, H_S is defined as

$$H_S = \frac{1}{2\mu\rho^2}\left[-\frac{4\hbar^2}{\sin\omega_\alpha}\left(\frac{\partial^2}{\partial\omega_\alpha^2} + 1\right)\sin\omega_\alpha + \frac{(\mathbf{J} - \mathbf{j}_\alpha)^2}{\cos^2(\omega_\alpha/2)} + \frac{\mathbf{j}_\alpha^2}{\sin^2(\omega_\alpha/2)}\right] + V(\rho, \omega_\alpha, \gamma_\alpha) \tag{5.6}$$

where $\mathbf{J}$ is the total angular momentum of the system and $\mathbf{j}_\alpha$ is the angular momentum of the diatom in the α arrangement.

Now there is a single scattering coordinate ρ that extends from the origin to infinity, just as in inelastic scattering. The various asymptotic arrangements are encompassed through the sweeping of the hyperspherical angles. In the hyperspherical coordinate approach to reactive scattering, one usually starts by solving for the multidimensional "surface" eigenfunctions

$$H_S^i \Phi_n^i = \epsilon_n^i \Phi_n^i \tag{5.7}$$

for fixed values of ρ_i. The full scattering wavefunction is then expanded in terms of these quasi-adiabatic surface functions and the coupled radial equation is propagated in the scattering coordinate ρ from the origin outward to match the asymptotic boundary conditions for various arrangements. Since the hyperspherical coordinates are not the correct coordinates to represent the asymptotic wavefunction, the matching to asymptotic states is usually accompanied by a coordinate transformation from the hyperspherical coordinates to the appropriate Jacobi coordinates. Detailed numerical implementation of the hyperspherical coordinates approach to atom-diatom reactive scattering is not presented here.

5.4 Variational Algebraic Approach

Although the use of hyperspherical coordinates *formally* transforms a reactive scattering problem into a nonreactive-like scattering problem in order to apply the standard inelastic computational techniques for calculation, it pays a heavy price for it. First, the Schrödinger equation expressed in hyperspherical coordinates is very complicated. Secondly, the hyperspherical surface functions that are used in coupled channel expansion of the scattering wavefunction are very difficult to calculate. This is because the hyperspherical surface functions are nonlocal bound state wavefunctions that span both the reactant and product arrangement valleys. In fact, the majority of the computational cost in the hyperspherical coordinate approach is in the calculation of hyperspherical surface functions that have to be calculated for many values of ρ_i.

A quite different approach to reactive scattering is the variational algebraic approach which provides a physically intuitive treatment for reactive scattering problems and is completely general, at least in formalism. The following section is devoted to the discussion of the variational algebraic approach.

5.4.1 Multi-arrangement Expansion of Wavefunction

At the heart of algebraic approach to reactive scattering is the multi-arrangement expansion of the full scattering wavefunction using nonorthogonal basis sets. In view of the asymptotic expansions of the full scattering wavefunction given in Eq. (4.271), we can expand the full scattering wavefunction Ψ_α in arrangement component wavefunctions $\phi_{\gamma\alpha}$ [56, 57]

$$\Psi_\alpha = \sum_\gamma \psi_{\gamma\alpha} \tag{5.8}$$

where the superscript α labels the initial arrangement. These arrangement wavefunctions are not orthogonal to each other

$$<\psi_{\gamma\alpha}|\psi_{\beta\alpha}> \neq 0 \qquad (\gamma \neq \beta) \tag{5.9}$$

In fact it is precisely this nonorthogonal property of the basis that gives rise to reactions.

The physical picture behind the multi-arrangement expansion in Eq. (5.8) is the idea of linear combination of atomic orbitals (LCAO) which forms the basis for modern electronic structure calculations. Here, the arrangement channel basis functions are like atomic orbitals in electron systems and the delocalized scattering wavefunction is like the molecular orbital. In view of this similarity, we might call the arrangement component wavefunction $\phi_{\gamma\alpha}$ "arrangement orbitals". We can therefore interpret the reactive scattering event as an atomic exchange process resulting from the overlap of arrangement orbitals. The comparison of expansion (5.8) to LCAO provides not only a clear and physically intuitive picture for reactive scattering processes, but also a practical means for numerical computation of reactive scattering problems.

5.4.2 Coupled Arrangement Integral Equation

Using the ansatz of (5.8), we can derive a coupled arrangement integral equation for reactive scattering, which is a generalization of the Lippmann-Schwinger equation for single arrangement scattering. [58]. Using expansion (5.8) in the Schrödinger equation

$$(E - H)\sum_{\gamma} \psi_{\gamma\alpha} = 0 \tag{5.10}$$

we can rearrange the above equation to obtain

$$\begin{aligned}(E - H_\beta)\psi_{\beta\alpha} &= V_\beta \psi_{\beta\alpha} + \sum_{\gamma \neq \beta} (H - E)\psi_{\gamma\alpha} \\ &= \sum_{\gamma} U_{\beta\gamma}\psi_{\gamma\beta}\end{aligned} \tag{5.11}$$

The above equation can be integrated to yield the coupled arrangement integral equation for the arrangement component wavefunction

$$\psi_\beta^\alpha = \phi_\alpha \delta_{\beta\alpha} + \sum_{\gamma} G_\beta \mathbf{U}_{\beta\gamma} \psi_\gamma^\alpha \tag{5.12}$$

Here the generalized arrangement-coupled potential matrix operator U is defined as

$$\mathbf{U}_{\beta\gamma} = V_\beta \delta_{\beta\gamma} + (H - E)(1 - \delta_{\beta\gamma}) \tag{5.13}$$

which can be written in matrix form

$$\mathbf{U} = \begin{bmatrix} V_1 & (H-E) & \cdots \\ (H-E) & V_2 & \cdots \\ \vdots & \vdots & \ddots \end{bmatrix} \tag{5.14}$$

It is easy to see from the matrix definition in Eq. (5.14) that **U** is hermitian.

Similarly we can define the asymptotic arrangement Hamiltonian written in matrix form

$$\mathbf{H}_0 = \begin{bmatrix} H_1 & 0 & \cdots \\ 0 & H_2 & \cdots \\ \vdots & \vdots & \ddots \end{bmatrix} \tag{5.15}$$

We can also define the asymptotic arrangement Green's function as

$$\mathbf{G}_0^+ = [E\mathbf{I} - \mathbf{H}_0 + i\epsilon]^{-1} \tag{5.16}$$

which is written in matrix form

$$\mathbf{G}_0^+ = \begin{bmatrix} G_1^+ & 0 & \cdots \\ 0 & G_2^+ & \cdots \\ \vdots & \vdots & \ddots \end{bmatrix}. \tag{5.17}$$

where

$$G_\alpha^+ = [E - H_\alpha + i\epsilon]^{-1} \tag{5.18}$$

Finally, we can define the matrix full scattering wavefunction

$$\psi = \begin{bmatrix} \psi_{11} & \psi_{12} & \cdots & \psi_{1N} \\ \psi_{21} & \psi_{22} & \cdots & \psi_{2N} \\ \vdots & \vdots & \ddots & \vdots \\ \psi_{N1} & \psi_{N2} & \cdots & \psi_{NN} \end{bmatrix} \tag{5.19}$$

and similarly define the matrix asymptotic wavefunction

$$\phi = \begin{bmatrix} \phi_1 & 0 & \cdots \\ 0 & \psi_2 & \cdots \\ \vdots & \vdots & \ddots \end{bmatrix} \tag{5.20}$$

Using above matrix definitions, the coupled arrangement integral equation of (5.12) can be cast in the form of a matrix LS equation

$$\boxed{\mathbf{\Psi} = \mathbf{\Phi} + \mathbf{G}_0\mathbf{U}\mathbf{\Psi}} \tag{5.21}$$

We can also define the matrix full Green's function as

$$\mathbf{G}^+ = [E\mathbf{I} - \mathbf{H}_0 - \mathbf{U} + i\epsilon]^{-1} \tag{5.22}$$

and rewrite Eq. (5.21) in terms of the matrix full Green's function

$$\boxed{\mathbf{\Psi} = \mathbf{\Phi} + \mathbf{G}\mathbf{U}\mathbf{\Psi}} \tag{5.23}$$

Equation (5.21) is a multi-arrangement generalization of the single arrangement (inelastic) Lippmann-Schwinger equation for reactive scattering. The generalized arrangement coupled potential $\mathbf{U}$ has the following properties:

1) it couples different arrangements symmetrically;

2) it is energy-dependent.

Because of the formal similarity of Eq. (5.21) to the standard LS equation for inelastic scattering, essentially all the results discussed for the single arrangement LS equation can be easily generalized to the case of multi-arrangements. For example, the T matrix operator is now defined in matrix form

$$\mathbf{T\Phi} = \mathbf{U\Psi} \tag{5.24}$$

from which one can derive the LS equation for the T matrix operator

$$\mathbf{T} = \mathbf{U} + \mathbf{U}\mathbf{G}^+\mathbf{U} \tag{5.25}$$

It is easy to verify that the definition of Eq. (5.24) yields the correct reactive T matrix element as given in Eq. (4.266). From Eq. (5.13), we have

$$<\phi_\beta|\mathbf{T}_{\beta\alpha}|\phi_\alpha> = \sum_\gamma <\phi_\beta|\mathbf{U}_{\beta\gamma}|\psi_{\gamma\alpha}>$$

$$= \sum_{\gamma} <(H-E)\phi_\beta|\psi_{\gamma\alpha}>$$

$$= <V_\beta\phi_\beta|\sum_{\gamma}\psi_{\gamma\alpha}>$$

$$= \sum_{\gamma} <\phi_\beta|V_\beta|\Psi_\alpha> \tag{5.26}$$

where we have used the result

$$\mathbf{U}_{\beta\gamma}|\phi_\beta>= (H-E)|\phi_\beta>= V_\beta|\phi_\beta> \tag{5.27}$$

Similar definitions and results hold for the K matrix as well. Thus with the matrix generalizations of $\Psi \to \mathbf{\Psi}$, $V \to \mathbf{U}$, and $G \to \mathbf{G}$, we can formally treat the reactive scattering problem as a generalized inelastic scattering problem by solving the matrix version of the LS equation.

5.4.3 Algebraic Method

Because of formal similarity of the matrix LS equation for reactive scattering to that of inelastic scattering, we do not need to explicitly specify whether we are dealing with reactive or inelastic scattering in the following discussion of algebraic methods. Although we make specific references to inelastic scattering, the results can be easily generalized to the case of reactive scattering. Starting from the standard LS equation for nonreactive scattering

$$\Psi_i = \Phi_i + G_0 V \Psi_i, \tag{5.28}$$

where i is the label for the initial channel and G_0 can be either G_0^+ or G_0^p, we multiply the above equation by the potential V from the left to obtain the equation for the amplitude density [57]

$$\xi_i = V\Phi_i + VG_0\xi_i \tag{5.29}$$

where the amplitude density is defined as $\xi_i = V\Psi_i$.. The K (or T) matrix can be directly calculated from the amplitude density

$$K_{fi} =<\Phi_f|V|\Psi_i>=<\Phi_f|\xi_i> \tag{5.30}$$

The amplitude density ξ_i behaves like a bound state wavefunction because it vanishes asymptotically where the interaction potential V vanishes although the scattering wavefunction Ψ_i does not. Thus it is convenient to directly expand ξ in any $\mathcal{L}^2$ basis set $u_n (n = 1, 2, \cdots)$,

$$\xi_i = \sum_n u_n a_{ni} \tag{5.31}$$

By substituting the above expansion for ξ_i into Eq. (5.29) and integrating over the basis functions u_n, we obtain linear algebraic equations for the amplitude density

$$\mathbf{Oa} = \mathbf{B} + \mathbf{C}'\mathbf{a} \tag{5.32}$$

where the various matrices are defined as

$$\mathbf{O}_{mn} = <u_m|u_n> \tag{5.33}$$

$$\mathbf{B}_{mi} = <u_m|V\Phi_i> \tag{5.34}$$

$$\mathbf{C}'_{mn} = <u_m|VG_0|u_n> \tag{5.35}$$

Here the basis functions u_n are assumed to be nonorthogonal for purpose of generality.

The linear algebraic Eq. (5.32) can be solved by simple matrix inversion to yield the expansion coefficients a_{ni}

$$\begin{aligned} \mathbf{a} &= [\mathbf{O} - \mathbf{C}']^{-1}\mathbf{B} \\ &= \mathbf{C}^{-1}\mathbf{B} \end{aligned} \tag{5.36}$$

where we define a new matrix

$$\mathbf{C} = \mathbf{O} - \mathbf{C}' \tag{5.37}$$

After the expansion coefficients $\mathbf{a}$ are obtained, the K matrix in Eq. (5.30) can be calculated by

$$\begin{aligned} \mathbf{K}_{fi} &= \sum_n <\Phi_f|u_n> a_{ni} \\ &= \boxed{\mathbf{D}^T\mathbf{C}^{-1}\mathbf{B}} \end{aligned} \tag{5.38}$$

where the matrix $\mathbf{D}$ is defined as

$$\mathbf{D}^T_{fn} = <\Phi_f|u_n> \tag{5.39}$$

After the K matrix is computed, we can obtain the T or S matrices through standard relations given in Chapter 4. Since we calculate the K matrix directly, G_0 is the principal value Green's function and all quantities can be chosen to be real. We note that Eq. (5.38) is not explicitly symmetric. If the numerical calculation is accurate, however, the K matrix should be symmetric or very nearly symmetric.

In the above equations for various matrix definitions, the labels of various matrices should be classified into two different types. The summation index n in Eq. (5.31) is a composite index that includes labels for basis functions of all degrees of freedom, or more clearly, it includes both internal channel functions and radial basis functions. Thus the total number of m or n values can be very large. On the other hand, the index i is a label for asymptotic open channels which does not include closed channels and radial basis functions. In general, the total number of i values is much smaller than the total number of n values by typically an order of magnitude. The largest matrix to be computed is the $\mathbf{C}$ matrix.

Since this algebraic method is very general, it can be straightforwardly applied to solve the coupled arrangement integral equation (5.21) for reactive scattering without any modification. Using this algebraic method, the calculation of scattering problems, reactive or inelastic, is very much like that for bound state problems: one chooses a basis set to expand the unknown scattering amplitude (or wavefunction), evaluating the necessary matrix elements of $\mathbf{O}, \mathbf{B}, \mathbf{C}, \mathbf{D}$, solving the linear algebraic equations to obtain the expansion coefficient matrix $\mathbf{a}$, and finally evaluates the K matrix through a simple matrix multiplication in Eq. (5.38). The basic procedures of the method are the *same* whether one solves for an inelastic scattering problem or a reactive scattering problem. The difference, however, is in the details of choosing the appropriate basis functions and the complexity of calculating the various Hamiltonian matrices.

The main numerical complication in solving the reactive scattering problem using the algebraic method is the calculation of "exchange" matrix elements. In the case of reactive scattering, the expansion of the amplitude density in Eq. (5.31) will include basis functions defined in all energetically accessible arrangements, viz.,

$$\xi_{\alpha i} = \sum_{\gamma n} u_{\gamma n} a_{\gamma n, \alpha i} \tag{5.40}$$

As is discussed in Sec. 5.4.1, the arrangement channel basis functions (like atomic orbitals in electronic structure calculations) are nonorthogonal. If we use $u_{\alpha m}$ and $u_{\beta n}$ to denote basis functions defined in the α and β arrangement, respectively, the overlap integral is given by

$$\mathbf{O}_{\alpha m, \beta n} = <u_{\alpha m}|u_{\beta n}> \neq \delta_{\alpha\beta}\delta_{mn} \tag{5.41}$$

Similarly one also needs to evaluate "exchange" matrix elements of the type

$$\mathbf{C}_{\alpha m,\beta n} = <u_{\alpha m}|\mathbf{U}_{\alpha\beta}G_\beta|u_{\beta n}> \tag{5.42}$$

Since for different γ, the basis functions $u_{\gamma n}$ are defined with respect to Jacobi coordinates of different arrangements, the above integrals are multidimensional and their evaluation requires a multidimensional coordinate transformation between Jacobi coordinates of different arrangements. This is a major source of computational cost in reactive scattering calculations.

5.4.4 Schwinger Variational Method

Since the numerically calculated scattering wavefunction Ψ or the amplitude density ξ inevitably has small errors $\delta\Psi$, it is desirable to improve the accuracy of the calculated scattering matrix T or S which contains all the scattering information. A mathematically elegant approach is to apply variational methods to evaluate the scattering matrices. The rationale of using the variational principle to evaluate the scattering matrix is to reduce the error in the calculated scattering matrix to second order in $\delta\Psi$. For example, since the K matrix has different mathematical expressions

$$K_{fi} = <\Psi_f|V|\Phi> = <\Phi_f|V|\Psi_i> = <\Psi_f|V - VG_0^pV|\Psi_i> \tag{5.43}$$

we can use the *variational* expression to evaluate the K matrix

$$\boxed{K_{fi} = <\Psi_f|V|\Phi_i> + <\Phi_f|V|\Psi_i> - <\Psi_f|V - VG_0^pV|\Psi_i>} \tag{5.44}$$

instead of the direct expression $K_{fi} = <\Phi_f|V|\Psi_i>$. Equation (5.44) is the Schwinger variational expression for the K matrix whose first order variation with respect to the variation of the scattering wavefunction Ψ can be shown to vanish, i.e.,

$$\delta K_{fi} = 0 \tag{5.45}$$

This can be proved by carrying out the explicit variation in Eq. (5.44) with respect to $\delta\Psi$.

$$\begin{aligned}\Delta K_{fi} &= <\delta\Psi_f|V|\Phi_i> + <\Phi_f|V|\delta\Psi_i> - <\delta\Psi_f|V - VG_0^pV|\Psi_i> \\ &\quad - <\Psi_f|V - VG_0^pV|\delta\Psi_i> \\ &= <\delta\Psi_f|V\ \ [|\Phi_i> - |\Psi_i> + G_0^pV|\Psi_i>] \\ &\quad + [<\Phi_f| - <\Psi_f| + <\Psi_f|VG_0^pV]\ \ |\delta\Psi_i>\end{aligned} \tag{5.46}$$

Since the terms in the brackets vanish via the LS equation, this proves that ΔK_{fi} vanishes to first order in $\delta\Psi$. Thus, if the error of the numerically calculated wavefunction is $\delta\Psi$, the corresponding error in the K matrix evaluated via the variational expression (5.44) is of the order $(\delta\Psi)^2$. This variational property generally helps converge the calculated scattering matrices in numerical computation.

There is also a fractional form of the Schwinger variational method

$$\boxed{K_{fi} = \frac{<\Psi_f|V|\Phi_i><\Phi_f|V|\Psi_i>}{<\Psi_f|V - VG_0^p V|\Psi_i>}} \tag{5.47}$$

which is essentially a normalized version of Eq. (5.44), and is independent of the normalization of the scattering wavefunction Ψ. It can be shown easily that this fractional expression is variational as well.

To obtain linear algebraic equations from variational expressions, one expands the scattering wavefunction Ψ in an $\mathcal{L}^2$ basis as in Eq. (5.31)

$$\Psi_i = \sum_n u_n a_{ni} \tag{5.48}$$

and uses the variational condition (5.45) to determine the expansion coefficients a_{ni}. The reason that one can use an $\mathcal{L}^2$ basis to expand the continuous wavefunction Ψ is due to the presence of V in front of every Ψ in Eq. (5.44) that kills off the contribution from the continuum part of Ψ to the integral. As a result, the scattering wavefunction Ψ can be treated like an $\mathcal{L}^2$ integrable quantity. This may seem strange but it is true because the correct asymptotic boundary condition of Ψ is actually enforced by the Green's function G_0^p in the equation.

Using the wavefunction expansion of (5.48), Eq. (5.44) can be written in matrix form

$$\mathbf{K} = \mathbf{a}^T\mathbf{B} - \mathbf{B}^T\mathbf{a} - \mathbf{a}^T\mathbf{C}\mathbf{a} \tag{5.49}$$

where the matrices $\mathbf{B}$ and $\mathbf{C}$ are defined as

$$\mathbf{B}_{ni} = <u_n|V|\Phi_i> \tag{5.50}$$

and

$$\mathbf{C}_{mn} = <u_n|V - VG_0^p V|u_n> \tag{5.51}$$

By setting the first derivative of $\mathbf{K}$ with respect to $\mathbf{a}^T$ equal to zero in the variational expression (5.49), one obtains the solution for the expansion coefficient

$$\mathbf{a} = \mathbf{C}^{-1}\mathbf{B} \tag{5.52}$$

A more rigorous derivation with respect to variation of individual coefficients a_{ni} will lead to the same result. Substituting Eq. (5.52) in Eq. (5.49), one obtains the variational result for the K matrix

$$\boxed{K_{fi} = \mathbf{B}^T\mathbf{C}^{-1}\mathbf{B}} \tag{5.53}$$

Although the variational result of Eq. (5.53) looks very similar to the nonvariational result of Eq. (5.38), Eq. (5.53) is numerically superior to Eq. (5.38). Firstly, the numerical error of the variational result is of second order in the error of the wavefunction $\delta\Psi$ while the error of the nonvariational result is in first order. Thus we expect the variational expression of Eq. (5.53) to give a more accurate result than the nonvariational expression of Eq. (5.38) for a given finite basis set. Secondly, Eq. (5.53) is explicitly symmetric and therefore guarantees the unitarity of the S matrix while Eq. (5.38) does not. Thus numerically, it is preferable to use the variational method to improve the accuracy of the numerical calculation. Besides these differences, the calculational procedures of the variational method for scattering (reactive or inelastic) are essentially the same as the nonvariational method discussed in the previous subsection, and are therefore not repeated here.

For reactive scattering, the expansion of the wavefunction can be explicitly written to include the arrangement label γ

$$\Psi_{\alpha i} = \sum_{\gamma n} u_{\gamma n} a_{\gamma n,\alpha i} \tag{5.54}$$

and one needs to evaluate similar exchange integrals as in the nonvariational method discussed in the previous section. However, the variational formalism remains completely general for either reactive or nonreactive scattering.

5.4.5 Kohn Variational Method

The Kohn variational principle can be applied to any of the scattering matrices such as the S and K matrices. There is, however, an important difference between the K matrix and the S matrix variational method. Although, the K matrix variational method involves only real functions, it is numerically less stable than the S matrix version due to the so called "Kohn anomalies" [59, 60]. For this reason, it is preferable to use the S matrix version of the Kohn variational principle for numerical calculations. If we denote $\tilde{\Psi}_i^+$ as a trial scattering wavefunction with outgoing boundary condition and $\tilde{\Psi}_i^-$ as its complex conjugate, the Kohn variational method gives the following expression for the S matrix element

$$S_{fi} = \tilde{S}_{fi} + \frac{i}{\hbar} <\tilde{\Psi}_f^-|H - E|\tilde{\Psi}_i^+> \tag{5.55}$$

where $\tilde{S}_{fi}$ is the S matrix element extracted from the asymptotic expression of the trial wavefunction $\tilde{\Psi}_i^+$. The correctly normalized $\tilde{\Psi}_i^+$ corresponding to Eq. (5.55) has the asymptotic boundary condition

$$\begin{aligned}\tilde{\Psi}_i^+(r \to \infty) \quad &\sim \quad \sum_n \left[-\frac{e^{-k_i r}}{\sqrt{v_i}} \varphi_i \delta_{ni} + \frac{e^{ik_n r}}{\sqrt{v_n}} \varphi_n \tilde{S}_{ni} \right] \\ &= -u_{0i}(r)\varphi_i + \sum_n u_{1n}(r)\varphi_n \tilde{S}_{ni} \end{aligned} \tag{5.56}$$

where the asymptotic functions are defined as

$$u_{1n}(r) = u_{0n}^*(r) = \frac{e^{ik_n r}}{\sqrt{v_n}} \tag{5.57}$$

with $v_n = \hbar k_n / m$.

Nonhermiticity of the kinetic operator

To prove the stationary property of Eq. (5.55), we first need to clarify the fact that the Hamiltonian operator $H = T + V$ is *not* hermitian with respect to two scattering (continuum) wavefunctions Ψ_1 and Ψ_2, i.e.,

$$\begin{aligned}<\Psi_2|H^\dagger|\Psi_1> &= <H\Psi_2|\Psi_1> \\ &= <\Psi_1|H|\Psi_2>^* \\ &\neq \ <\Psi_2|H|\Psi_1> \end{aligned} \tag{5.58}$$

This is because the kinetic energy operator T gives rise to a surface integral

$$\begin{aligned}<\Psi_2|T^\dagger|\Psi_1> &= <\Psi_1|T|\Psi_2>^* \\ &= -\frac{\hbar^2}{2m} \left[\int_0^\infty \Psi_1^*(r) \Psi_2^{''}(r) dr \right]^* \\ &= -\frac{\hbar^2}{2m} \left[\Psi_1 \Psi_2^{*'} - \Psi_1^{'} \Psi_2^* \right] \Bigg|_{r=0}^{r=\infty} + <\Psi_2|T|\Psi_1> \end{aligned} \tag{5.59}$$

Although the scattering wavefunctions vanish at the origin, they give a non-vanishing contribution at $r = \infty$ to the first term in Eq. (5.59), which is in general nonzero when both Ψ_1 and Ψ_2 are continuous wavefunctions. Thus the Hamiltonian operator is not hermitian in this sense, and we should be aware of the inequality in (5.58). However, if at least one of these two wavefunctions is of finite range, then the Hamiltonian operator is hermitian because the surface integral in Eq. (5.59) vanishes in this case. Thus great care has to be exercised when dealing with integrals of this type.

Stationary condition

Following the preceding discussion on hermiticity of the Hamiltonian, we can evaluate the variation of the Kohn integral with respect to variation of the wavefunction $\delta\Psi = \tilde{\Psi} - \Psi$

$$\begin{aligned}\delta I_{fi} &= \frac{i}{\hbar}\left[<\delta\Psi_f^-|H-E|\Psi_i^+> + <\Psi_f^-|H-E|\delta\Psi_i^+>\right] \\ &= \frac{i}{\hbar}<\Psi_f^-|H-E|\delta\Psi_i^+>\end{aligned} \tag{5.60}$$

where the Schrödinger equation $(E-H)\Psi = 0$ satisfied by the exact wavefunction Ψ has been employed to eliminate the first term. If the Hamiltonian is written as the kinetic energy operator for the radial coordinate r plus H' for the remaining coordinates $\mathbf{q}$

$$H = -\frac{\hbar^2}{2m}\frac{\partial^2}{\partial r^2} + H'(\mathbf{q}), \tag{5.61}$$

then upon partial integration, Eq. (5.60) becomes

$$\delta I_{fi} = \frac{i}{\hbar}(-\frac{\hbar^2}{2m})\int d\mathbf{q}\left[\Psi_f^{-*}\frac{\partial\delta\Psi_i^+}{\partial r} - \frac{\partial\Psi_f^{-*}}{\partial r}\delta\Psi_i^+\right]\Bigg|_{r=0}^{r=\infty} \tag{5.62}$$

Using the regular condition for the scattering wavefunction at $r=0$

$$\Psi(r=0) = \delta\Psi(0) = 0, \tag{5.63}$$

and the asymptotic condition from (5.56) for Ψ_f^- and Ψ_i^+

$$\Psi_f^-(r\to\infty) \sim -u_{0f}^*\varphi_f + \sum_n u_{1n}^*\varphi_n\tilde{S}_{nf}^* \tag{5.64}$$

$$\delta\Psi_i^+(r\to\infty) \sim \sum_n u_n(r)\varphi_n(\mathbf{q})\delta\tilde{S}_{ni} \tag{5.65}$$

we derive the variational condition

$$\begin{aligned}\delta I_{fi} &= -\frac{i\hbar}{2m}\left[-u_{0f}(r)u'_{1i}(r) + u_{0f}(r)u_{1i}(r)\right]\Big|_{r=\infty} \delta\tilde{S}_{fi} \\ &= -\delta\tilde{S}_{fi}\end{aligned} \tag{5.66}$$

where the explicit expressions for u_{0n} and u_{1n} in Eq. (5.57) have been used. Hence, the variation δI_{fi} cancels the first term in the Kohn expression of Eq. (5.55). Thus we have proved that the Kohn expression for the S matrix in Eq. (5.55) is stationary with respect to small changes of the scattering wavefunction from the exact value, viz.,

$$\delta S_{fi} = 0 \tag{5.67}$$

Proof for reactive scattering

The same proof is also valid for reactive scattering. In this case, the scattering wavefunction is a superposition of arrangement wavefunctions

$$\Psi_{\alpha i} = \sum_{\gamma} \psi_{\gamma}^{\alpha i} \tag{5.68}$$

We can write the variation of the integral as

$$\begin{aligned}\delta I_{\beta f,\alpha i} &= \frac{i}{\hbar} <\Psi_{\beta f}^{-}|H - E|\Psi_{\alpha i}^{+}> \\ &= \frac{i}{\hbar}\sum_{\gamma} <\psi_{\gamma}^{\beta f}|H_{\gamma} + V_{\gamma} - E|\delta\psi_{\gamma}^{\alpha i}> \\ &\quad + \frac{i}{\hbar}\sum_{\alpha' \neq \beta'} <\psi_{\beta'}^{\beta f}|H - E|\delta\psi_{\alpha'}^{\alpha i}>\end{aligned} \tag{5.69}$$

Using the asymptotic condition for the reactive wavefunction in Eq. (4.271), it can be shown that the surface integral resulting from partial integration in the first term of Eq. (5.69) vanishes unless $\gamma = \beta$. Therefore the first term in Eq. (5.69) becomes

$$\begin{aligned}\sum_{\gamma} <\psi_{\gamma}^{\beta f}|H - E|\delta\psi_{\gamma}^{\alpha i}> &= i\hbar\tilde{S}_{\beta f,\alpha i} \\ &\quad + \sum_{\gamma} <\delta\psi_{\gamma}^{\alpha i}|H - E|\psi_{\gamma}^{\beta f}>^{*}\end{aligned} \tag{5.70}$$

Since the exchange integral is of short range, the Hamiltonian operator is hermitian in the second term of Eq. (5.69) which therefore vanishes

$$\sum_{\alpha'\neq\beta'} <\psi_{\beta'}^{\beta f}|H-E|\delta\psi_{\alpha'}^{\alpha i}> = \sum_{\alpha'\neq\beta'} <\delta\psi_{\beta'}^{\alpha i}|H-E|\psi_{\alpha'}^{\beta f}>^* \tag{5.71}$$

Thus, Eq. (5.69) simplifies to

$$\delta I_{\beta f,\alpha i} = -\tilde{S}_{\beta f,\alpha i} + <\delta\Psi^{\alpha i}|H-E|\Psi^{\beta f}>^* = -\tilde{S}_{\beta f,\alpha i} \tag{5.72}$$

which proves the variational property of Eq. (5.55) for reactive scattering.

Variational algebraic expression for the S matrix

Taking into consideration the regular boundary condition at the origin and the asymptotic outgoing wave boundary condition of the scattering wave-function, we can express Ψ_i^+ in the form

$$\Psi_i^+ = -u_{0i}(r)\varphi_i(\mathbf{q}) + \sum_{n=1}^{N_0} u_{1n}(r)\varphi_n(\mathbf{q})S_{ni} + \Gamma_i \tag{5.73}$$

where we require

$$u_{1n} = u_{0i}^* \begin{cases} \overset{r\to 0}{\longrightarrow} 0 \\ \overset{r\to\infty}{\longrightarrow} \dfrac{e^{ik_n r}}{\sqrt{v_n}} \end{cases} \tag{5.74}$$

and $\Gamma_i(r=0) = \Gamma_i(r=\infty) = 0$. We can thus expand Γ_i in an $\mathcal{L}^2$ basis

$$\Gamma_i = \sum_{t>1,n} u_{tn}(r)\varphi_n(\mathbf{q})a_{tn,i} \tag{5.75}$$

where $u_t(r)$ are any $\mathcal{L}^2$ basis functions for the radial coordinate and the label for u_t is deliberately set to start from $t = 2, 3, \cdots$ for convenience. Equation (5.73) can now be written as

$$\Psi_i^+ = -u_{0i}(r)\varphi_i(\mathbf{q}) + \sum_{t=1,n} u_{tn}\varphi_n(\mathbf{q})a_{tn,i} \tag{5.76}$$

with the prescription $u_{tn} = u_t$ for $t > 1$ and $a_{1n,i} = \tilde{S}_{ni}$.

Since the variational condition $\delta S_{fi} = 0$ in Eq. (5.55) simply yields the stationary Schrödinger equation $(E-H)\Psi_{i,f} = 0$, by substituting the expansion of (5.76) for Ψ_i^+ in the Schrödinger equation, we obtain the linear algebraic equations for the expansion coefficients

$$\mathbf{Ca} = \mathbf{B} \tag{5.77}$$

where the matrices are defined as

$$\mathbf{B}_{tm,n} = <u_{tn}^* \varphi_n | H - E | u_{0n} \varphi_n> \tag{5.78}$$

and

$$\mathbf{C}_{tm,t'n} = <u_{tm}^* \varphi_n | H - E | u_{t'n} \varphi_n> \tag{5.79}$$

for $(t, t' = 1, 2, \cdots)$. The variational expression (5.55) for the S matrix can be shown to give rise to the S matrix

$$\boxed{\mathbf{S} = \frac{i}{\hbar}\left(\mathbf{C}_{00} - \mathbf{B}^T \mathbf{C}^{-1} \mathbf{B}\right)} \tag{5.80}$$

where $\mathbf{C}_{00}$ is defined as

$$[\mathbf{C}_{00}]_{mn} = <u_{0m}^* \varphi_m | H - E | u_{0n} \varphi_n> \tag{5.81}$$

If we define a "super" matrix $\mathbf{M}$ by

$$\mathbf{M}_{tm,t'n} = <u_{tm}^* \varphi_n | H - E | u_{t'n} \varphi_n> \qquad (t, t' = 0, 1, 2, \cdots) \tag{5.82}$$

which can be written in block matrix form

$$\mathbf{M} = \begin{bmatrix} \mathbf{C}_{00} & \mathbf{B} \\ \mathbf{B}^T & \mathbf{C} \end{bmatrix} \tag{5.83}$$

then we can rewrite Eq. (5.80) for the S matrix in a compact form

$$\boxed{\mathbf{S} = \frac{i}{\hbar}\left[(\mathbf{M}^{-1})_{00}\right]^{-1}} \tag{5.84}$$

Here the well known Löwdin-Feshbach partitioning of the matrix is employed [61].

For reactive scattering, all the matrices $\mathbf{B}, \mathbf{C}, \mathbf{M}$, etc., have the following common structure

$$\mathbf{X} = \begin{bmatrix} X_{11} & X_{12} & \cdots \\ X_{21} & X_{22} & \cdots \\ \vdots & \vdots & \ddots \end{bmatrix} \tag{5.85}$$

where the diagonal matrices $X_{\alpha\alpha}$ are "direct" Hamiltonian matrices constructed from basis functions defined in the same arrangement. The off diagonal matrices $X_{\beta\alpha}(\beta \neq \alpha)$ are "exchange" Hamiltonian matrices constructed from basis functions from different arrangements. The calculation of these direct and exchange matrices will be discussed in the following section.

Selection of basis functions

In numerical applications of the Kohn variational method, one needs to select proper basis functions to expand the scattering wavefunction. The internal functions φ_n are normally chosen to be eigenfunctions of a reference Hamiltonian with the scattering coordinate fixed at some value. If φ_n are chosen to be eigenfunctions of the asymptotic Hamiltonian, only open channels of φ_n that are associated with the free radial functions $u_{1n} = u_{0n}^*$ are needed in the expansion of the scattering wavefunction. However, both open and closed channels φ_n that are associated with the $\mathcal{L}^2$ functions $u_{tn}(t > 1)$ are needed. To satisfy the boundary condition (5.74), one can simply multiply the asymptotic free functions by a cutoff (or switching) function, e.g.,

$$u_{0n} = f_{cut}\frac{e^{-ik_nR}}{\sqrt{v_n}} \tag{5.86}$$

where the cutoff function f_{cut} has the behavior

$$\begin{cases} f_{cut}(r \to 0) \longrightarrow 0 \\ f_{cut}(r \to \infty) \longrightarrow 1 \end{cases}$$

A good cutoff function is

$$f_{cut} = exp[-(r_0/r)^p] \tag{5.87}$$

which is essentially a smoothed step function. The parameters r_0 and p can be optimized in actual numerical calculations.

For reactive scattering, the expansion of the wavefunction in Eq. (5.76) can be written explicitly to include arrangement labels

$$\Psi_{\alpha i}^+ = -u_{0\alpha i}(r_\alpha)\varphi_i(\mathbf{q}_\alpha) + \sum_{\gamma tn} u_{\gamma tn}(r_\gamma)\varphi_{\gamma n}(\mathbf{q}_\gamma)a_{\gamma tn,\alpha i} \tag{5.88}$$

and will result in the following exchange integrals between basis functions of different arrangements

$$\mathbf{M}_{\alpha tm,\beta t'n} = <u_{\alpha tm}^*\varphi_{\alpha m}|H - E|u_{\beta t'n}\varphi_{\beta n}> \tag{5.89}$$

in addition to the direct integrals between basis functions of the same arrangement.

5.5 Atom-Diatom Reactive Scattering

Background

The ability to accurately predict the outcome of a chemical reaction in detail based on first principles has long been a holy grail for theoretical chemists since the discovery of quantum mechanics in the 1920s. It has been known that chemical reactions are the results of molecular collisions that can be rigorously described, at least in principle, by quantum reactive scattering theory. In principle, the whole scenario seems rather straightforward. First, one performs an *ab initio* quantum chemistry calculation to generate electronic energies at various nuclear configurations and fit them to a global potential energy surface. Secondly, one performs a quantum reactive scattering calculation to obtain detailed dynamics information such as reaction probabilities, cross sections, rate constants, etc.. In practice, however, such a theoretical endeavor is a formidable computational task at best because quantitatively accurate calculations for the majority of chemical reactions are enormously complex due to the inherent mathematical difficulties in solving many-body Schrödinger equations.

Besides the huge computational cost required in electronic structure calculations to generate accurate potential energy surfaces, which is outside the topic of the present work, the quantum reactive scattering calculation itself presents a major challenge to theoretical dynamicists in the study of reaction dynamics. From the very beginning, the theory of quantum reactive scattering has focused on the development of computational methodologies for the atom-diatom reaction A + BC → AB + C, AC + B which is the simplest possible chemical reaction system. In particular, H + H_2 and its isotopically substituted reactions have dominated theoretical and computational study of reactive scattering for several decades [52,62]. After the first report of a three dimensional quantum calculation for the H + H_2 reaction on the empirical PK2 potential energy surface in 1976 [52], it took another decade or so for theoretical chemists to fully develop general and powerful numerical methods to compute atom-diatom reactions. Thanks to the development of new computational methods [63–71] and the rapid increase in the speed of modern computers, tremendous progress has been made during the past decade in the theory and computation of atom-diatom reactive scattering. Rigorous and detailed quantum dynamics calculations for H + H_2 and other simple triatomic reactions in three-dimensional space have been reported and some excellent agreement between exact quantum dynamics calculations and experiments has been obtained for $H + H_2$ and its isotopic reactions using algebraic variational methods and hyperspherical coordinate methods as reviewed in Ref. [62]. Recent dynamics calculations for $H + H_2$ [72] and $F + H_2$ [73] show that the remaining discrepancies between the dynamics calculation and experiment for these two systems appear to be entirely attributable to deficiencies on the part of the potential energy surface (PES) or its proper treatment in the dynamics calculation, rather than the dynamics calculation itself.

Variational algebraic methods, as well as hyperspherical coordinate methods, have enjoyed tremendous success for the past ten years in treating atom-diatom reactive scattering problems [63–65, 71]. The use of multi-arrangement basis functions (LCAO approach) to expand the reactive scattering wavefunction or solve coupled arrangement Schrödinger equations has played a key role in the success of the methods. Some detailed accounts on the $H + H_2$ reaction can be found in a review [62]. In this section, we provide a specific treatment of atom-diatom reactive scattering by using the S matrix Kohn variational method to show the flavor of the variational algebraic method.

5.5.1 Hamiltonian and Basis Set

Since there are three arrangements for a triatomic system (excluding breakup), the Hamiltonian can be written in any of the three sets of *mass-scaled* Jacobi coordinates shown in Fig. 4.2

$$H = -\frac{\hbar^2}{2\mu}\frac{1}{R_\alpha}\frac{\partial^2}{\partial R_\alpha^2}R_\alpha + \frac{\mathbf{L}_\alpha^2}{2\mu R_\alpha^2} - \frac{\hbar^2}{2\mu}\frac{1}{r_\alpha}\frac{\partial^2}{\partial r_\alpha^2}r_\alpha + \frac{\mathbf{j}_\alpha^2}{2\mu r_\alpha^2}$$
$$+V^\alpha(\mathbf{R}_\alpha, \mathbf{r}_\alpha) + V_\alpha(r_\alpha) \qquad (\alpha = 1, 2, 3) \qquad (5.90)$$

where V^α is the interaction potential and $V_\alpha(r_\alpha)$ is the asymptotic diatomic potential, $\mathbf{L}_\alpha$ and $\mathbf{j}_\alpha$ are, respectively, the orbital and diatomic angular momentum operators, and μ is the system reduced mass defined by

$$\mu = \sqrt{\frac{m_A m_B m_C}{m_A + m_B + m_C}} \qquad (5.91)$$

If we choose a body-fixed (BF) angular momentum basis, the arrangement channel basis set for the γ arrangement can be denoted

$$|\varphi_{\gamma n}\rangle = |\gamma J M v j K\rangle \qquad (5.92)$$

which in coordinate representation takes the explicit form

$$
\begin{aligned}
<\hat{\mathbf{R}}_\gamma \mathbf{r}_\gamma|\gamma JMvjK> &= \mathcal{Y}_{jK}^{JM}(\hat{\mathbf{R}}_\gamma \hat{\mathbf{r}}_\gamma)\chi_{vj}(r_\gamma) \\
&= \widetilde{D}_{KM}^{J} P_{jK}\chi_{vj} \qquad (5.93)
\end{aligned}
$$

where $\mathcal{Y}_{jK}^{JM}$ is the BF angular momentum eigenfunction defined in Eq. (D.28), $\widetilde{D}_{KM}^{J}$ is the parity-adapted Wigner rotation matrix, and P_{jK} are normalized associated Legendre polynomials. These quantities are all defined in the appendix. The diatomic vibrational eigenfunctions χ_{vj} are obtained by solving the coupled vibrational eigenvalue equation

$$
\left[-\frac{\hbar^2}{2\mu}\frac{d^2}{dr_\gamma^2} + \frac{\mathbf{j}_\gamma^2}{2\mu r_\gamma^2} + V_\gamma(r_\gamma)\right]\chi_{vj}(r_\gamma) = \epsilon_{\gamma vj}\chi_{vj}(r_\gamma) \qquad (5.94)
$$

We denote $u_{\gamma t}(R_\gamma)(t > 1)$ as the $\mathcal{L}^2$ basis functions for the translational coordinate which can be chosen to be local Gaussian functions. The expansion of the scattering wavefunction starting from an initial arrangement α in the Kohn variational approach in Eq. (5.73) is now explicitly written as

$$
\begin{aligned}
\Psi_{v_0 j_0 K_0}^{JM\alpha} &= -u_{\alpha 0 v_0 j_0 K_0}(R_\alpha)\mathcal{Y}_{jK_0}^{JM}\chi_{\alpha v_0 j_0}(r_\alpha) + \sum_{\gamma tvjK} u_{\gamma tvjK}(R_\gamma) \\
&\times \mathcal{Y}_{jK}^{JM}\chi_{\gamma vj}(r_\gamma)c_{\gamma tvjK,\alpha v_0 j_0 K_0}^{JM} \qquad (5.95)
\end{aligned}
$$

The free functions u_0 can be chosen as

$$
u_{\alpha 0 vjK}(R_\gamma) = u_{1\gamma vjK}^{*}(R_\gamma) = f_{cut}(R_\gamma)\frac{e^{-ik_{vj}R_\gamma}}{\sqrt{v_{vj}}} \qquad (5.96)
$$

where the cutoff function f_{cut} is defined in Eq. (5.87).

5.5.2 Calculation of Direct Matrix Elements

As mentioned in previous sections, the reactive scattering application of the algebraic methods discussed before requires one to calculate two types of matrix elements: direct matrix elements and exchange matrix elements. Since the direct matrix elements are calculated between basis functions of the same arrangement, we drop the arrangement label in the expression. These matrix elements contain the information about inelastic scattering and are relatively easy to calculate. For example, the direct Hamiltonian matrix elements are calculated by the following integrations

$$
[\mathbf{H}_d^J]_{t'v'j'K',tvjK} = <JMt'v'j'K'|H|JMtvjK>
$$

$$= <u_{t'vjK}| - \frac{\hbar^2}{2\mu}\frac{d^2}{dR^2} + \epsilon_{vj}|u_{tvjK}> \delta_{v'v}\delta_{j'j}\delta_{K'K}$$

$$+ <u_{t'v'j'K}|V^K_{v'j',vj}|u_{tvjK}> \delta_{K'K} + \frac{\hbar^2}{2\mu}W^{Jj}_{K',K}$$

$$\times <u_{t'v'jK'}|\frac{1}{R^2}|u_{tvjK}> \delta_{v'v}\delta_{j'j} \quad (5.97)$$

where the potential matrix elements are defined as

$$[V^K]_{v'j',vj}(R) = <v'j'K|V|vjK>$$

$$= \int_0^\pi \sin\theta \, d\theta \, P_{j'K}(\theta) <\chi_{v'j'}|V|\chi_{vj}> P_{jK}(\theta) \quad (5.98)$$

and $W^{Jj}_{K',K}$ is the tridiagonal centrifugal coupling matrix defined in Eq. (3.69). For convenience we have defined a reduced rovibration eigenstate $|vjK>$

$$<r\theta|vjK>= P_{jK}(\theta,0)\chi_{vj}(r) \quad (5.99)$$

If we expand the potential in Legendre polynomials

$$V(R,r,\theta) = \sum_n V_n(R,r)P_n(\theta) \quad (5.100)$$

with the expansion coefficients determined by

$$V_n(R,r) = \frac{2l+1}{2}\int_0^\pi \sin\theta d\theta V(R,r,\theta)P_n(\theta), \quad (5.101)$$

Eq. (5.98) can be simplified to yield

$$[V^{JK}]_{v'j',vj} = \sum_n <\chi_{v'j'}|V_n\chi_{vj}> \sqrt{\frac{2j+1}{2j'+1}}$$

$$\times <jKn0|j'K><j0n0|j'0> \quad (5.102)$$

where the integration formula (A.30) involving three spherical harmonics in the appendix has been used.

Finally, the overlap matrix is trivially obtained by calculating the overlap integral of the radial basis functions (which are assumed to be nonorthogonal)

$$[\mathbf{O}^J_d]_{t'v'j'K',tvjK} =<u_{t'vjK}|u_{tvjK}> \delta_{v'v}\delta_{j'j}\delta_{K'K} \quad (5.103)$$

Thus the direct part of the "super" matrix defined in Eq. (5.82) is simply assembled as follows

$$[\mathbf{M}_d^J]_{t'v'j'K',tvjK} = [\mathbf{H}_d^J]_{t'v'j'K',tvjK} - E[\mathbf{O}_d^J]_{t'v'j'K',tvjK} \qquad (5.104)$$

for $(t, t' = 0, 1, 2, \cdots)$.

5.5.3 Calculation of Exchange Matrix Elements

Transformation of rotation marix between different arrangements

The computation of the exchange matrix is much more computationally expensive than that of the direct matrix. In calculating the exchange matrix elements between different arrangements, we need to use an important result for the rotation matrix. Let $R^J(\Omega_\alpha)$ denote the rotation operator that rotates the SF frame to a BF frame through Euler angles Ω_α such that in the BF frame the z_α axis points in the direction of the vector $\mathbf{R}_\alpha$ in the α arrangement. A similar definition of $R^J(\Omega_\beta)$ holds for the β arrangement in which the z_β axis points along the direction of the vector $\mathbf{R}_\beta$. As shown in Fig. 4.2, the pairs of vectors $(\mathbf{R}_\alpha, \mathbf{r}_\alpha)$ and $(\mathbf{R}_\beta, \mathbf{r}_\beta)$ are in the same plane with a common y axis perpendicular to the plane. Thus the β BF frame can be simply obtained from the α BF frame through a rotation by an angle between the vectors $\mathbf{R}_\alpha$ and $\mathbf{R}_\beta$ around the y axis. Thus, the angular momentum eigenstate $|JM\beta>$ in the β BF frame is related to that of $|JM\alpha>$ in the α BF frame by a unitary transformation

$$|JK\alpha> = e^{-i\Gamma_{\beta\alpha}J_y}|JK\beta> = \sum_{K'} d^J_{K'K}(\Gamma_{\beta\alpha})|JK'\beta> \qquad (5.105)$$

where $d^J_{K'K}(\Gamma_{\beta\alpha})$ is the reduced rotation matrix

$$d^J_{K'K}(\theta) = <JK'|e^{-i\theta J_y}|JK> \qquad (5.106)$$

and the rotation angle is defined as

$$\cos(\Gamma_{\beta\alpha}) = (\hat{\mathbf{R}}_\beta \cdot \hat{\mathbf{R}}_\alpha) \qquad (5.107)$$

with the positive rotation from the vector $\mathbf{R}_\alpha$ to $\mathbf{R}_\beta$.

Since the total angular momentum eigenstates $|JM>$ in the SF frame are related to those in the BF frame through the rotation matrix

$$|JM> = \sum_K D^J_{KM}(\Omega_\beta)|JK\beta> = \sum_K D^J_{KM}(\Omega_\alpha)|JK\alpha> \qquad (5.108)$$

where Ω_α and Ω_β denote the corresponding Euler angles, we can use Eq. (5.105) to directly obtain the equation

$$\sum_K D^J_{KM}(\Omega_\beta)|JK\beta> = \sum_{K'} D^J_{K'M}(\Omega_\alpha)|JK'\alpha>$$
$$= \sum_K \left[\sum_{K'} D^J_{K'M}(\alpha) d^J_{KK'}(\Gamma_{\beta\alpha})\right] |JK\beta> \quad (5.109)$$

Equation (5.109) yields the transformation relation between $D^J_{KM}(\Omega_\beta)$ and $D^J_{KM}(\Omega_\alpha)$

$$\boxed{D^J_{KM}(\Omega_\beta) = \sum_{K'} d^J_{KK'}(\Gamma_{\beta\alpha}) D^J_{K'M}(\Omega_\alpha)} \quad (5.110)$$

Using the orthogonality property of the rotation matrix

$$\sum_K d^J_{K'K}(\theta) d^J_{K''K}(\theta) = \delta_{K'K''} \quad (5.111)$$

we can invert Eq. (5.110) to yield

$$D^J_{KM}(\Omega_\alpha) = \sum_{K'} d^J_{K'K}(\Gamma_{\beta\alpha})$$
$$= \sum_{K'} d^J_{KK'}(\Gamma_{\alpha\beta}) D^J_{K'M}(\Omega_\beta) \quad (5.112)$$

where the angle $\Gamma_{\alpha\beta} = -\Gamma_{\beta\alpha}$.

If we use the normalized rotation matrix defined in Eq. (D.29) in the appendix, we can write

$$\widetilde{D}^J_{KM}(\Omega_\beta) = \sum_{K'} d^J_{KK'}(\Gamma_{\beta\alpha}) \widetilde{D}^J_{K'M}(\Omega_\alpha) \quad (5.113)$$

We can thus obtain the overlap integral over Euler angles

$$<\widetilde{D}^J_{K'M}(\beta)|\widetilde{D}^J_{KM}(\alpha)> = d^J_{KK'}(\Gamma_{\alpha\beta}) = d^J_{K'K}(\Gamma_{\beta\alpha}) \quad (5.114)$$

Integrals of exchange matrix

Using the transformation property in Eq. (5.113) and symmetry properties of the rotation matrix given in Appendix D.2, the integration over Euler angles between angular momentum eigenfunctions of different arrangements can be carried out as in Eq. (5.114) to yield the result

$$
\begin{aligned}
\langle \mathcal{Y}^{JMp}_{\beta j'K'} | \mathcal{Y}^{JMp}_{\alpha jK} \rangle &= \frac{1}{\sqrt{2(1+\delta_{K0})}} \frac{1}{\sqrt{2(1+\delta_{K0})}} \\
&= \langle \left[\widetilde{D}^J_{K'M}(\beta) + (-1)^{P+K'} \widetilde{D}^J_{-K'M}(\beta) \right] P_{j'K'}(\beta) | \\
&\quad \left[\widetilde{D}^J_{KM}(\alpha) + (-1)^{P+K} \widetilde{D}^J_{-KM}(\alpha) \right] P_{jK}(\alpha) \rangle \\
&= \langle P_{j'K'}(\beta) | \tilde{d}^J_{K'K}(\Gamma_{\beta\alpha}) | P_{jK}(\alpha) \rangle
\end{aligned}
\tag{5.115}
$$

where the new reduced rotation matrix is defined as

$$
\tilde{d}^J_{KK'} = \sqrt{\frac{1}{(1+\delta_{K0})(1+\delta_{K'0})}} \left[d^J_{KK'} + (-1)^{P+K} d^J_{-KK'} \right] \tag{5.116}
$$

for $(K, K' \geq 0)$ and P is the total parity defined in the appendix.

We are now in a position to compute the exchange integral for the Hamiltonian matrix

$$
\begin{aligned}
[\mathbf{H}^J_{ex}]_{\beta,\alpha} &= [\mathbf{H}^J_{ex}]_{\beta t'v'j'K',\alpha tvjK} \\
&= \langle \beta JMt'v'j'K' | H | \alpha JMtvjK \rangle \\
&= \langle \beta t'v'j'K' | \bar{d}^J_{KK'}(\Gamma_{\beta\alpha}) \left[-\frac{\hbar^2}{2\mu} \frac{d^2}{dR_\alpha^2} + \epsilon_{\alpha vj} + V \right] | \alpha tvjK \rangle \\
&\quad + \frac{\hbar^2}{2\mu} \sum_{K''} \langle \beta t'v'j'K' | \bar{d}^J_{K''K'}(\Gamma_{\beta\alpha}) W^{Jj}_{K'',K} \frac{1}{R_\alpha^2} | \alpha tvjK \rangle
\end{aligned}
\tag{5.117}
$$

where the matrix $W^{Jj}_{K'',K}$ is defined in Eq. (3.68). Similarly, the exchange integral of the overlap matrix is given by

$$
[\mathbf{O}^J_{ex}]_{\beta t'v'j'K',\alpha tvjK} = \langle \beta t'v'j'K' | \bar{d}^J_{KK'}(\Gamma_{\beta\alpha}) | \alpha tvjK \rangle \tag{5.118}
$$

which can be written explicitly as a three dimensional integral

$$
\begin{aligned}
[\mathbf{O}^J_{ex}]_{\beta,\alpha} &= [\mathbf{O}^J_{ex}]_{\beta t'v'j'K',\alpha tvjK} \\
&= \iiint d\tau\, u_{\beta t'v'j'K'}(R_\beta) \chi_{v'j'}(r_\beta) P_{j'K'}(\theta_\beta) \\
&\quad \times \bar{d}^J_{KK'}(\Gamma_{\beta\alpha}) u_{\alpha tvjK}(R_\alpha) \chi_{vj}(r_\alpha) P_{j'K'}(\theta_\alpha)
\end{aligned}
\tag{5.119}
$$

The 3D volume integral can be done by using any of the following sets of coordinates or mixed combinations of them with appropriate Jacobian factors

$$d\tau = R_\alpha^2 r_\alpha^2 \sin\theta_\alpha dR_\alpha dr_\alpha d\theta_\alpha$$
$$= R_\beta^2 r_\beta^2 \sin\theta_\beta dR_\beta dr_\beta d\theta \qquad (5.120)$$

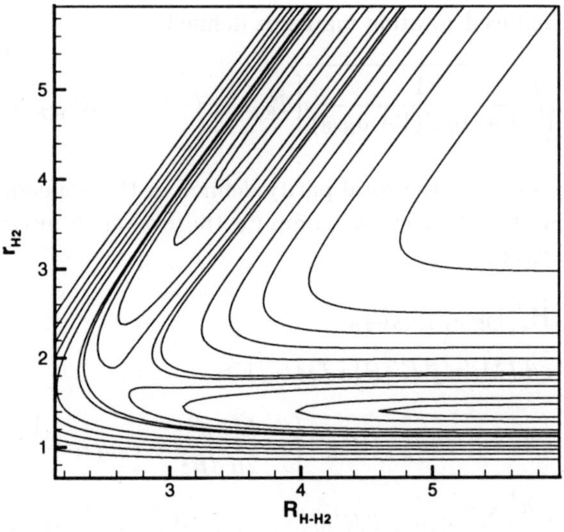

Figure 5.1: Contour plot of the LSTH potential energy surface for H + H_2 reaction in collinear geometry. Distances are in bohrs

After the Hamiltonian and overlap matrices have been constructed, it is then straightforward to carry out matrix computations to obtain the S matrix via the Kohn variational expression of (5.80).

5.5.4 H + H_2 Reaction

Background

The reaction of H + H_2 and its isotopic variants has played a pivotal role in the history of molecular reaction dynamics. Theoretically, it is the simplest chemical reaction system with the three lightest atoms and only three electrons. Thus high quality *ab initio* calculations are practical for this simple system. The global potential energy surface for dynamics claculation has been constructed by fitting *ab initio* points. The most widely used PES for H + H_2 is the LSTH potential surface [74]. H + H_2 is a collinearly dominated reaction with the minimum energy path lying along the collinear geometry. Figure 5.1 shows a contour plot of the LSTH potential energy surface (PES) for this system. The LSTH potential is purely repulsive and has a barrier of 0.36 eV above the asymptote at the saddle point.

After the first accurate three-dimensional reactive scattering calculation for the H + H_2 reaction at low collision energies on an empirical potential energy surface in 1976 [52], it took another decade or so to fully solve the reactive scattering problem for this system using the more general and powerful methods discussed in this chapter. Many of the recent theoretical dynamics calculations for the H + H_2 reaction have been summarized in a recent review [62]. In the following, we show some benchmark results from quantum dynamics calculations for H + H_2 and D + H_2 using the S matrix Kohn variational method. The technical details of the numerical calculations are omitted here.

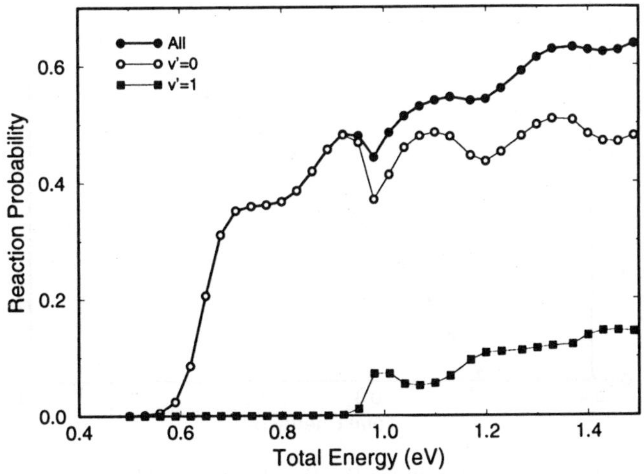

Figure 5.2: Total reaction probabilities for the reaction H + H_2 → H + $H_2(v')$ from the ground state of the H_2 reagent for total angular momentum J=0

Reaction probability

Figure 5.2 shows the calculated product vibration-specific reaction probability $P_{v\leftarrow 00}$ from the ground state of H + H_2 at fixed total angular momentum J=0 on the LSTH PES. Here the hydrogen atom is treated as distinguishable in the dynamics calculation. In practice, of course, the reactant and product are the same species and therefore indistinguishable. However, the symmetry relation tells us that if the product H_2 is of different rotation symmetry than the reagent H_2, then it must come from reactive collisions. For example, if the reagent H_2 is in an even rotational state (j=even), then only reactive collisions can produce odd rotational states of H_2 and vice versa. But if the product H_2 is also in an even rotational state, one cannot determine whether it is from a reactive or inelastic collision. In fact in this case, the result is in general a combination of both reactive and inelastic collisions. Since the H_3 system belongs to the D_{3h} symmetry group, one can utilize the group symmetry relations to reduce the dynamics calculation into those for different symmetry separately in order to realize computational savings and distinguish different symmetries from numerical calculations [75]

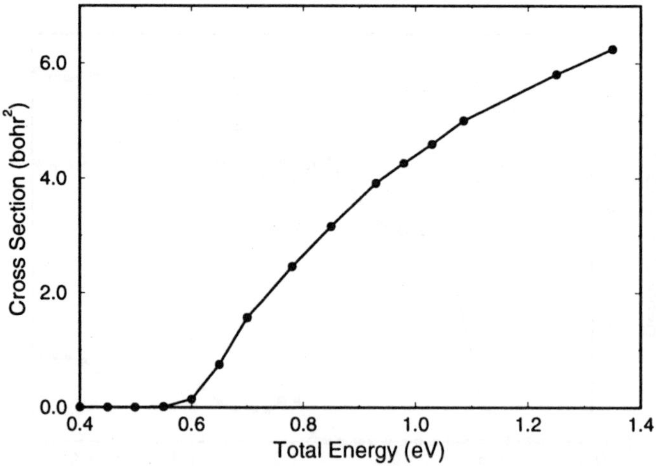

Figure 5.3: Integral cross section for the D + H_2 reaction from the ground state of H_2

The exact quantum mechanical result in Fig. 5.2 displays a number of important quantum effects that are absent in classical trajectory calculations. First, the probability curve as a function of energy rises smoothly from zero as the collision energy increases from below to above the barrier. This is a clear indication of quantum tunneling effects. In classical dynamics calculations, the reaction probability typically rises steeply from zero like a step function with a clear classical threshold as the collision energy increases from below to above the reaction barrier. Thus the exact quantum calculation, even just for J=0, provides an accurate determination of the reaction threshold which is very important in determining the thermal rate constant at room temperatures. Secondly, the probability curve has oscillatory structures as a function of energy that have been well known to be dynamical or Feshbach resonances. The phenomenon of dynamical resonance is interesting because it shows that resonances can exist in molecular reactions even on a purely repulsive potential energy surface like H + H_2! In a dynamical or Feshbach resonance, the translational energy of the collision system is temporarily transferred to other bound degrees of freedom and the system becomes temporarily trapped. This can occur even for multidimensional systems on a purely repulsive potential energy surface. Very similar probability curves are seen for the isotopic reaction D + H_2.

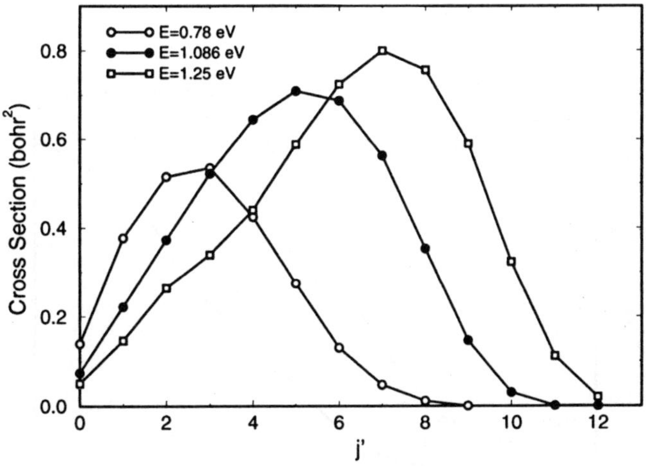

Figure 5.4: Product rotational state distribution of the integral cross section for the reaction D + H_2 → H + DH(j′) from the ground state of H_2

Reaction cross section

Unfortunately, the observable dynamical quantities in collisional experiments are reaction cross sections, not microscopic reaction probabilities shown in Fig. 5.2. Since the cross section involves a summation over all values of the total angular momentum J or impact parameters given by Eqs. (4.276) and (4.277), one needs to perform reactive scattering calculations for all contributing partial waves or total angular momentum J. Since the number of channels or quantum states increases linearly with J, exact dynamics calculations for higher J are extremely demanding computationally. Figure 5.3 shows the exact integral cross sections for the isotope reaction $D + H_2 \rightarrow H + HD$ as a function of total scattering energy from the work of Ref. [75]. The converged cross sections were obtained by including reaction probabilities calculated for fixed total angular momentum up to J=32. The integral cross section shown in Fig. 5.3 is a monotonically increasing smooth function of scattering energy without any visible trace of resonances that are displayed in the energy-dependence of microscopic reaction probabilities at fixed total angular momentum J. Thus the resonances are quenched as a result of summation over J in the evaluation of integral cross sections. In many situations, the integral cross section generally does not provide much information on reaction resonances.

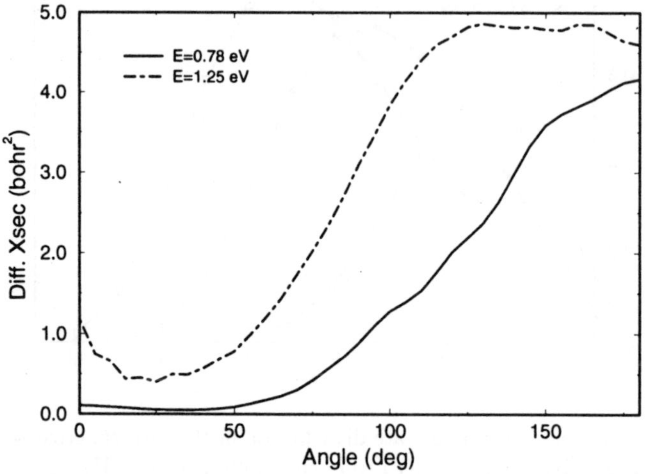

Figure 5.5: Differential cross section for the $D + H_2$ reaction from the ground state of H_2

The exact quantum reactive scattering calculation provides complete state to state dynamics information. Figure 5.4 shows product rotational state distributions for the reaction D + H_2 from the ground state of the H_2 reagent at three scattering energies [75]. These results show Boltzmann-like distribution of the rotational states of the product HD with the maximum peak shifting to higher values of j as the scattering energy increases. These theoretical results were found to be in good agreement with experiments reported shortly thereafter [76].

The angular distribution or differential cross section for the D + H_2 reaction has also been calculated in the work of Ref. [75]. Figure 5.5 shows the calculated differential cross section at two scattering energies of 0.78 and 1.25 eV. The product HD is predominantly backward scattered in the center-of-mass system. This is typical of collinearly dominated reactions with repulsive potentials in which the main contribution to the scattering angle is coming from the impact at small partial waves (low values of J). At the high energy of 1.25 eV, however, there is some forward component characteristic of pickup reaction in which the contribution to the forward scattering mainly comes from impact at high partial waves. These angular distributions were also in good agreement with results from molecular beam experiments, although small discrepancies exist [77].

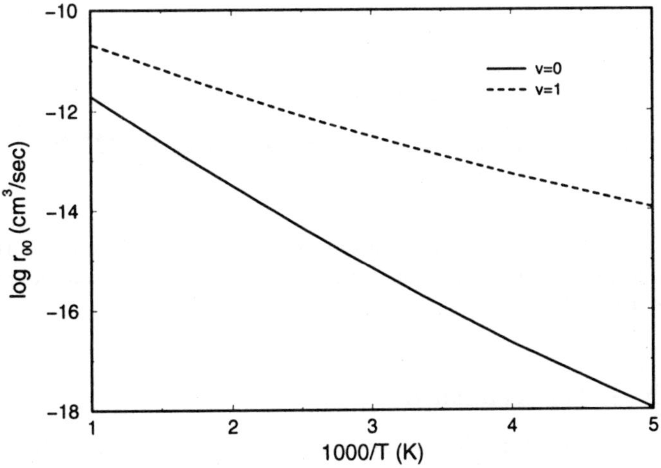

Figure 5.6: Rate constant for the D + H_2 reaction from the ground state of H_2

By thermally averaging the collision energy for the total reactive flux, we obtain the initial-state specific rate constant

$$r_i = <v\sigma_i>$$
$$= (\frac{8kT}{\pi\mu})^{1/2}(kT)^{-2}\int_0^\infty dE_t\, E_t\, exp(-E_t/kT)\sigma_i \qquad (5.121)$$

Figure 5.6 plots $\log r_i$ as a function of inverse temperature for the initial states $v = 0$ and $v = 1$. As can be seen, there is a significant enhancement of reaction rates when the reagent H_2 is vibrationally excited.

Recently, it was proposed that the geometric phase effect may play a role in the reaction of D + H_2 at higher collision energies where discrepancies between theoretical and experimental cross sections exist. Reference [78] gives a detailed account of this phenomenon.

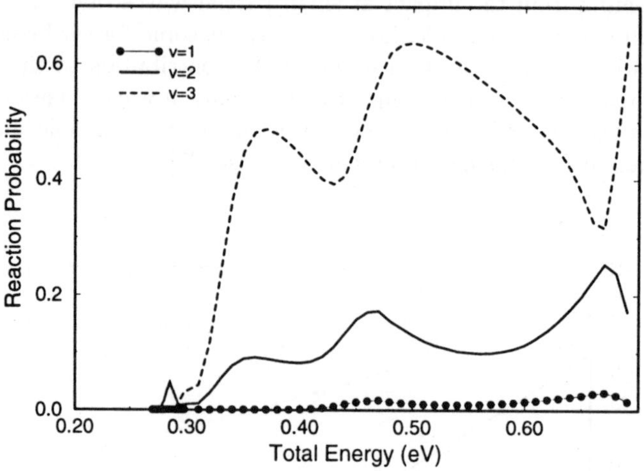

Figure 5.7: Product vibrational state specific reaction probabilities for the F + H_2 reaction from the ground rotational state of $H_2(v)$ for total angular momentum J=0 on the T5A PES

5.5.5 F + H_2 Reaction

Another important atom-diatom system is the reaction of F + H_2 → H + HF which has been extensively studied both experimentally and theoretically. This reaction is highly exoergic (by about 1.4 eV) and has a well in the interaction region. Consequently, the reaction dynamics is more complicated than the H + H_2 reaction. Due to its high exoergicity, there are

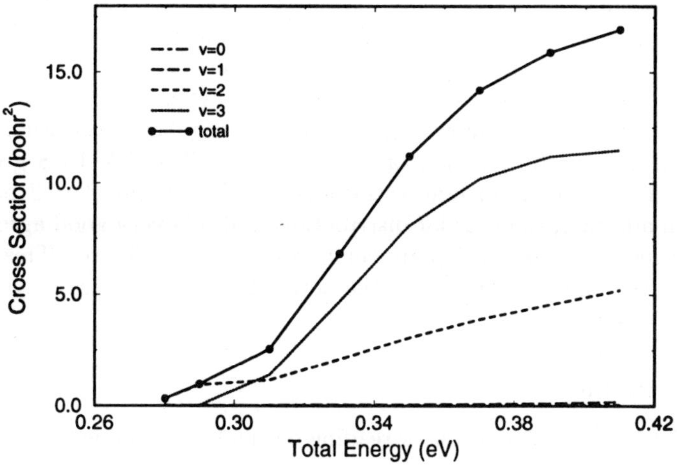

Figure 5.8: Product vibrational state specific integral cross sections for the F + H_2 reaction from the ground rotational state of $H_2(v)$ on the T5A PES

four vibrational states of the product HF that are energetically open at normal scattering energies. The vibrational population of the product HF was found to be inverted.

One of the interesting aspects from experimental investigation is the discovery of possible dynamical resonances in crossed molecular beam experiments [79]. Rigorous three dimensional quantum calculations for this reaction have been reported by various researchers [80–86]. Figure 5.7 shows the calculated reaction probabilities from the ground state of H_2 to specific product vibrational states on the T5A PES [88]. Here we see that the resonance structure is more pronounced than in the H + H_2 reaction and the population is completely inverted! The population of v=0 and v=1 products are negligible and the dominant population is in v=3 vibrational state of HF.

The integral cross sections from [81] on the T5A PES are also shown in Fig. 5.8. However, the results in Fig. 5.7 and Fig. 5.8 are not in good agreement with experimental results. Experiments indicated that v=2 state of HF is the most dominant product of this reaction while the theoretical calculation on T5A surface shows that the v=3 state should be the dominant product. Since quantum dynamics calculations are exact in nature, the discrepancy must lie in errors in the potential energy surface.

This is indeed the case.

Recently, a new *ab initio* potential surface (SW) [89] has become available for the F + H_2 reaction which is significantly improved over all of its predecessors. Quantum dynamics calculations on this new PES have been performed by Castillo *et al.* using the hyperspherical coordinate method [73]. Figure 5.9 shows their calculated angular distribution of the HF product from the reaction of F + para-H_2. The SW PES gives essentially the correct branching ratio between $v' = 2$ and $v' = 3$ product states. In addition, the angular distribution is also in very good agreement with the molecular beam experiment as shown in the figure. This result highlights the important role that the potential energy surface plays in the dynamics calculation.

Further reading

Refs. [90–94] are good sources for reading on the topic of molecular collision dynamics.

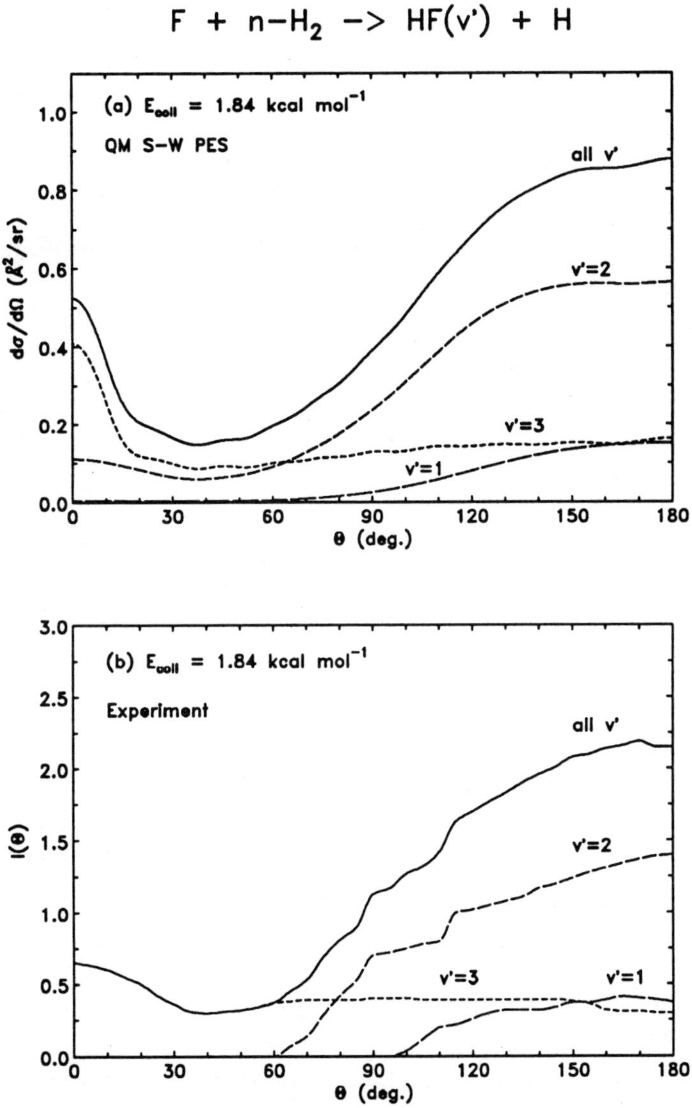

Figure 5.9: Comparison of calculated and experimental angular distributions for the F + n-H_2 →HF(v') + H reaction at collision energy E_{coll}=1.84 kcal/mol. (a) Quantum results of Ref. [73]. (b) Expreimental result of Ref. [79]. (From Ref. [73] with permission, Courtesy of Jesus F. Castillo and David E. Manolopoulos)

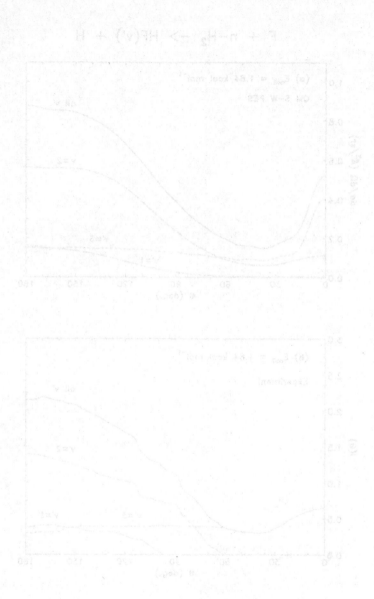

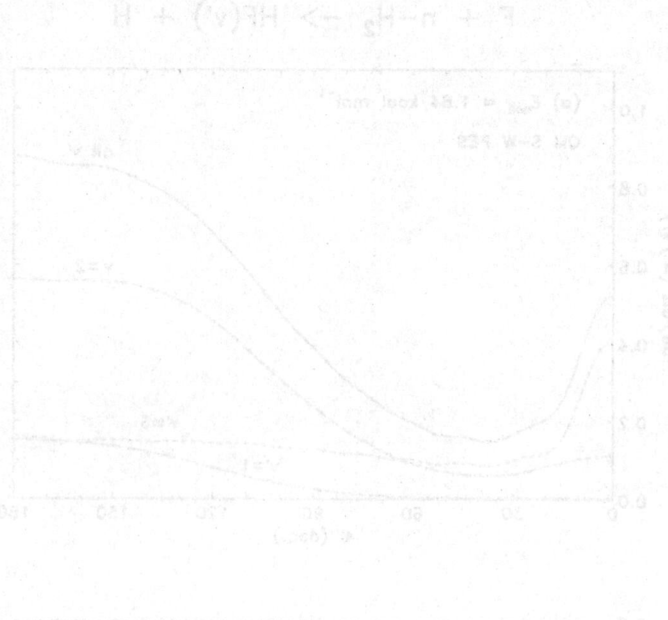

Chapter 6

Time-Dependent Approach to Reactive Scattering

6.1 Introduction

Time-independent methods such as variational algebraic methods and hyperspherical coordinate methods have had great success in theoretical calculation for atom-diatom reactions during the past decade. However, there are significant difficulties in extending these methods to polyatomic reactions involving more than three atoms. The fundamental reason is that the computational cost of these coupled channel TI methods scales as N^3 with the number of basis functions N. Therefore in order to extend the reactive scattering calculations to polyatomic reactions, one has to use methods whose computational cost scales much slower than N^3. Recently, the time-dependent (TD) wave packet approach has provided some hope for accurate quantum reactive scattering calculations of chemical reactions involving more than three atoms [95]. In the time-dependent approach, one can calculate the total reaction probability by evaluating the reactive flux at a dividing surface and simply use an absorbing potential to absorb the wavefunction beyond the transition state region [96]. In particular, it is convenient to choose the Jacobi coordinates corresponding to the reactant arrangement to propagate the time-dependent wavefunction up to the absorbing region.

6.2 Representations

The starting point for discussion is the time-dependent (TD) Schrödinger equation

$$i\hbar\frac{\partial}{\partial t}|\Psi_S(t)>= H|\Psi_S(t)> \tag{6.1}$$

where H is the Hamiltonian operator and Ψ_S is the TD wavefunction. The central task of this section is to find the numerical solution for Ψ_S which contains all the necessary dynamics information of the system. For most of our discussions, we will assume that the Hamiltonian H is time-independent, which is usually the case for a closed system, unless otherwise specified explicitly.

The wavefunction Ψ_S satisfying Eq. (6.1) is in the Schrödinger representation (SR), and has the formal solution (assuming H independent of time),

$$|\Psi_S(t)>= e^{-\frac{i}{\hbar}Ht}|\Psi_S(0)> \tag{6.2}$$

where the time evolution operator is unitary since H is Hermitian. In the Schrödinger representation, the wavefunction $\Psi_S(t)$ is time-dependent as in (6.1) but the operators are time-independent. Often it is useful to employ the interaction representation (IR) by splitting the Hamiltonian H into two parts,

$$H = H_0 + V \tag{6.3}$$

where H_0 is a reference Hamiltonian and V is the residual interaction potential. The reference Hamiltonian H_0 is often chosen to be the asymptotic limit of H for scattering problems but the best choice generally depends on the specific problem at hand. One can then define an interaction representation which is related to the Schrödinger representation by a unitary transformation

$$|\Psi_I(t)>= e^{\frac{i}{\hbar}H_0t}|\Psi_S(t)> \tag{6.4}$$

It is relatively straightforward to derive the equation for $\Psi_I(t)$,

$$i\hbar\frac{\partial}{\partial t}|\Psi_I>= V_I(t)|\Psi_I(t)>, \tag{6.5}$$

where $V_I(t)$ is the generalized interaction operator in the interaction representation (IR) defined as,

$$V_I(t)>= e^{\frac{i}{\hbar}H_0t}Ve^{-\frac{i}{\hbar}H_0t} \tag{6.6}$$

In general, one can regard the IR as a general representation while both Schrödinger and Heisenberg representations are special cases corresponding, respectively, to the choice of $H_0 = 0$ for the former and V=0 for the latter in (6.3). There are certain advantages to using the IR wavefunction for some applications as shown in refs. [97–101].

6.3 Methods of Time Propagation

6.3.1 Finite Difference Method

Solving Eq. (6.1) or (6.5) for a given initial wavefunction $\Psi(0)$ constitutes a time propagation of the wavefunction, which is carried out by integrating the wavefunction in time. The propagation can be accomplished by using a variety of integration methods. The most straightforward approach is based on finite difference schemes including the Runga-Kutta method, second order difference (SOD) method, or higher order difference methods. At present, however, more sophisticated methods, such as the split-operator (SP) method [102], Chebychev polynomial method [103], short iterative Lanczos method [104–107] as well as other methods, are often used in practical applications. Here, we briefly describe three commonly used propagators, namely, the SOD, SP, and Chebychev polynomial methods.

In the SOD method, one approximates the time-derivative of the wavefunction by second-order finite differencing,

$$\frac{\partial}{\partial t}\Psi(t) = \frac{\Psi(t+\Delta) - \Psi(t-\Delta)}{2\Delta} + O(\Delta^2) \tag{6.7}$$

which results in the following iterative formula for the wavefunction

$$\boxed{\Psi(t+\Delta) = \Psi(t-\Delta) - 2\frac{i}{\hbar}\Delta H\Psi(t).} \tag{6.8}$$

The wavefunction obtained by the SOD method of (6.8) is correct to second order $O(\Delta^2)$. Since the iterative procedure in Eq. (6.8) requires wavefunctions at two prior times, Ψ_0 and Ψ_1, the SOD method requires the use of a self-starting scheme such as the Runga-Kutta method to generate Ψ_1 first. The SOD method is stable with respect to the time increment Δ below a certain critical value Δ_{max}. If the step size Δ is greater than Δ_{max}, the solution becomes unstable and increases exponentially (blow-up). The SOD method is extremely easy to use but is not very efficient for large scale calculations because it generally requires the use of a small time step in order to obtain stable and accurate solutions.

6.3.2 Split-Operator Method

The split-operator method is a popular method and has been widely used in many practical applications. It approximates the short time propagator by the equation,

$$e^{-\frac{i}{\hbar}H\Delta} = e^{-\frac{i}{\hbar}H_0\Delta/2}e^{-\frac{i}{\hbar}V\Delta}e^{-\frac{i}{\hbar}H_0\Delta/2} + O(\Delta^3) \tag{6.9}$$

where the Hamiltonian H is split into two parts as in (6.3) and the error term is due to non-commutivity of H_0 and V. The wavefunction is propagated by the formula

$$\boxed{\Psi(t+\Delta) = e^{-\frac{i}{\hbar}H_0\Delta/2}e^{-\frac{i}{\hbar}V\Delta}e^{-\frac{i}{\hbar}H_0\Delta/2}\Psi(t)} \tag{6.10}$$

The split operator propagation of $\Psi(t)$ is explicitly unitary, which is a main factor contributing to the numerical stability of the SP method. The original derivation of Eq. (6.9) is given by Fleck *et al* in Ref. [102]. However, since the split-operator scheme is closely related to the interaction representation, we give here a simple derivation of the split-operator method by utilizing the interaction representation.

The solution to Eq. (6.5) in the time interval $[t, t+\Delta]$ can formally be written as

$$\Psi_I(t+\Delta) = \frac{1}{i\hbar}\int_t^{t+\Delta} V_I(t')\Psi_I(t')dt' \tag{6.11}$$

which can be solved iteratively to generate a solution to second order accuracy in Δ

$$\Psi_I(t+\Delta) = \left[1 + \frac{1}{i\hbar}\int_t^{t+\Delta} V_I(t')dt' + \left(\frac{1}{i\hbar}\right)^2 \int_t^{t+\Delta} V_I(t')dt' \int_t^{t'} V_I(t'')dt''\right] \times\Psi_I(t) + O(\Delta^3) \tag{6.12}$$

Now, we expand the interaction $V_I(t)$ around the middle point $\bar{t} = t + \Delta/2$

$$V_I(t') = V_I(\bar{t}) + (t-\bar{t})V_I'(\bar{t}) + O(\Delta^2) \tag{6.13}$$

where V' denotes derivative of the potential. Substituting the expansion of Eq. (6.13) into Eq. (6.12) and keeping terms of less than third order in Δ, we obtain the solution

$$\Psi_I(t+\Delta) \approx \left[1 + \frac{\Delta}{i\hbar}V_I(\bar{t}) + \frac{1}{2}\left(\frac{\Delta}{i\hbar}\right)^2 V_I^2(\bar{t})\right]\Psi_I(t) \tag{6.14}$$

with errors in third order $O(\Delta^3)$. In deriving the above equation, the fact that

$$\int_t^{t+\Delta} (t' - \bar{t})dt' = 0 \tag{6.15}$$

was utilized. Equation (6.14) can be replaced by (still in third order)

$$\begin{aligned}\Psi_I(t+\Delta) &\approx e^{-\frac{i}{\hbar}V_I(\bar{t})\Delta}\Psi_I(t) \\ &= e^{\frac{i}{\hbar}H_0\bar{t}}e^{-\frac{i}{\hbar}V(\bar{t})\Delta}e^{-\frac{i}{\hbar}H_0\bar{t}}\Psi_I(t)\end{aligned} \tag{6.16}$$

Equation (6.16) is the split-operator formula in the interaction representation. By transforming the wavefunction to the Schrödinger representation, we obtain the standard split-operator method

$$\begin{aligned}\Psi_S(t+\Delta) &= e^{-\frac{i}{\hbar}H_0(t+\Delta)}\Psi_I(t+\Delta) \\ &\approx e^{-\frac{i}{\hbar}H_0\Delta/2}e^{-\frac{i}{\hbar}V(\bar{t})\Delta}e^{-\frac{i}{\hbar}H_0\Delta/2}\Psi_S(t)\end{aligned} \tag{6.17}$$

Since the SP method is a short time propagator like the SOD, it can handle complicated Hamiltonians including time-dependent Hamiltonians and complex Hamiltonians. A particularly attractive feature of the SP method is its numerical stability with respect to time step Δ in numerical integration because the propagator is explicitly unitary and therefore conserves the normalization of the wavefunction. The price to pay is that we need to deal with exponential operators which may require diagonalization of some smaller-sized matrices.

6.3.3 Chebychev Polynomial Method

Another widely used propagation scheme is the Chebychev polynomial method introduced by Kosloff and Kosloff [103]. This is a global propagator in the sense that it expands the propagator $e^{-\frac{i}{\hbar}Ht}$ in the interval [0, t]. The method is based on the Chebychev expansion relation for the function exp(iRX) ($X \in [-1, 1]$) [108],

$$e^{iRX} = \sum_n A_n(R)T_n(X) \tag{6.18}$$

where the coefficient A_n is given by

$$A_n(R) = (2 - \delta_{n0})i^n J_n(R) \tag{6.19}$$

and J_n is the Bessel function of the first kind of order n. Using Eq. (6.18), one obtains the expansion relation for the propagator,

$$\begin{aligned} e^{-\frac{i}{\hbar}Ht} &= e^{-\frac{i}{\hbar}(E_{max}+E_{min})t/2}e^{iR\hat{X}} \\ &= e^{-\frac{i}{\hbar}(E_{max}+E_{min})t/2}\sum_n A_n(R)T_n(\hat{X}) \end{aligned} \tag{6.20}$$

where

$$R = (E_{max} - E_{min})t/2 \tag{6.21}$$

$$\hat{X} = (E_{max} + E_{min} - 2H)/(E_{max} - E_{min}) \tag{6.22}$$

and E_{max} and E_{min} are, respectively, the maximum and minimum eigenvalues of the Hamiltonian operator. Thus eigenvalues of the normalized Hamiltonian operator $\hat{X}$ fall between $[-1, 1]$. The time propagation is usually carried out by using the recurrence relation for Chebychev polynomials [108]

$$T_{n+1}(X) = 2XT_n(X) - T_{n-1}(X) \tag{6.23}$$

with T_0=1 and $T_1 = X$.

The Chebychev method converges exponentially with the number of expansion terms N for a given step size Δ and is particularly advantageous and efficient when Δ is large. In actual computations, one first needs to determine or estimate the spectrum range $[E_{min}, E_{max}]$. Since, the order of expansion is proportional to the spectrum range, one wants to choose the smallest possible spectrum range needed to enclose all accessible eigenvalues of the Hamiltonian. For a given spectrum, the order of expansion is proportional to the time step Δ. Unlike short-time propagators such as SOD or SP, the Chebychev method with a large time steps is not directly applicable to time-dependent or non-hermitian Hamiltonians. However, since each iteration of Eq. (6.23) in fact is physically equivalent to a short time propagation of the wavefunction, it is possible to modify the Chebychev propagator by including appropriate terms in each iteration [109].

Whatever propagation method is used, it is necessary to evaluate the action of the Hamiltonian operator on the wavefunction $\Psi(t)$. This is normally carried out by expanding $\Psi(t)$ in a suitable basis set and then evaluating the operator action on basis functions. One can use FFT (fast Fourier transform) techniques [102,110], discrete variable representation (DVR) [11] techniques, or simply calculate matrix elements of the operator in a given basis set.

6.3.4 Gaussian Wavepackets

In this section we discuss the basic properties of Gaussian wavepackets. The initial wavepacket $\psi(0)>$ employed in time-dependent scattering calculations is often chosen to be a Gaussian function and has the form

$$\psi(x,0) = (\frac{1}{2\pi\delta^2})^{1/4} \exp[-(x-x_0)^2/4\delta^2]e^{ik_0x}, \tag{6.24}$$

which travels toward the positive direction of x. The Gaussian packet $\psi(x,0)$ can be written as a superposition of plane waves

$$\psi(x,0) = \frac{1}{\sqrt{2\pi}} \int_{-\infty}^{\infty} \phi(k)e^{ikx}\, dk \tag{6.25}$$

with its momentum representation given by

$$\phi(k) = \left(\frac{2\delta^2}{\pi}\right)^{1/4} e^{-\delta^2(k-k_0)^2 - i(k-k_0)x_0} \tag{6.26}$$

Free propagation

Under the propagation of a free particle Hamiltonian (V=0)

$$H_0 = -\frac{\hbar^2}{2m}\frac{d^2}{dx^2} \tag{6.27}$$

the evolution of a Gaussian wavepacket can be calculated analytically

$$\begin{aligned}\psi(x,t) &= e^{-\frac{i}{\hbar}H_0}\psi(x,0) \\ &= \frac{1}{\sqrt{2\pi}} \int_{-\infty}^{\infty} \phi(k)e^{ikx} \exp\left(-\frac{i}{\hbar}\frac{\hbar^2 k^2}{2m}\right) dk \\ &= \left(\frac{\delta^2}{2\pi\gamma^2(t)}\right)^{1/4} \exp\left[-\frac{(x-x_0-v_0t)^2}{4\gamma(t)} + ik_0x - i\frac{k_0^2\hbar t}{2m}\right]\end{aligned} \tag{6.28}$$

where $v_0 = \hbar k_0/m$ and

$$\gamma(t) = \delta^2 + i\frac{t\hbar}{2m} \tag{6.29}$$

The absolute square of the wavepacket is given by

$$|\psi(x,t)|^2 = \left(\frac{1}{2\pi\delta^2 s(t)}\right)^{1/2} \exp\left[-\frac{(x-x_0-v_0t)^2}{2\delta^2 s(t)}\right] \tag{6.30}$$

where the spreading parameter is

$$s(t) = 1 + \frac{\hbar^2 t^2}{4m^2\delta^4} \tag{6.31}$$

Thus a Gaussian packet remains Gaussian with its center traveling at the classical speed and the width of the packet spreads even in the absence of an interaction potential. The Gaussian packet defined in Eq. (6.24) has the minimum width allowed by the uncertainty principle.

Semiclassical propagation

As is well known for Hamiltonians with quadratic potentials, the initial Gaussian packet remains Gaussian under time propagation. For a general potential, however, the wavepacket will not remain Gaussian during propagation. If the wavepacket is expected to remain compact and the distortion of the potential from quadratic form is relatively small over the range of the wavepacket, then it may be reasonable to approximate the potential by its local quadratic expansion around the center of the wavepacket. Thus a Gaussian packet will remain a Gaussian in this approximate quadratic potential. This is the idea of "Gaussian wavepacket propagation" introduced by Heller [111]. Since the validity condition of this approach is similar to that of the semiclassical approximation, it is essentially a semiclassical wavepacket approach.

The *ansatz* of Gaussian wavepacket propagation is the assumed Gaussian form of the wavefunction

$$\psi(x,t) = \left(\frac{2\alpha(0)}{\pi}\right)^{1/4} \exp\left[-\alpha(t)[x - x_c(t)]^2 + \frac{i}{\hbar}p(t)[x - x_c(t)] + \frac{\beta(t)}{\hbar}\right] \tag{6.32}$$

where $\alpha(t), x_c(t), p(t), \beta(t)$ are time-dependent parameters to be determined. By using quadratic approximation for the potential at the center of the wavepacket $x = x_c(t)$

$$V(x) \simeq V(x_c) + V'(x_c)(x - x_c) + \frac{1}{2}V''(x - x_c)^2 \tag{6.33}$$

we can determine these time-dependent parameters by substituting Eqs. (6.32) and (6.32) into the Schrödinger equation

$$i\hbar\frac{\partial}{\partial t}\psi(t) = \left(-\frac{\hbar^2}{2m}\frac{d^2}{dx^2} + V\right)\psi(t) \tag{6.34}$$

By equating various terms in like powers of x on both sides of the above equation, we can derive equations of motion for the center of the packet

$$\dot{x}_c = p/m, \quad \dot{p} = -V'(x_c), \tag{6.35}$$

and for the coefficients

$$\dot{\alpha} = \frac{i}{2\hbar}\left(V''(x_c) - \frac{4\alpha\hbar^2}{m}\right) \tag{6.36}$$

$$\dot{\beta} = i\left(\frac{m\dot{x}_c^2}{2} - V(x_c) - \frac{\alpha\hbar^2}{m}\right) \tag{6.37}$$

Using the method of Gaussian propagation, the wavepacket remains Gaussian and the motion of its center is governed by classical mechanics. It has some useful applications but is also limited by its simplicity. Because it is a classical wavepacket, it does not apply to classically forbidden processes. Also, since the wavepacket does not change its shape, it cannot describe a process in which the wavepacket splits such as in a reaction process. It is possible, however, to use superposition of several Gaussian wavepackets to describe such processes and some useful applications have been reported.

6.4 Application to Reactive Scattering

The time-dependent wave packet approach provides an attractive alternative to the standard time-independent close-coupling (CC) approach for scattering problems. Although both approaches are formally equivalent, they are technically different and the TD approach has some computational advantages for large-scale scattering problems. The most important difference is that the TD approach solves for one column of the S matrix at a time while the TI CC approach solves for the whole S matrix in a single TD calculation. As a result, the computational time of each TD calculation scales less than N^2 versus the N^3 scaling of the TI CC calculation, where N is the total number of basis functions. In addition, a single wavepacket propagation enables one to obtain scattering information for all the energies contained in the initial wavepacket. These features make the TD approach increasingly attractive for large-scale computational calculations in chemical dynamics.

The most straightforward application of the TD approach to scattering problems is to launch an initial wavepacket in a specific internal state from the asymptotic region with a positive momentum toward the interaction region. In practical applications, one may wish to use a shrinking wavepacket instead of the minimum width packet of Eq. (6.24). This can be accomplished by simply employing $\psi(x, -t_0)$ as the initial wavepacket $\psi(0)$. The propagation of $\psi(x, -t_0)$ forward in time will therefore produce a minimum width wavepacket at $t = t_0$ as illustrated in Fig. 6.1.

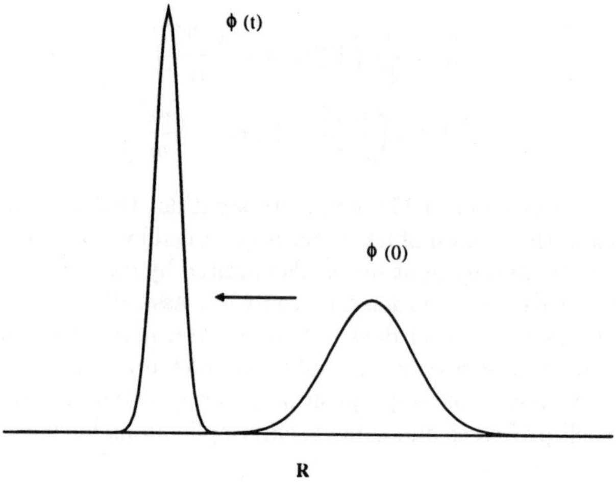

Figure 6.1: Illustration of a shrinking wavepacket.

6.4.1 Extraction of Scattering Information

As the TD wavefunction $\Psi(t) = e^{-\frac{i}{\hbar}Ht}\Psi(0)$ is propagated, one can obtain the stationary scattering wavefunction $\psi(E)$ from $\Psi(t)$ by a time→energy Fourier transform as discussed in Chapter 4

$$\begin{aligned} |\psi^+(E)\rangle &= \delta(E-H)|\Psi(0)\rangle\ /a(E) \\ &= \frac{1}{2\pi\hbar a(E)}\int_{-\infty}^{\infty} e^{\frac{i}{\hbar}Et}\Psi(t)dt \end{aligned} \tag{6.38}$$

where the energy-dependent coefficient is given by

$$a(E) = [\langle\psi(0)|\delta(E-H)|\psi(0)\rangle]^{-1/2} \tag{6.39}$$

which normalizes the stationary solution to the δ-function in energy, viz.,

$$\langle\psi^+(E)|\psi^+(E')\rangle = \delta(E-E') \tag{6.40}$$

Using the properties of the Mϕller operator discussed in Chapter 4, one can also express the normalizing coefficient as

$$a(E) = [<\Phi_{in}|\delta(E - H_0)|\Phi_{in}>]^{-1/2} \tag{6.41}$$

where Φ_{in} is defined by

$$|\Psi(0)>= \Omega_+|\Phi_{in} \tag{6.42}$$

Using the normalization of (6.40) for $\psi^+(E)$, the coefficient $a(E)$ can also be calculated as follows

$$\begin{aligned} a(E) &= <\psi^+(E)|\Psi(0)> \\ &= <\phi(E)|\Omega_+^\dagger|\Psi(0)> \\ &= <\phi(E)|\Omega_+^\dagger\Omega_+|\Phi_{in}> \\ &= <\phi(E)|\Phi_{in}> \end{aligned} \tag{6.43}$$

where $\phi(E)$ is the energy-normalized free or asymptotic function. If the initial wavepacket is chosen to be localized in the asymptotic region of the reagent and with the incoming wave only, then

$$\begin{aligned} \Phi_{in} &= \Omega_+^\dagger\Psi(0) \\ &= \lim_{t\to-\infty} e^{\frac{i}{\hbar}H_0t}e^{-\frac{i}{\hbar}Ht}\Psi(0) \\ &= \Psi(0) \end{aligned} \tag{6.44}$$

because the back propagation of $\Psi(0)$ by the full propagator is completely canceled by the forward propagation.

Once the stationary solution $\psi(E)$ is obtained, one can employ asymptotic boundary conditions to extract the scattering matrix. However, if only total reaction probabilities are needed, the calculation can be greatly simplified by evaluating the reactive flux at any fixed hypersurface (preferably close to the transition state) without the need to compute the state-to-state S matrix.

6.4.2 Reactive Flux and Total Reaction Probability

The conservation relation corresponding to the time-dependent Schrödinger equation $i\hbar\frac{\partial}{\partial t}\Psi = H\Psi$ can be written as a continuity equation

$$\frac{\partial\rho}{\partial t} + \nabla\cdot\mathbf{J} = 0 \tag{6.45}$$

where the divergence operator is defined in the $N-1$-dimensional hypersurface. Here the density is given by $\rho = |\Psi(t)|^2$ and the flux is defined by the equation

$$\nabla \cdot \mathbf{J} = \frac{i}{\hbar}[\Psi^* H\Psi - (H\Psi)^*\Psi] \tag{6.46}$$

For any stationary wavefunction ψ, ρ is independent of time, so $\nabla \cdot \mathbf{J} = 0$. This means that the flux is constant across any fixed hypersurface. If the Hamiltonian H can be expressed as the sum of a kinetic energy operator for the coordinate S and a reduced Hamiltonian for the remaining N-1 degrees of freedom

$$H = \frac{p_S^2}{2m_S} + H_S \tag{6.47}$$

where H_S is the reduced Hamiltonian, then we can evaluate the flux at a fixed surface at $S = S_0$ by integrating over the remaining $N-1$ coordinates in Eq. (6.46)

$$\boxed{J_S = <\psi|\mathbf{F}|\psi>} \tag{6.48}$$

where the flux operator is defined as

$$\begin{aligned} \mathbf{F} &= \frac{1}{2}\left[\delta(S-S_0)\frac{p_S}{m_S} + \frac{p_S}{m_S}\delta(S-S_0)\right] \\ &= \text{Im}\left[\frac{\hbar}{m_S}\delta(S-S_0)\frac{\partial}{\partial S}\right] \end{aligned} \tag{6.49}$$

Since the flux J_S is a constant and therefore independent of the position of the surface to calculate, we can of course evaluate the reactive flux at a fixed surface in the asymptotic region of the product. By using the S matrix asymptotic boundary condition for the reactive scattering wavefunction

$$\psi_{\alpha i}^+ \overset{R_\beta \to \infty}{\longrightarrow} \sum_n \frac{e^{i(k_{\beta n} R_\beta - L_{\beta n}\pi/2)}}{\sqrt{v_{\beta n}}} \varphi_{\beta n} S_{\beta n, \alpha i} \tag{6.50}$$

for $\alpha \neq \beta$, we can calculate the flux at a surface with a fixed value of $S = R_\beta$ to obtain

$$J_S = \sum_n |S_{\beta n, \alpha i}|^2 \tag{6.51}$$

Thus the reactive flux gives the total reaction probability

$$P_{\alpha i} = J_S = <\psi_{\alpha i}|\hat{F}|\psi_{\alpha i}> \tag{6.52}$$

where $P_{\alpha i}$ is the total $\alpha(i) \rightarrow \beta(all)$ reaction probability. In TD calculations, however, it is preferable to evaluate the reactive flux at a location near the transition state because this will generally shorten the propagation time needed to converge the flux.

6.4.3 Use of Absorbing Potentials

One common difficulty in time-dependent wavepacket calculations for scattering problems is the artificial boundary reflection from the end of the numerical grid. This is because the Schrödinger wavefunction for any scattering problem is unbound while the basis set or numerical grid used to represent the Schrödinger wavefunction is of finite size. Thus the Schrödinger wavefunction is forced to reflect back after reaching the end of the grid as if there were an infinite wall at the boundary. This artificial reflection will distort and ruin the correct dynamics. As a result, one often has to use a large enough numerical grid to minimize the artificial reflection. For some applications, it is possible to directly solve the TD wavefunction in the interaction representation which is often localized and bounded [97].

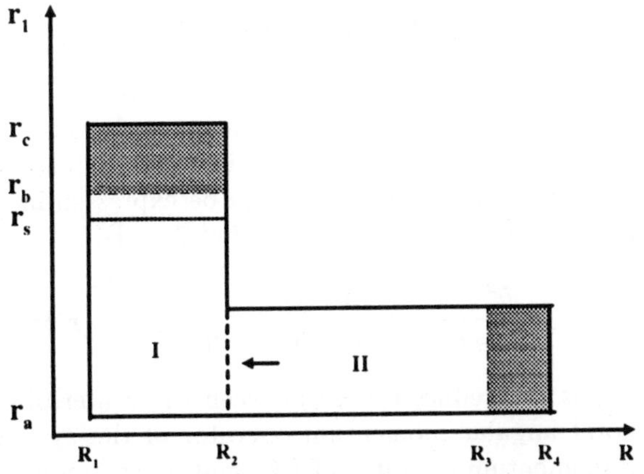

Figure 6.2: Illustrative drawing of the configuration space for atom-diatom reactive scattering. R is the radial coordinate between the atom and the center-of-mass of the diatom, and r is the vibrational coordinate of the diatom. Shaded regions represent absorbing potentials.

However, a robust, albeit empirical, approach is to employ an absorbing potential or optical potential to absorb the wavefunction near the grid boundary to effectively eliminate artificial boundary reflections. The optical potential has long been used in various applications in physics and chemistry. Its recent successful application to reactive scattering has made its use a standard approach in time-dependent scattering calculations. The most widely used absorbing potential is negative imaginary (also called NIP) defined near the boundary of the numerical grid

$$V_{abs}(x) = -i\alpha(x - x_0)^n \tag{6.53}$$

for $x_0 \le x \le x_0 + L$ and is zero for $x < x_0$. The general criterion for $V_{abs}(x)$ is that it is smooth enough to avoid reflection and at the same time rises rapidly to efficiently absorb the wavefunction within the absorbing region. The index n is generally chosen to have a value around 2. Of course, other functional forms of $V_{abs}(x)$ are also widely used. Various studies indicate that the absorption length L should be comparable or greater than the de Broglie wavelength λ of the wavefunction in order to achieve good absorption. In the following section, we discuss specific applications of TD wavepacket methods for studying dynamics of molecular reactions.

6.5 Triatomic Reaction: A + BC

6.5.1 Hamiltonian and Wavepacket Propagation

For an atom-diatom reaction A + BC, there are two possible products (neglecting the three body fragmentation or breakup channel): B + AC and C + AB as shown in Fig. 6.3. For a fixed total angular momentum J, the Hamiltonian of the system can be expressed in terms of the Jacobi coordinates of the reactant arrangement A + BC,

$$H = -\frac{\hbar^2}{2\mu_R}\frac{\partial^2}{\partial R^2} + \frac{(\mathbf{J}-\mathbf{j})^2}{2\mu_R R^2} + \frac{\mathbf{j}^2}{2\mu_r r^2} + V(\mathbf{r},\mathbf{R}) + h(r), \tag{6.54}$$

where μ_R is the reduced mass between the center-of-mass of A and BC, $\mathbf{J}$ the total angular momentum operator of the system, $\mathbf{j}$ the rotational angular momentum operator of BC, and μ_r the reduced mass of BC. The standard substitution for the wavefunction $\psi = \psi/(Rr)$ has been implicitly assumed. The diatomic reference Hamiltonian $h(r)$ is defined as

$$h(r) = -\frac{\hbar^2}{2\mu_r}\frac{\partial^2}{\partial r^2} + V_r(r), \tag{6.55}$$

where V_r is a diatomic reference potential (usually chosen as an asymptotic diatomic potential).

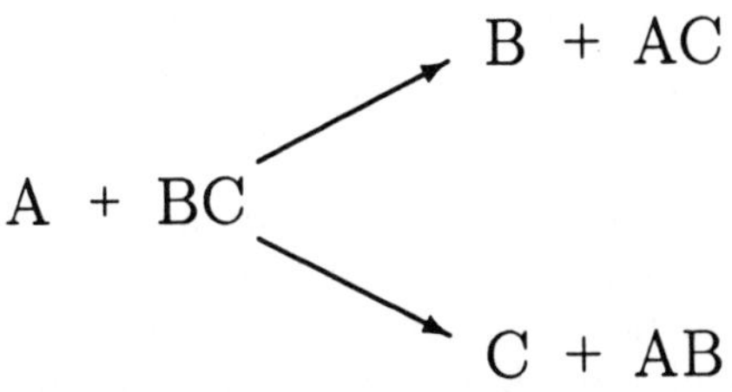

Figure 6.3: Reaction of A + BC to produce B + AC and C +AB.

The time-dependent wavefunction Ψ_r satisfying the absorbing boundary condition can be expanded in terms of the BF (body-fixed) translational-vibrational-rotational basis $\{u_n^v(R)\phi_v(r)\mathcal{Y}_{jK}^{JMp}(\hat{R},\hat{r})\}$ as

$$\psi_{r,v_0j_0K_0}^{JMp}(\mathbf{R},\mathbf{r},t) = \sum_{n,v,j,K} F_{nvjK,v_0j_0K_0}^{JMp}(t)u_n^v(R)\phi_v(r)\mathcal{Y}_{jK}^{JMp}(\hat{R},\hat{r}), \quad (6.56)$$

where n is the translational basis label, M is the projection quantum number of J on the space-fixed z axis, (v_0, j_0, K_0) denotes the initial rovibrational state, and p is the system parity.

The functions $\phi_v(r)$ are eigenfunctions of the diatomic Hamiltonian of Eq. (6.55). The definition of the nondirect product basis functions $u_n^v(R)$ is given in Sec. 6.6. For the sake of conciseness, we omit the labels $v_0j_0K_0$ and JMp in the following discussion.

The split-operator method for wavepacket propagation is employed to propagate the wavefunction

$$\Psi_r(\mathbf{R},\mathbf{r},t+\Delta) = e^{-iH_0\Delta/2}e^{-iU\Delta}e^{-iH_0\Delta/2}\Psi_r(\mathbf{R},\mathbf{r},t), \quad (6.57)$$

where the reference Hamiltonian H_0 is defined as,

$$H_0 = -\frac{\hbar^2}{2\mu_R}\frac{\partial^2}{\partial R^2} + h(r), \quad (6.58)$$

and the effective potential operator U is defined as

$$U = \frac{(\mathbf{J}-\mathbf{j})^2}{2\mu_R R^2} + \frac{\mathbf{j}^2}{2\mu_r r^2} + V(R,r,\theta) = V_{rot} + V \quad (6.59)$$

The matrix version of Eq. (6.57) for the expansion coefficient vector $\mathbf{F}$ is then given by

$$\boxed{\mathbf{F}(t+\Delta) = e^{-i\mathbf{H}_0\Delta/2}e^{-i\mathbf{U}\Delta}e^{-i\mathbf{H}_0\Delta/2}\mathbf{F}(t)} \quad (6.60)$$

and the operator $e^{-i\mathbf{U}\Delta}$ is further split as

$$e^{-i\mathbf{U}\Delta} = e^{-i\mathbf{V}_{rot}\Delta} e^{-i\mathbf{V}\Delta} e^{-i\mathbf{V}_{rot}\Delta} \tag{6.61}$$

where V_{rot} is diagonal in the angular momentum basis representation and V is diagonal in the coordinate representation

The initial wavefunction is chosen as the product of a specific rovibrational eigenfunction and a localized translational wave packet,

$$\Psi_i(0) = g_{k_0}(R)\phi_{v_0 j_0}(r)\mathcal{Y}^{JMp}_{j_0 K_0}(\hat{R}, \hat{r}) \tag{6.62}$$

where the wavepacket $\varphi_{k_0}(R)$ is chosen to be a standard Gaussian function

$$\varphi_{k_0}(R) = (\frac{1}{\pi\delta^2})^{1/4} \exp[-(R - R_0)^2/2\delta^2] e^{-ik_0 R}, \tag{6.63}$$

The exact rovibrational function $\phi_{v_0 j_0}(r)$ of BC is expanded in terms of the reference vibrational functions $\phi_v(r)$.

6.5.2 Reaction of H + O_2

Background

The endothermic reaction $H(^2S) + O_2(^3\Sigma_g^-) \rightarrow OH(^2\Pi) + O(^3P)$ plays a central role in combustion chemistry. It is responsible for the chain-branching in the oxidation of hydrogen and is a dominant molecular oxygen consuming step in hydrogen-oxygen and methane-oxygen combustion mechanisms. For these reasons, it is named "the most important combustion reaction" [112]. The reaction is endothermic by 0.71 eV and can only proceed in high temperature flames. In addition, through third body collision, the reaction can form the intermediate species: the HO_2 radical, which is an important intermediate for many chemical reactions in atmospheric and other areas of chemistry. A considerable amount of experimental work has been carried out to measure the reaction rate coefficient and absolute reaction cross sections. Very recently, three dimensional quantum dynamics calculations have been reported for the H + O_2 reaction [34,37,39,113–115]. From a dynamical point of view, the H + O_2 reaction presents a real challenge to accurate quantum dynamics calculations. Current *ab initio* electronic structure calculations show no barrier along the minimum energy path but there is a deep well of about –2.38 eV relative to the minimum of the H + O_2 asymptotic potential. Figure 6.4 shows the energy profile and the equilibrium geometry of the HO_2 complex. The contour plot of the DMBE IV PES [116] is shown in Fig. 6.5. The deep well, which corresponds to the hydroperoxyl radical HO_2, supports hundreds of bound states as discussed in Sec. 3.3.3.

Because of long-lived resonances, the time-dependent wavepacket dynamics calculation requires wavepacket propagation up to about the lifetime of the resonances. Figure 6.6 shows the calculated total reaction probabilities from the ground state of O_2(v=0,j=1) in the energy range of 0.81–0.89 eV computed at different propagation times [37]. The threshold energy for the H + O_2 → HO + O reaction reaction is 0.8115 eV. As shown in the Figure, the wave packet propagation up to t=15,000 a.u. yields essentially zero probabilities at energies below 0.89 eV. Some broad peaks appeared at t=15,000 a.u. and more peaks are observed at t=50,000 a.u. Thus the resonance structures grow more pronounced and more peaks appear as the propagation time increases.

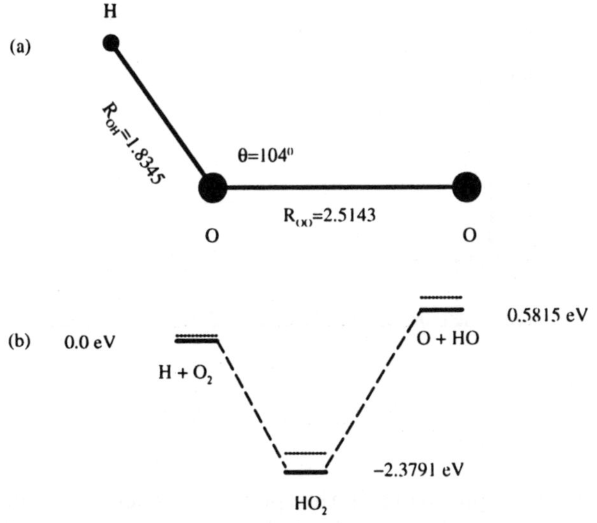

Figure 6.4: (a) The equilibrium geometry of HO_2 where distances are in bohrs. (b) The energy profile of the reaction $H(^2S) + O_2(^3\Sigma_g^-) \rightarrow HO_2(^2A'') \rightarrow OH(^2\Pi) + O(^3P)$. The dotted lines above the solid lines represent the zero point energy levels.

The reaction probabilities at high energies are converged within relatively short propagation times. Most high-energy results are converged around t=50,000 a.u.. Similar convergence is also seen for reaction probabilities out of vibrationally excited initial states. Figure 6.7 shows converged reaction probabilities from several excited vibrational states of O_2(v,1) for v=0–3. The trend is that vibrational excitation gradually decreases the amplitude of probability. It should be pointed out, however, that the plotted energy is the total scattering energy (internal energy of O_2 plus translational energy). Thus it does not imply that the vibrational excitation decreases the reaction probability at fixed kinetic energies. It is also clear from the Figures that resonances persist all the way to high energies without any significant broadening of the widths.

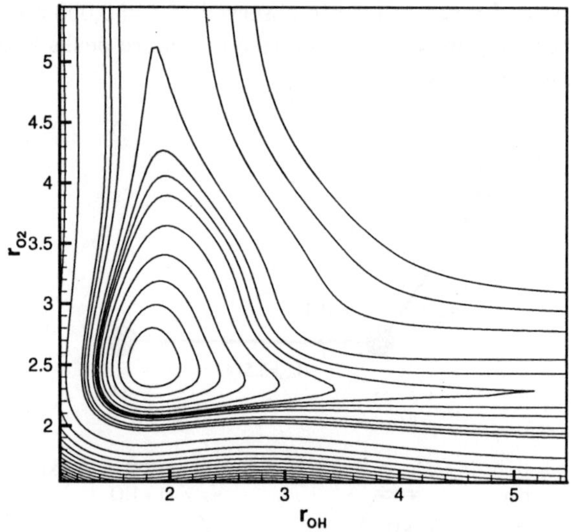

Figure 6.5: Contour plot of the H + O_2 potential energy surface with the bending angle fixed at γ=132 degrees.

It is worth emphasizing at this point a particular feature of the TD wave packet method. The complicated energy-dependences of the reaction probabilities shown in Fig. 6.6 and 6.7 are mapped out in a single wave packet propagation for hundreds of energies at only a fractional cost of the total wave packet propagation. This is extremely attractive for systems like H + O_2 for which several hundreds of close-coupling (CC) calculations need to be repeated in order to map out the resonance structures in the standard time-independent approach. Such energy-dependent propagation could be very expensive computationally as the CPU time in each CC propagation scales as N^3.

The reaction probabilities shown above are calculated for total angular

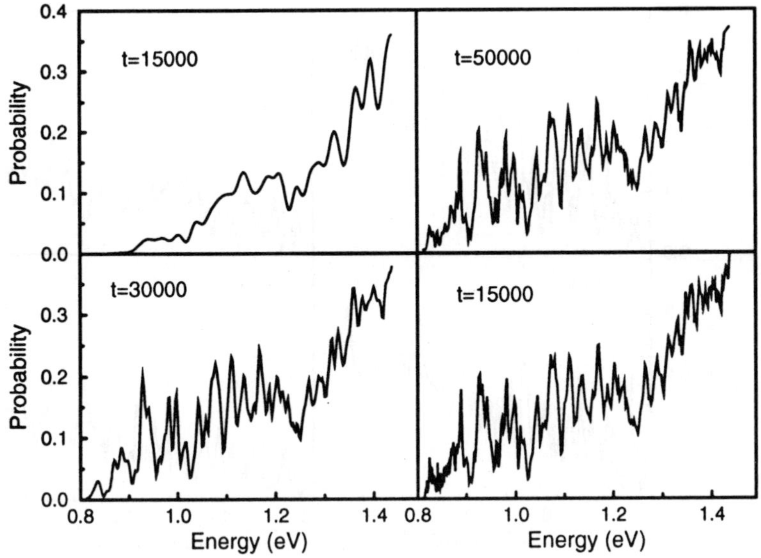

Figure 6.6: Total reaction probabilities from the ground state $H + O_2(v = 0, j = 1)$ for total angular momentum J=0 as a function of scattering energy computed at different propagation times t.

momentum J=0. Cumulative reaction probabilities for J=0 were reported by the reactive flux method [113]. Recent $J > 0$ wavepacket calculations have been reported for the $H + O_2$ reaction as well [115].

6.6 Tetraatomic Reaction: AB + CD

6.6.1 Beyond Triatomic Systems

Of more practical interest to chemistry are polyatomic reactions that involve more than three atoms. But going beyond the atom-diatom to polyatomic reactive scattering presents a new grand challenge to quantum dynamicists. The major challenge in theoretical treatment is how to handle the exponential increase of computational cost due to the increase of mathematical dimensionalities when the number of atoms in the system increases. For example, the dimensionality (internal degrees of freedom) increases from three for a triatomic system to six for a tetraatomic system—a two-fold increase in dimensionality! Since the addition of each atom adds

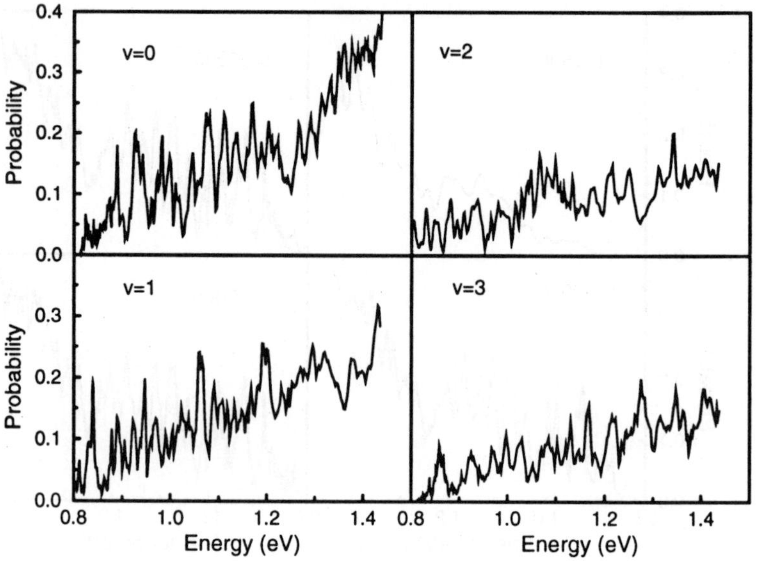

Figure 6.7: Total reaction probabilities from initial O_2 vibrational states of v=0, 1, 2, and 3.

three internal degrees of freedom to the system, it is interesting to note that the transition from triatomic to tetraatomic systems causes the maximum relative increase in dimensionality (100%). Thus the rigorous dynamical treatment for tetraatomic reactions is hardly a trivial extension of the previous treatments for triatomic reactions and its success is a major advance in reaction dynamics. In fact, a number of computational methodologies that work effectively for simple triatomic systems prove difficult or even impossible to apply at present to polyatomic reactions due to the requirement of impractically large computational resources. For example, in the algebraic variational approach, one is required to invert the Hamiltonian matrix to solve linear algebraic equations. Even for a simple tetraatomic reaction like H_2 + OH, the size of the Hamiltonian matrix is prohibitively large to be inverted directly on today's computers. It is therefore necessary to utilize alternative methods such as iterative methods to solve linear algebraic equations due to their lower computational scaling than matrix inversion. Thus the critical measure of the applicability of any method to polyatomic reactions is the scaling of its computational cost with respect to the number of basis functions or degrees of freedom. Many standard time-independent scattering methods such as variational methods or propagation methods scale as N^3 with the number of basis functions N, and are thus difficult to extend to large systems. Until a few years ago, the reduced dimensionality approach (RDA) [117,118] provided the only viable means for tackling the four-atom reactive scattering problem. In the RDA approach a four-atom reaction system is reduced to an effective atom-diatom system through elimination of three internal coordinates, either by applying adiabatic approximation for three internal angular variables [117] or by restricting the system to certain geometric configurations [118]. Although the RDA methods are computationally simple to apply and can often give reasonably good results when all the missing degrees of freedom are properly accounted for, they generally do not give definitive results and/or predictions of the dynamics of the reactive scattering problem for a given potential energy surface. The status is similar for other dynamically approximate methods including the IOSA method [119], the mixed quantum/classical method [120] and full-dimensional planar models [121,122,152]. These approximate methods are nevertheless very useful for studying the dynamics of complex reactions for which rigorous dynamical methods are not available.

The ultimate goal in quantum reaction dynamics is to develop rigorous quantum methods that can provide definitive results and/or predictions for the dynamics of polyatomic reactions for given potential energy surfaces. Significant progress has been made in that direction during the past few years and the above goal has been at least partially realized for a few important benchmark tetraatomic reactions. Rigorous quantum reactive scattering calculations in full-dimensional space have been reported for reactions of H_2 + OH [124–127], DH + OH [128], D2 + OH [129], and HO + CO [130], including calculations of initial state specific cross sections for H_2 + OH and its isotope reactions. Most recently, quantum state-to-state calculations have become available for the H_2 + OH reaction [131,132] and its reverse reaction H + H_2O [133,134]. In the following, we present the TD theoretical treatment for tetraatomic reactive scattering.

6.6.2 Hamiltonian and Basis Functions

In this section, we are primarily interested in the calculation of total (final state summed) reaction probabilities while the discussion of complete state-to-state reactive scattering calculations will be given in Chapter 7. As is well known to the reactive scattering community, the choice of suitable coordinates as well as the basis functions associated with them is at the heart of any reactive scattering problem. The Jacobi coordinates are natural coordinates for describing wavefunctions that are confined primarily to the

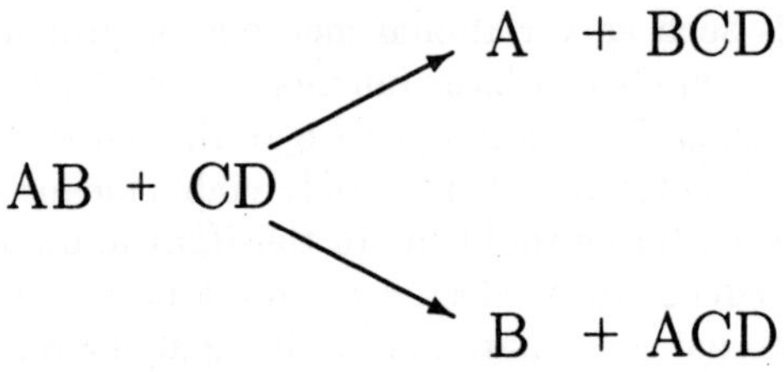

Figure 6.8: Reaction of AB + CD to produce A + BCD and B +ACD.

corresponding arrangement channel space. Thus, the Jacobi coordinates of the reactant arrangement are generally a good choice for calculating initial state selected but final state summed reaction probabilities because we only need to propagate the wavefunction to just beyond the transition state region. Specifically for a diatom-diatom reaction AB + CD to produce atom-triatom products A + BCD and/or B + ACD as illustrated in Fig. 6.8, the Hamiltonian expressed in reactant Jacobi coordinates defined in Fig. 3.2 in full dimensions (6D) can be written as

$$H = -\frac{\hbar^2}{2\mu}\frac{\partial^2}{\partial R^2} + \frac{(\mathbf{J}-\mathbf{j}_{12})^2}{2\mu R^2} + h_1(r_1) + h_2(r_2) + \frac{\mathbf{j}_1^2}{2\mu_1 r_1^2} + \frac{\mathbf{j}_2^2}{2\mu_2 r_2^2} + V(\mathbf{r}_1, \mathbf{r}_2, \mathbf{R}) \tag{6.64}$$

where μ is the reduced mass between the center-of-mass of AB and CD, $\mathbf{J}$ the total angular momentum operator, and $\mathbf{j}_1$ and $\mathbf{j}_2$ the rotational angular momentum operators of AB and CD, which are coupled to form $\mathbf{j}_{12}$. The reference diatomic vibrational Hamiltonian $h_i(r_i)$ (i=1,2) is defined as

$$h_i(r_i) = -\frac{\hbar^2}{2\mu_i}\frac{\partial^2}{\partial r_i^2} + V_i(r_i), \tag{6.65}$$

whose eigenfunctions and eigenenergies are ϕ_{v_i} and ε_{v_i}, respectively, and V_i is a reference diatomic vibrational potential. The expression for the Hamiltonian given in terms of the Jacobi coordinates for the A + BCD arrangement is very similar.

Before the numerical solution for the TD wavefunction can be started, one needs to find a suitable basis set to expand the TD wavefunction. For a general diatom-diatom reaction of the type AB + CD, one can expand the TD wavefunction in terms of body-fixed (BF) rovibrational eigenfunctions defined in terms of the reagent Jacobi coordinates

$$\Psi^{JMp}_{v_0j_0K_0}(\mathbf{R},\mathbf{r}_1,\mathbf{r}_2,t) = \sum_{n,v,j,K} F^{JMp}_{nvjK,v_0j_0K_0}(t)u^{v_1}_n(R)\phi_{v_1}(r_1)\phi_{v_2}(r_2) \times \mathcal{Y}^{JMp}_{jK}(\hat{R},\hat{r}_1,\hat{r}_2), \tag{6.66}$$

where n is the translational basis label, v denotes (v_1, v_2), j denotes (j_1, j_2, j_{12}), (v_0, j_0) is the label for the initial rovibrational state, and p is the system parity defined as $\epsilon = (-1)^{j_1+j_2+L}$ with L being the orbital angular momentum quantum number. The determination of the TD coefficient $F^{JM\epsilon}_{nvjK,v_0j_0K_0}(t)$ gives the solution of the TD Schrödinger equation. In order to save computational cost, we separate the interaction region from the asymptotic region in the dynamics calculation [135,136,199]. A simple way to implement this is to use nondirect product basis functions and define normalized translational basis functions as [125],

$$u^{v_1}_n(R) = \begin{cases} \sqrt{\dfrac{2}{R_4 - R_1}} \sin \dfrac{n\pi(R - R_1)}{R_4 - R_1} & v_1 \le v_{asy} \\[2ex] \sqrt{\dfrac{2}{R_2 - R_1}} \sin \dfrac{n\pi(R - R_1)}{R_2 - R_1} & v_1 > v_{asy} \end{cases},$$

where R_2 and R_4 define, respectively, the interaction and asymptotic grid (cf. Fig. 6.2), and v_{asy} is the number of energetically open vibrational states plus a few closed vibrational states of the reactive AB diatom. The use of a nondirect product basis makes it simple to separate the asymptotic region from the interaction region, and thus a substantial amount of computational savings can be realized.

The coupled total angular momentum eigenfunctions $\mathcal{Y}^{JMp}_{jK}$ in Eq. (6.66) are defined in Eq. (D.34) in the appendix

$$\begin{aligned} \mathcal{Y}^{JMp}_{jK} &= \frac{1}{\sqrt{2(1+\delta_{K0})}}\left[\mathcal{Y}^{JM}_{jK} + \hat{p}\mathcal{Y}^{JM}_{jK}\right] \\ &= \frac{1}{\sqrt{2(1+\delta_{K0})}}\left[\mathcal{Y}^{JM}_{jK} + (-1)^P\mathcal{Y}^{JM}_{j-K}\right] \end{aligned} \tag{6.67}$$

where $\mathcal{Y}^{JM}_{jK} = \widetilde{D}^J_{KM}\mathcal{Y}^{j_{12}K}_{j_1j_2}$ and $\mathcal{Y}^{j_{12}K}_{j_1j_2}$ is the angular momentum eigenfunction of j_{12},

$$\mathcal{Y}^{j_{12}K}_{j_1j_2}(\theta_1,\theta_2,\phi) = \sum_{m_1} \langle j_1m_1j_2K - m_1|j_{12}K\rangle P_{j_1m_1}(\theta_1) \times Y_{j_2K-m_1}(\theta_2,\phi), \tag{6.68}$$

where P_{jm} are normalized associated Legendre polynomials and Y_{jm} are spherical harmonics. Note in Eq. (6.67) that if $K = 0$, then the allowed quantum numbers are restricted by the requirement

$$(-1)^{P+j_1+j_2+j_{12}} = 1 \tag{6.69}$$

which limits the allowable values of j_{12}.

Time propagation of wavefunction

The split-operator propagator is used to carry out the time propagation of the wavepacket,

$$\Psi^{JM\epsilon}(\mathbf{R}, \mathbf{r}_1, \mathbf{r}_2, t+\Delta) = e^{-iH_0\Delta/2} e^{-iU\Delta} e^{-iH_0\Delta/2} \Psi^{JM\epsilon}(\mathbf{R}, \mathbf{r}_1, \mathbf{r}_2, t), \tag{6.70}$$

where the reference Hamiltonian $\mathrm{H_0}$ is defined as,

$$H_0 = -\frac{\hbar^2}{2\mu}\frac{\partial^2}{\partial R^2} + h_1(r_1) + h_2(r_2) \tag{6.71}$$

and the effective potential operator U in Eq. (6.70) is defined as

$$\begin{aligned} U &= \frac{(\mathbf{J}-\mathbf{j}_{12})^2}{2\mu R^2} + \frac{\mathbf{j}_1^2}{2\mu_1 r_1^2} + \frac{\mathbf{j}_1^2}{2\mu_2 r_2^2} + V(\mathbf{r}_1, \mathbf{r}_2, \mathbf{R}) \\ &= V_{rot} + V(\mathbf{r}_1, \mathbf{r}_2, \mathbf{R}). \end{aligned} \tag{6.72}$$

Utilizing the split-operator scheme again, we can split $e^{-iU\Delta}$ as

$$e^{-iU\Delta} = e^{-iV_{rot}\Delta/2} e^{-iV\Delta} e^{-iV_{rot}\Delta/2}, \tag{6.73}$$

where V_{rot} and V are defined in Eq. (6.72). The action of the operator $e^{-iV_{rot}\Delta/2}$ on the wavefunction is straightforward because it is diagonal in the coupled angular momentum representation. The matrix version of Eq. (6.70) for the expansion coefficient vector $\mathbf{F}$ is then given by

$$\mathbf{F}(t+\Delta) = \exp(-i\mathbf{H}_0\Delta/2)\exp(-i\mathbf{U}\Delta)\exp(-i\mathbf{H}_0\Delta/2)\mathbf{F}(t), \tag{6.74}$$

where $\mathbf{H}_0$ is the diagonal matrix defined in Ref. [125].

At a given radial quadrature point (R_m, r_{1n}, r_{2l}), the standard method for handling the potential operator $e^{-iU\Delta}$ is to diagonalize the potential matrix $\mathbf{U}$ in the angular basis $\mathcal{Y}_{jK}^{JMp}$. This approach preserves the unitarity of the operator $e^{-iV\Delta}$ and is efficient when the size of the angular basis Y_{jK}^{JMp} is relatively small. However, if the coupled angular basis is large, this approach can become computationally expensive because one needs to calculate and store all the transformation matrices that diagonalize the potential matrices **U** at all the radial grid points. Thus for large systems, the matrix diagonalization method will require a large-memory computer. We can use a normalized quadrature scheme to treat angular quadratures which avoids explicit matrix diagonalization and therefore does not require large computer memory for matrix storage.

The exponential potential operator $e^{-iV\Delta}$ is now treated by quadrature approximation for which we define a transformation matrix **Q** by

$$Q^{jK}_{i\,k\,l} = \sqrt{W_{1i}W_{2k}W_{3l}}\; <\theta_{1i}\,\theta_{2k}\,\phi_l|\mathcal{Y}^{JMp}_{jK}>, \tag{6.75}$$

where $(\theta_{1i}\,\theta_{2k}\,\phi_{3l})$ are angular quadratures and (W_{1i}, W_{2k}, W_l) are the corresponding angular weights. Thus Eq. (6.73) is approximated by the angular quadrature

$$e^{-iU\Delta} = e^{-iV_{rot}\Delta/2}\mathbf{Q}^{+}e^{-iV\Delta}\mathbf{Q}e^{-iV_{rot}\Delta/2} \tag{6.76}$$

where $\mathbf{Q}^{+}$ is the complex conjugate of **Q**. In general, one does not need to store the large matrix **Q** but only a few small sub-matrices contained in Eq. (6.75). Because the transformation matrix **Q**, which transforms from the coupled angular momentum representation to the grid representation, is not unitary, we therefore use a simple method to retain the unitarity of the propagator. Specifically, we renormalize the wavefunction after it has been propagated by the operator $e^{-iV\Delta}$. The detailed procedure has been described in Ref. [128] and [130].

6.6.3 Treatment for Spectator Bonds

For a polyatomic reaction with a nonreactive bond, such as the OH bond in the H_2 + OH reaction, one often does not need to treat the nonreactive OH bond coordinate explicitly in the dynamics calculation. The PA5D model treats the vibration of the nonreactive CD bond diabatically which results in an effective 5D model in which the effective 5D potential is simply obtained by averaging the original PES over the vibrational function of the nonreactive CD (or OH) bond. Specifically, the interaction potential $V(\mathbf{r}_1, \hat{r}_2, \mathbf{R})$ in Eq. (6.72) is obtained by averaging the 6D potential over the vibrational function of the nonreactive CD [125], i.e.,

$$\boxed{V(\mathbf{r}_1, \hat{r}_2, \mathbf{R}) = <\phi_{v_2}|V(\mathbf{r}_1, \mathbf{r}_2, \mathbf{R})|\phi_{v_2}>} \tag{6.77}$$

and the rotation constant for the nonreactive diatom CD is given by $B_{v_2} = <\phi_{v_2}|1/2\mu_2 r_2^2|\phi_{v_2}>$. The study in Ref. [125] showed that the PA5D treatment is significantly better than the simpler rigid-bond treatment with fixed bond length and it gives reaction probabilities that are essentially indistinguishable from those of the full 6D calculation for the OH + H_2 reaction. This is very encouraging for polyatomic reactions because it demonstrates the practicality of eliminating spectator bond lengths from explicit dynamics calculations.

Extraction of dynamics information

From the propagation of an initial wavepacket $|\chi_i(0)\rangle$, the time-independent (TI) wavefunction $\psi_i^+(E)$ can be obtained by Fourier transforming the TD wavefunction as will be discussed in Sec. 7.2

$$\psi_i^+(E) = \frac{1}{2\pi\hbar a_i(E)} \int_{-\infty}^{\infty} e^{\frac{i}{\hbar}(E-H)t} \chi_i(0) dt, \tag{6.78}$$

and similarly for the derivative of the wavefunction $\psi_i^{'+}(E)$. The coefficient $a_i(E)$ is easily evaluated from the free energy-normalized asymptotic function $\phi_i(E)$ as $a_i(E) = \langle\phi_i(E)|\chi_i(0)\rangle$ [125]. The total reaction probability from a given initial state i can be calculated by using the flux formula discussed in Sec. 6.4.2

$$P_i^R(E) = \sum_f |S_{fi}^R|^2 = \langle\psi_i^+(E)|\hat{F}|\psi_i^+(E)\rangle \tag{6.79}$$

The initial wavepacket $|\chi_i(0)\rangle$ is usually chosen to be a Gaussian function with an average momentum k_0 traveling toward the $-R$ direction (interaction region)

$$\phi_{k_0}(R) = (\frac{1}{\pi\delta^2})^{1/4} \exp[-(R-R_0)^2/2\delta^2] e^{-ik_0R}, \tag{6.80}$$

multiplied by the internal function $|\varphi_i\rangle$ in eq. (6.66). In actual propagation, the TD wavefunction is absorbed at the edges of the grid to avoid boundary reflection as discussed in the previous section.

6.6.4 Reaction of H_2 + OH

Background

The H_2 + OH is perhaps the "simplest" prototype four-atom reaction system and has enjoyed much theoretical and experimental attention, much like the status of the H + H_2 reaction for triatomic systems. However, besides theoretical interest, H_2 + OH is an important reaction in combustion and is the main source of water in typical hydrocarbon/air flames at atmospheric pressure. The H_2 + OH reaction is exothermic by ~15 kcal/mol and has a near T-shaped transition state geometry with the O-H-H atoms lying close to a linear geometry. The potential barrier to reaction is ~6 kcal/mol [140]. Extensive experimental studies have been reported including the investigation of the effect of vibrational excitation of OH and H_2 on reaction rate, as well as studies of the reverse reaction H + H_2O. Experiments have also been reported recently on the measurement of differential cross sections and absolute reaction cross sections for the OH + H_2/D_2 reaction [141]. Extensive theoretical studies have been reported for this reaction. Many of the theoretical investigations for the H_2 + OH reaction have been discussed in a recent article [95].

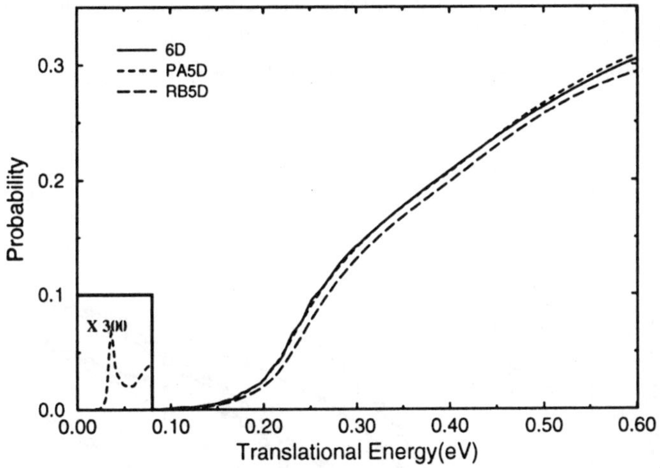

Figure 6.9: Total reaction probability as a function of translational energy for the reaction $H_2(00) + OH(00) \rightarrow H + H_2O$ for total angular momentum J=0.

Reaction probability

In this subsection, we show some results of application of the TD approach described in the previous section for calculations of total (final state summed) reaction probabilities for H_2 + OH. The details of the numerical calculation can be found in Ref. [125]. The potential energy surface used for the calculation of the H_2 + OH reaction is the Walch-Dunning-Schatz-Elgersma PES [140] slightly modified by Clary [118]. Figure 6.9

shows the energy-dependence of the calculated total (final state summed) reaction probabilities from the initial ground state of the reagents H_2 + OH. The long-dashed line is the result of the RB5D (rigid bond) calculation in which the OH bond distance is fixed at the equilibrium distance of free OH. Thus the dynamics calculation for the RB5D model includes 5 internal coordinates. There is a small but noticeable difference between the RB5D result and the exact 6D result as shown in Fig. 6.9. On the other hand, the potential-averaged 5D (PA5D) treatment with only one vibrational basis described in Sec. 6.6.3 gives essentially identical reaction probabilities to those obtained from the 6D calculation. This indicates that the OH bond is a spectator bond whose coordinate does not need to be explicitly included in the dynamics calculation. One only needs to treat its vibration by averaging the 6D potential to obtain an effective 5D potential and carry out an effective 5D dynamics calculation. We expect this conclusion to be generally true for similar reactions involving nonreactive or spectator bonds. This should result in significant savings in computation for polyatomic reactions with many spectator bonds.

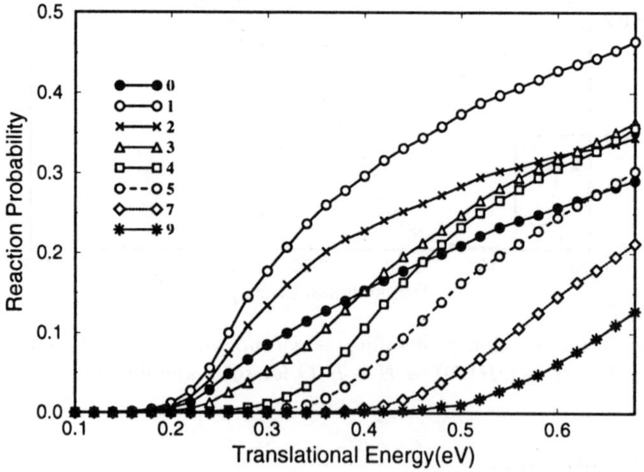

Figure 6.10: Total reaction probability as a function of translational energy for the reaction $D_2(0j) + OH(00) \to D + HOD$ for different values of j.

It is also of interest to note the sharp peak inside the rectangular box at the lower left corner in Fig. 6.9. This is the reaction probability at very low kinetic energies and the result is enlarged by a factor of 300 to make it visible in the figure. Although the reaction probability is only on the order of 10^{-4} at the top of this peak, it has a very narrow width corresponding to a resonance lifetime longer than 100 femtoseconds. it is believed to be an artifact of the Schatz-Elgersma PES which is known to have an unphysical shallow well near the entrance channel. This peak does not have any significant effect on the reaction cross sections at energies above 0.1 eV. However, its effect is significant on the calculated reaction cross sections for vibrationally excited H_2 as discussed in detail in [125]. This demonstrates that exact quantum dynamics calculation provides a very sensitive probe of the global potential energy surface.

Unlike the spectator bond of OH, the vibrational motion of H_2 is found to significantly enhance the reaction probability as expected for this reaction [125]. Also of interest is the rotational state dependence of the reaction probability which shows a strong steric effect for H_2 + OH and its isotopic reactions. Figure 6.10 shows the rotational state dependence of total reaction probabilities for the isotopic D_2 + OH reaction at total angular momentum J=0. As shown in the figure, the reaction probability is quite sensitive to the initial rotational states of D_2 or H_2 [128, 129]. In particular, the maximum reaction probability always shows up for the j=1 state of $H(D)_2$. This is believed to be a general phenomenon for collinearly dominated reactions at zero total angular momentum due to spatial orientation of the rotational wavefunction as explained in [128].

Reaction cross section

The reaction cross section from a specific initial state is obtained by summing the reaction probabilities $P^{J\epsilon}_{v_0 j_0 K_0}$ over all contributing partial waves (total angular momentum J),

$$\sigma_{v_0 j_0}(E) = \frac{1}{(2j_1+1)(2j_2+1)} \frac{\pi}{k^2_{v_0 j_0}} \sum_{J K_0 p} (2J+1) P^{Jp}_{v_0 j_0 K_0}(E), \qquad (6.81)$$

where p is the parity and K_0 denotes the initial rotation projection quantum numbers on the BF z axis. Since the exact close-coupling calculation for J>0 is extremely expensive computationally, the standard CS approximation is used in calculations for J>0. The CS approximation is expected to give good results for collinearly dominated reactions such as H_2 + OH. Figure 6.11 shows the calculated integral cross sections as a function of translational energy.

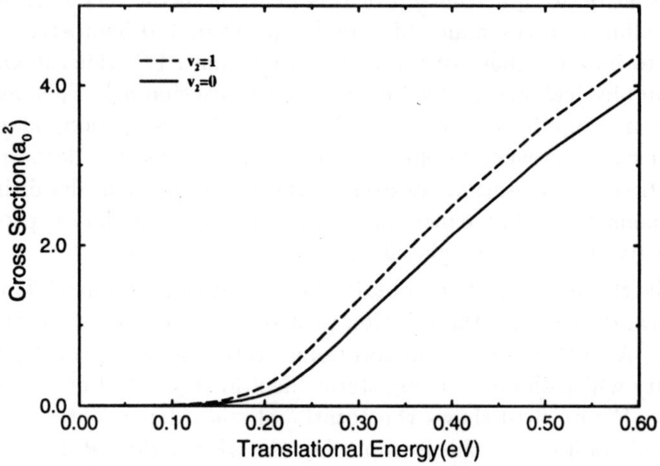

Figure 6.11: Integral cross sections for the reaction $H_2(00)$ + $OH(v_2 0)$. Solid line is the result for v_2=0 and the long-dashed line for v_2=1.

Rate constant

The initial state-specific rate constant is obtained by thermally averaging the collision energy of the reactive flux as given in Eq. (5.121)

$$r_{v_0 j_0}(T) = \left(\frac{1}{2}\right)\sqrt{\frac{8kT}{\pi\mu}}\frac{1}{(kT)^2}\int_0^\infty dE_t\, E_t \exp(-E_t/kT)\sigma_{v_0 j_0}(E_t) \quad (6.82)$$

where E_t is the translational energy. For H_2 + OH, an extra factor of 1/2 has been included in Eq. (6.82) to account for the fact that only half of the reagent $H_2(^1\Sigma)$ + $OH(^2\pi)$ collisions access the $^2A'$ surface which correlates with the products $H_2O(^1A')$ + $H(^2A')$ [140]. Figure 6.11 shows the calculated rate constant for temperatures below 1000 K. As shown, the rate constant has the familiar Arrhenius form. We note, however, that the rate constant defined in Eq. (6.82) involves only Boltzmann averaging over the translational energy but not the rovibrational energy. This is not the standard definition of the thermal rate constant for which all the rovibrational states are Boltzmann averaged.

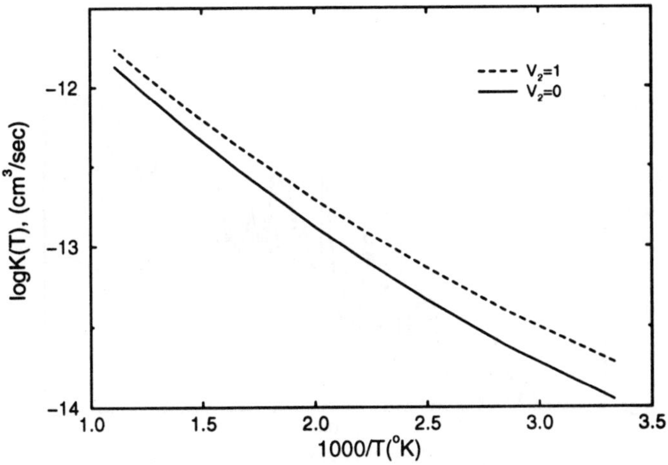

Figure 6.12: The rate constant for the reaction $H_2(00) + OH(v_2 0)$ as a function of temperature. Solid line is the result for v_2=0 and the dashed line v_2=1.

6.6.5 Reaction of HO + CO

Background

The reaction of HO + CO to produce H + CO_2 plays a very important role in combustion processes. It is named "the second most important combustion reaction" [112]. However, quantum dynamics calculation for the HO + CO reaction is much more challenging computationally than the direct H_2 + OH reaction. First, the HOCO system has only one hydrogen atom and is thus much heavier than the three-hydrogen H_2OH system. Secondly, the PES for the HO + CO system has deep wells of about 1.6 eV which support stable species of *trans-* and *cis-*HOCO radicals. Thus the reaction is strongly influenced by dynamical resonances with lifetimes on the order of one picosecond. These features have imposed serious challenges to rigorous quantum reactive scattering calculations for the HO + CO reaction.

The HO + CO reaction has been the subject of many experimental studies [142–145] including a recent crossed molecular beam study by Alagia *et al* [147]. The reaction rate is found to be nearly independent of temperature below 500 K [142]. The molecular beam experiment has shown strong peaking both in the forward and backward directions, indicating the existence of intermediate species [147]. In the theoretical front, *ab initio*

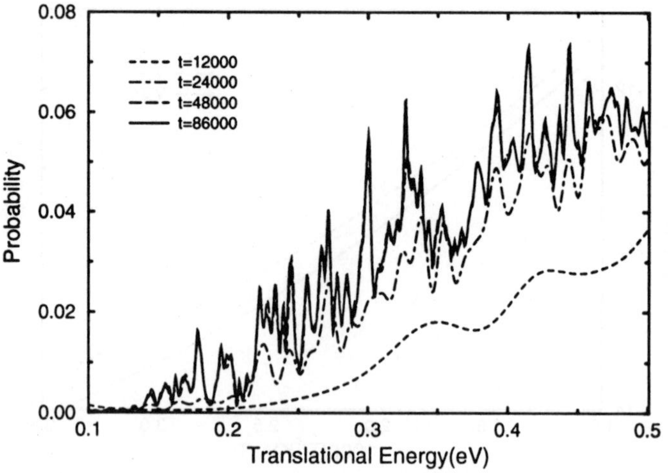

Figure 6.13: Total reaction probabilities for the reaction HO + CO in the ground state with total angular momentum J=0 calculated through wavepacket propagation at different times in atomic units.

calculations have been performed and the calculated energy points have been fitted to a global potential surface by Schatz *et al* [146]. Quasiclassical trajectory calculations have been performed in full-dimensions and state-selected reaction cross sections and resonance lifetimes have been reported [148]. Reduced dimensionality quantum dynamics calculations have also been reported [149, 150, 152, 153]. In the following, we show results of a rigorous dynamics calculation of Ref. [130] which is the first quantum calculation in full physical dimensions using the PA5D treatment for the nonreactive CO bond. As is shown for the H_2 + OH reaction, the PA5D treatment for the spectator bond is an excellent approximation and the result is essentially indistinguishable from that of the exact 6D calculation.

Results of theoretical calculation

Because of resonances, the TD calculation for HO + CO requires one to propagate the wavefunction for more than 1 picosecond in order to uncover the resonance structures [130]. Figure 6.13 shows the calculated total reaction probabilities from the ground state of the reagents that are obtained from different propagation times. As shown in the figure, the reso-

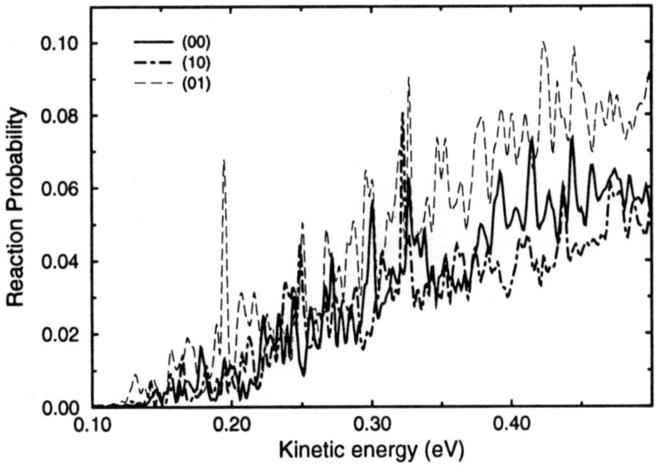

Figure 6.14: Total reaction probabilities for the reaction $HO(v = 0, j_1)$ + $CO(v = 0, j_2)$ for different initial states of (j_1, j_2).

nance structure does not show up until after propagating the wavepacket to about 48,000 a.u.. The energy-dependence of converged reaction probabilities shows many narrow but overlapping resonances. Although the HO + CO reaction is exothermic by about 0.97 eV and there is no barrier along the minimum energy path in the entrance channel, the reaction probability is quite small (less than 10% in Fig. 6.13). The system does not seem to follow the reaction path and therefore produces very little product. From the propagation time needed to converge the reaction probabilities, the lifetimes of most resonances are less than 1 ps with only a couple of them lasting a little longer than that. Figure 6.14 shows reaction probabilities from different initial rotational states of the reagents as a function of kinetic energy. The quantum dynamics calculation shows that both steric and resonance effects play important roles in the HO + CO reaction. The calculated reaction probability decreases as the dimensionality of the model increases [130].

Chapter 7

State-to-State Reactive Scattering

7.1 Introduction

Much progress has been made in the theoretical and experimental study of elementary chemical reactions during the past decade. Experimentally, complete state-to-state integral cross sections have been measured for some simple reactions and the measurement of state-to-state differential cross sections now appears on the horizon. These experiments will require accurate theoretical calculations of state-to-state dynamics for comparison and interpretation in order to provide detailed dynamical information for chemical reactions at complete state-to-state levels. In the full scale time-independent hyperspherical coordinate or algebraic variational approaches to reactive scattering, the complete state-to-state scattering matrix is obtained from a single calculation. Because these are N^3 methods, the computational cost is very high and one either obtains nothing or the complete state-to-state S matrix. In the TD wavepacket approach, however, one starts from a given initial state and solves for a column of the S matrix only. Since it is an N^2 approach, one obtains reaction product state distributions from a given initial state only from a single wavepacket calculation. If scattering information from different initial states is needed, it is necessary to carry out a new TD calculation for these initial states as desired. Because theoretical calculation at the state-to-state level is generally much more demanding computationally than the calculation of total (final state summed) reaction probabilities, it is important and practical to develop efficient approaches to compute state-to-state S matrix elements in the TD wavepacket approach. In this chapter, we discuss a number of theoretical methods designed specifically for the calculation of state-to-state S matrix elements or probability.

7.2 State-to-State S Matrix Elements

7.2.1 Time-independent Expressions of S Matrix Elements

In quantum scattering theory, there are many seemingly different but mathematically equivalent expressions for S matrix elements which are sometimes confusing to the readers. In this section, we derive a number of formally equivalent expressions for the state-to-state S matrix elements starting from the basic definition in order to give a clear picture and connections between the various expressions for the S matrix elements. Starting from the general definition of on-shell S matrix elements S_{fi} from Eq. (4.58) in Sec. 4.2.2, we can write generalized reactive S matrix elements

$$\delta(E_f - E_i)S_{fi} = <\psi^-_{\beta f}(E_f)|\psi^+_{\alpha i}(E_i)> \tag{7.1}$$

where $|\psi^+_{\alpha i}(E_i)>$ and $|\psi^-_{\beta f}(E_f)>$ are energy-normalized δ-function incoming and outgoing scattering wavefunctions, and $i(f)$ labels the initial (final) state. Here we use α and β as, respectively, the initial and final arrangement labels, and use γ to denote the intermediate arrangement label. The scattering wavefunctions $|\psi^+_{\alpha i}(E_i)>$ and $|\psi^-_{\beta f}(E_f)>$ satisfy the asymptotic boundary conditions of Eq. (4.271)

$$\begin{aligned} \psi^+_{\alpha i}(E) \stackrel{R_\gamma \to \infty}{\longrightarrow} & \sqrt{\frac{\mu_\gamma}{2\pi\hbar^2}} \Bigg[- \frac{e^{-ik_i R_\alpha}}{\sqrt{k_i}} |\varphi_{\alpha i}> \delta_{\alpha\gamma} \\ & + \sum_m S_{\gamma m, \alpha i} \frac{e^{ik_m R_\gamma}}{\sqrt{k_m}} |\varphi_{\gamma m}> \Bigg] \end{aligned} \tag{7.2}$$

and

$$\begin{aligned} \psi^-_{\beta f}(E) \stackrel{R_\gamma \to \infty}{\longrightarrow} & \sqrt{\frac{\mu_\gamma}{2\pi\hbar^2}} \Bigg[- \frac{e^{ik_f R_\beta}}{\sqrt{k_f}} |\varphi_{\beta f}> \delta_{\gamma\beta} \\ & + \sum_m S^*_{\gamma m, \beta f} \frac{e^{-ik_m R_\gamma}}{\sqrt{k_m}} |\varphi_{\gamma m}> \Bigg] \end{aligned} \tag{7.3}$$

where μ_γ is the reduced translational mass and $\varphi_{\gamma m}$ is mth internal channel function in the γ arrangement.

We can construct an $\mathcal{L}^2$ wavepacket $\chi^+_{\alpha i}$ which consists of incoming states only by the equation

$$|\chi^+_{\alpha i}> = \int a_{\alpha i}(E)|\psi^+_{\alpha i}(E)> dE \tag{7.4}$$

and similarly for another wavepacket $\chi^-_{\beta f}$ which consists of outgoing states only by

$$|\chi^-_{\beta f}> = \int a_{\beta f}(E)|\psi^-_{\beta f}(E)> dE. \tag{7.5}$$

where $a_{\alpha f}(E)$ and $a_{\beta f}(E)$ are coefficients of superposition. The $\chi^+_{\alpha i}$ and $\chi^-_{\beta f}$ so defined contain a range of scattering energies determined by the coefficients $a_{\alpha i}(E)$ and $a_{\beta f}(E)$. Equations (7.4) or (7.5) can be inverted to express $|\psi^+_{\alpha i}(E)>$ and $|\psi^-_{\beta f}(E)>$ in terms of the $\mathcal{L}^2$ wavepackets

$$|\psi^+_{\alpha i}(E)> = \frac{1}{a_{\alpha i}(E)}\delta(E-\hat{H})|\chi^+_{\alpha i}> \tag{7.6}$$

and

$$|\psi^-_{\beta f}(E)> = \frac{1}{a_{\beta f}(E)}\delta(E-\hat{H})|\chi^-_{\beta f}> . \tag{7.7}$$

Since $\psi^+_{\alpha i}$ and $\psi^-_{\beta f}$ are δ-function normalized in energy, the coefficients $a_{\alpha i}(E)$ and $a_{\beta f}(E)$ are then given by

$$\begin{aligned} a_{\alpha i}(E) &= <\psi^+_{\alpha i}(E)|\chi^+_{\alpha i}> \\ &= <\phi_{\alpha i}(E)|\Omega^\dagger_{\alpha +}|\chi^+_{\alpha i}> \end{aligned} \tag{7.8}$$

and

$$\begin{aligned} a_{\beta f}(E) &= <\psi^+_{\beta f}(E)|\chi^+_{\beta f}> \\ &= <\phi_{\beta f}(E)|\Omega^\dagger_{\beta -}|\chi^+_{\beta f}> \end{aligned} \tag{7.9}$$

where $\phi_{\alpha i}(E)$ and $\phi_{\beta f}(E)$ are the asymptotic functions in the α and β arrangements. Here the Mϕller operators are defined as

$$\Omega_{\alpha +} = \lim_{t\to -\infty} e^{\frac{i}{\hbar}Ht} e^{-\frac{i}{\hbar}H_\alpha t} \tag{7.10}$$

and

$$\Omega_{\beta-} = \lim_{t\to\infty} e^{\frac{i}{\hbar}Ht} e^{-\frac{i}{\hbar}H_\beta t} \tag{7.11}$$

Replacing the right hand side of Eq. (7.1) by Eqs. (7.6) and (7.7) and integrating over the energy E_f, one obtains the time-independent expression for the S matrix element

$$S_{fi}(E) = \frac{1}{a^*_{\beta f}(E)a_{\alpha i}(E)} <\chi^-_{\beta f}|\delta(E-\hat{H})|\chi^+_{\alpha i}> \tag{7.12}$$

Utilizing the relation

$$\delta(E-\hat{H}) = \frac{i}{2\pi}(G^+ - G^-) \tag{7.13}$$

and the fact that the term $<\chi^-_{\beta f}|G^-|\chi^+_{\alpha i}>$ vanishes because the wavepacket $\chi^+_{\alpha f}$ is composed of incoming waves only, we obtain a general expression for the S matrix element in terms of the Green's function [114, 154]

$$\boxed{S_{fi}(E) = \frac{i}{2\pi a^*_{\beta f}(E)a_{\alpha i}(E)} <\chi^-_{\beta f}|G^+(E)|\chi^+_{\alpha i}>} \tag{7.14}$$

Although we started from the assumption that $\chi^+_{\alpha i}$ is composed of incoming waves only, the final expression (7.14) for the S matrix element is more general and is valid even if $\chi^+_{\alpha i}$ includes outgoing components. For example, if we replace $\chi^+_{\alpha i}$ by $\chi^+_{\alpha i} + \chi^-_{\alpha i}$, Eq. (7.14) is still valid because the term $<\chi^-_{\beta f}|G^+(E)|\chi^-_{\alpha i}>$ vanishes. Similarly, Eq. (7.14) also holds if $\chi^-_{\beta f}$ contains an incoming component $\chi^+_{\beta f}$

7.2.2 Time-dependent Expressions of S Matrix Elements

Correlation function formalism for the S matrix elements

Eq. (7.14) is the most general expression for the S matrix element from which various forms of S matrix expressions can be derived straightforwardly. One of the useful expressions for the S matrix immediately arises by performing the standard energy $\to$ time Fourier transform for the Green's function in Eq. (7.14)

$$\begin{aligned} S_{fi}(E) &= \frac{1}{2\pi\hbar a^*_{\beta f}(E)a_{\alpha i}(E)} \int_0^\infty dt e^{\frac{i}{\hbar}Et} <\chi^-_{\beta f}|e^{-\frac{i}{\hbar}\hat{H}t}|\chi^+_{\alpha i}> \\ &= \frac{1}{2\pi\hbar a^*_{\beta f}(E)a_{\alpha i}(E)} \int_0^\infty dt e^{\frac{i}{\hbar}Et} C_{fi}(t) \end{aligned} \tag{7.15}$$

where $C_{fi}(t)$ is the correlation function defined by

$$C_{fi}(t) = <\chi^-_{\beta f}|e^{-\frac{i}{\hbar}\hat{H}t}|\chi^+_{\alpha i}> \quad (7.16)$$

Equation (7.15) is the correlation function expression for the S matrix derived by Tannor and Weeks following a somewhat different path [155]. The initial and final wavepackets are defined by the standard relation involving the Mϕller operator in Sec. 4.1.1

$$|\chi^+_{\alpha i}> = \Omega_{\alpha +}|\chi^0_{\alpha i}> \quad (7.17)$$

$$|\chi^-_{\beta f}> = \Omega_{\beta -}|\chi^0_{\beta f}> \quad (7.18)$$

where $\chi^0_{\alpha i}$ and $\chi^0_{\beta f}$ are the corresponding "free" arrangement wavepackets. For example, $\chi^0_{\alpha i}$ can be chosen to be the product of the internal eigenstate $\phi_{\alpha i}$ of the arrangement Hamiltonian H^{int}_α and a radial Gaussian function. If the free packet $\chi^0_{\alpha i}$ is chosen to be located in the asymptotic region of the α arrangement with incoming wave only, then $\chi^+_{\alpha i}$ is identical to $\chi^0_{\alpha i}$ via definitions of (7.17) and (7.10). Similarly, if $\chi^-_{\beta f}$ is chosen to be located in the asymptotic region of the β arrangement with outgoing wave only, then $\chi^-_{\beta f}$ is identical to $\chi^0_{\beta f}$.

One attractive feature of the expression (7.15) is that the time correlation function is independent of the scattering energy. Thus one can save the correlation function at discrete time steps for any desired final state and retrieve it later to extract the corresponding S matrix element at any desired energies contained in the energy range of the wavepackets. However, the calculation of the correlation function $C_{fi}(t)$ entails the calculation of overlap integrals which can be numerically complicated and expensive for reactive scattering. A crucial question is again the choice of the coordinates to carry out the TD propagation of the wavefunction. One can choose either the reactant or product Jacobi coordinates or any other complete set of coordinates to carry out the TD propagation. There are advantages and disadvantages associated with any set of coordinates depending on specific applications.

If we use the reactant Jacobi coordinates to propagate the initial wavepacket, the explicit calculation of the correlation function is given by

$$\begin{aligned} C_{fi}(t) &= <\chi^-_{\beta f}|\chi^+_{\alpha i}(t)> \\ &= \sum_n W_{\alpha n} <\chi^-_{\beta f}(X_{\alpha n})|\chi^+_{\alpha i}(X_{\alpha n}, t)> \end{aligned} \quad (7.19)$$

where $X_{\alpha n}$ and $W_{\alpha n}$ denote a set of quadrature points and weights defined in the initial (α) arrangement Jacobi coordinates. If computer memory is not a severe limitation in the calculation, one can store final-state wavepackets on these quadrature points and save them in memory in order to speed up the calculation of the correlation function in Eq. (7.19). This is not unrealistic because the final state wavepacket $\chi_{\beta f}^-$ can be chosen to be very localized in the coordinate space and thus it requires relatively fewer quadrature points to store in memory. Otherwise one needs to re-calculate the numerical values of the final state wavepackets at given quadrature points at each time step.

One could also use the final (β) arrangement Jacobi coordinates to carry out the wavepacket propagation. In this case, Eq. (7.19) is rewritten as

$$C_{fi}(t) = \sum_n W_{\beta n} <\chi_{\beta f}^-(X_{\beta n})|\chi_{\alpha i}^+(X_{\beta n}, t)> \tag{7.20}$$

where the quadrature points $X_{\beta n}$ and weights $W_{\beta n}$ are defined in the final (β) arrangement Jacobi coordinates. The use of final (product) Jacobi coordinates in the wavepacket propagation facilitates the extraction of final state information because the TD wavepacket is represented in the same coordinate system as the final state wavepackets. Thus the overlap integral in Eq. (7.20) is straightforward to compute without the need to perform coordinate transformations. However, one still needs to transform the initial wavepacket, which is defined in the reactant Jacobi coordinates, to the product Jacobi coordinates prior to the calculation of the correlation function.

Scattering amplitude formalism for the S matrix

If we choose the radial component of the product wavepacket $\chi_{\beta f}^-$ in Eq. (7.14) to be a δ-function times an outgoing asymptotic radial function such as the plane wave

$$<R_\beta|\chi_{\beta f}^->= \delta(R_\beta - R_\beta^\infty)\exp(ik_f R_\beta)|\varphi_{\beta f}> \tag{7.21}$$

where $\varphi_{\beta f}$ is an internal eigenfunction of H_β and R_β^∞ is a fixed radial coordinate in the asymptotic region of the β arrangement, the coefficient $a_{\beta f}(E)$ via Eq. (7.9) is calculated to be (cf. Eq. (4.113))

$$a_{\beta f}(E) = <-i\sqrt{\frac{\mu_\beta}{2\pi\hbar^2 k_f}}\exp(ik_f\hat{R})|\delta(\hat{R} - R_\beta^\infty)\exp(ik_f\hat{R})>$$

$$= i\sqrt{\frac{\mu_\beta}{2\pi\hbar^2 k_f}} \tag{7.22}$$

Eq. (7.14) then becomes

$$\boxed{S_{fi}(E) = \frac{i}{a_{\alpha i}(E)}\sqrt{\frac{k_f}{2\pi\mu_\beta}}e^{-ik_f R_\beta^\infty}\int_0^\infty dt e^{\frac{i}{\hbar}Et} A_{fi}^+(R_\beta^\infty, t).} \tag{7.23}$$

Equation (7.23) expresses the S matrix element in terms of the Fourier transform of the amplitude of the time-dependent wavefunction at a fixed radial coordinate [114, 156, 157]

$$A_{fi}^+(R_\beta^\infty, t) = <R_\beta^\infty| <\varphi_{\beta f}|\chi_{\alpha i}^+(t)> \tag{7.24}$$

More generally at a finite radial coordinate R_β^∞, the plane wave $e^{ik_f R_\infty}$ should be replaced by a more appropriate asymptotic radial function $\phi_f(R)$, such as the Ricatti-Hankel function that approaches the plane wave in the limit of $R \to \infty$.

Eq. (7.23) has an attractive numerical advantage in that one does not need to calculate any overlap integral to extract S matrix elements. Instead, one only needs to evaluate the radial component of the TD wavefunction $A_{fi}^+(t)$ at a large fixed radial coordinate R_β^∞ in the product asymptotic region for any desired open channel of the product. The amplitude $A_{fi}^+(t)$ can be stored on computer disk at each time step and then retrieved later to obtain the corresponding S matrix element S_{fi} at any desired energy via Eq. (7.23). However, one has to use a relatively large value of R_β^∞ at which the radial wavefunction can be represented by the appropriate asymptotic function such as a plane wave or Ricatti-Hankel function. The calculation for the S matrix element in Eq. (7.23) in reactive scattering is facilitated if the TD wavefunction $\chi_{\alpha i}^+(t)$ is represented in the product Jacobi coordinates.

Flux formula for state-to-state reaction probability

In both the correlation function formalism and scattering amplitude formalism discussed above, one needs to know the asymptotic radial function in order to calculate the S matrix element S_{fi}. This will require one to propagate the wavepacket into the product asymptotic space where the radial function $\phi_{\beta f}(R)$ is known. However, if one only needs the absolute value of the S matrix element $|S_{fi}|$ or the probability $P_{fi} = |S_{fi}|^2$, one can calculate the state-to-state reaction probabilities by evaluating the reactive flux (cf. Eq. (7.2))

$$\begin{aligned} |S_{fi}|^2 &= 2\pi\hbar Re\left[A^{+*}_{fi}(R_\beta,E)\hat{v}_\beta A^{+}_{fi}(R_\beta,E)\right]|_{R_\beta=R_L} \\ &= \frac{2\pi\hbar^2}{\mu_\beta} Im\left[A^{+*}_{fi}(R_\beta,E)\frac{d}{dR_\beta}A^{+}_{fi}(R_\beta,E)\right]\Big|_{R_\beta=R_L} \end{aligned} \tag{7.25}$$

where $A^{+}_{fi}(R_\beta)$ is given in terms of the Fourier transform of the time-dependent amplitude $A^{+}_{fi}(R_\beta,t)$ of Eq. (7.24)

$$\begin{aligned} A^{+}_{fi}(R_\beta,E) &= <R_\beta|\ <\varphi_{\beta f}|\psi^{+}_{i}(E_i)> \\ &= \frac{1}{2\pi\hbar a_{\alpha i}(E)}\int_0^\infty dt e^{iEt} A^{+}_{fi}(R_\beta,t). \end{aligned} \tag{7.26}$$

In Eq. (7.25), the flux is calculated at a surface defined by $R_\beta = R_L$ in the product asymptotic region beyond which the final state interaction is negligible and therefore the flux is invariant with respect to further increase of the distance. This state-to-state approach is also facilitated by the use of the product Jacobi coordinates to represent the TD wavefunction $\chi^{+}_{\alpha i}(t)$. As is discussed in Ref. [157], the flux formula (7.25) is valid in regions beyond which the inelastic scattering process is absent irrespective of the elastic scattering process. Therefore one does not need to know the exact asymptotic radial wavefunctions to extract the product state distribution. This will generally enable one to obtain converged state-to-state reaction probabilities at a relatively short radial distance, especially when a long-range elastic interaction is present in the specific product arrangement.

Other expressions for the S matrix element

If in the amplitude expression for the S matrix element in Eq. (7.23) the radial component of $\chi^{+}_{\alpha i}$ is also chosen to be a δ-function times the asymptotic radial function such as a plane wave

$$<R_\alpha|\chi^{+}_{\alpha i}>= \delta(R_\alpha - R^{\infty}_{\alpha})\exp(-ik_iR_\alpha)|\varphi_{\alpha i}> \tag{7.27}$$

the coefficient $a_{\alpha i}(E)$ is calculated to be

$$\begin{aligned} a_{\alpha i}(E) &= <i\sqrt{\frac{\mu_\alpha}{2\pi\hbar^2 k_i}}\exp(-ik_i\hat{R})|\delta(\hat{R}-R^{\infty}_{\alpha})\exp(ik_i\hat{R})> \\ &= -i\sqrt{\frac{\mu_\alpha}{2\pi\hbar^2 k_i}} \end{aligned} \tag{7.28}$$

We thus obtain another form of the S matrix element

$$S_{fi}(E) = (-i\hbar)\sqrt{v_{\beta f} v_{\alpha i}} e^{-ik_f R_\beta^\infty} G^+_{\beta f,\alpha i}(R_\beta^\infty | R_\alpha^\infty)(E) e^{-ik_i R_\alpha^\infty} \quad (7.29)$$

where $v = \hbar k/\mu$ is the velocity and $G^+_{fi}(R_\beta^\infty | R_\alpha^\infty)(E)$ is the matrix element of the Green's function

$$G^+_{fi}(R_\infty^\beta | R_\infty^\alpha)(E) = <R_\beta^\infty| <\varphi_{\beta f}|G^+(E)|\varphi_{\alpha i}> |R_\alpha^\infty> \quad (7.30)$$

Equation (7.29) is the reactive scattering version of the "classical" S matrix expression derived by Miller [158]. Equation (7.29) expresses the S matrix element in terms of the matrix element of the propagator (Green's function) which is very useful for introducing semiclassical approximations for the S matrix element. For example, we can express Eq. (7.29) in terms of the time-dependent propagator

$$S_{fi}(E) = -\sqrt{v_{\beta f} v_{\alpha i}} \int_0^\infty dt\, e^{-ik_f R_\beta^\infty} <R_\beta^\infty| <\varphi_{\beta f}|e^{-\frac{i}{\hbar}Ht}|\varphi_{\alpha i}> |R_\alpha^\infty> \quad (7.31)$$

7.2.3 State-to-State Reactive Scattering of H + O_2

As shown in Sec. 6.5, the reaction of H + O_2 is dominated by many overlapping resonances and the reaction probability shows a complicated structure as a function of scattering energy. For more detailed investigations of reaction dynamics, it is desirable to calculate state-to-state reaction probabilities. Although state-to-state reaction probabilities can be calculated, for example, by the hyperspherical coordinate approach [39], the computation becomes very expensive when probabilities at a large number of energies are needed, due to the N^3 computational cost for each energy [39]. The time-dependent approach described in Sec. 6.5 can efficiently produce reaction probabilities for many energies from a single wavepacket calculation but only produces total (final state summed) reaction probabilities. In this section we show state-to-state reaction probabilities for H + O_2 on the DMBE IV potential energy surface [116] using the time-dependent method described in the previous section [114]. The calculation is done for reagents in the ground state of O_2(j=1) and for total angular momentum J=0.

Figure 7.1 shows state-to-state reaction probabilities for the reaction of H + O_2(v=0,j=1) → HO(v′=0,j′=0) + O in the energy range from 0.81 to 1.0 eV. The energy E is defined as the internal energy of O_2(v=0,j=1) (0.81 eV) plus the translational energy. As is expected, the state-to-state reaction probability is dominated by narrow resonances. As shown in the figure, the TD state-to-state calculation agrees well with the hyperspherical

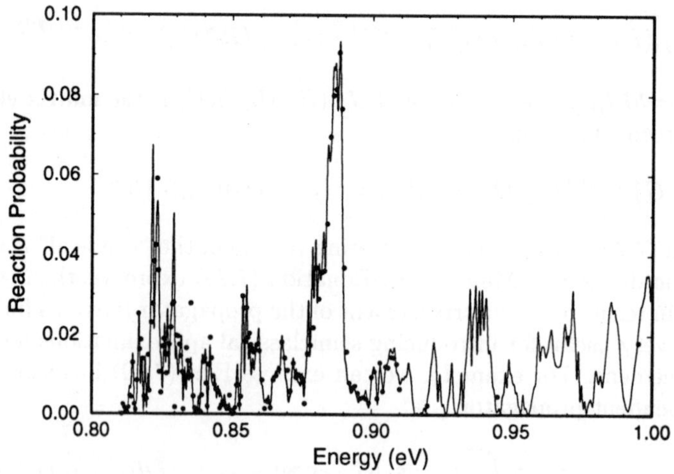

Figure 7.1: The reaction probability of H + $O_2(v = 0, j = 1) \to$ H + $HO(v' = j' = 0)$ as a function of total energy E defined as the rovibrational energy of O_2 + translational energy of H + O_2. The black dots are the results from the hyperspherical coordinate calculation of Ref. [39].

coordinate calculation of Ref. [39]. Obviously, the complete determination of the resonance structure in Fig. 7.1 requires calculations for numerous scattering energies, which is made easy by using the TD approach described in the previous section.

Reaction probabilities to various rotational states j′ of HO(v′=0) are shown in Fig. 7.2 for energies up to 1.5 eV. As is clear from Fig. 7.1, the reaction probability to every product rotational state is dominated by narrow and mostly overlapping resonances. The resonance structure of the state-to-state reaction probability is similar to that of the final state-summed reaction probabilities calculated in Ref. [37]. It it useful to note that the TD curves in these figures are generated from the calculated reaction probabilities at more than 5,000 scattering energies which are trivially obtained after the correlation function $C_{fi}(t)$ of Eq. (7.20) is computed.

Figure 7.3 plots the rotational state distribution of the product OH at 6 evenly spaced scattering energies of 0.92, 10.2, 1.12, 1.22, 1.32, and 1.42 eV. A brief inspection of Fig. 7.3 indicates that the reaction generally produces rotationally excited OH product and the detailed results are sensitive to the specific scattering energy.

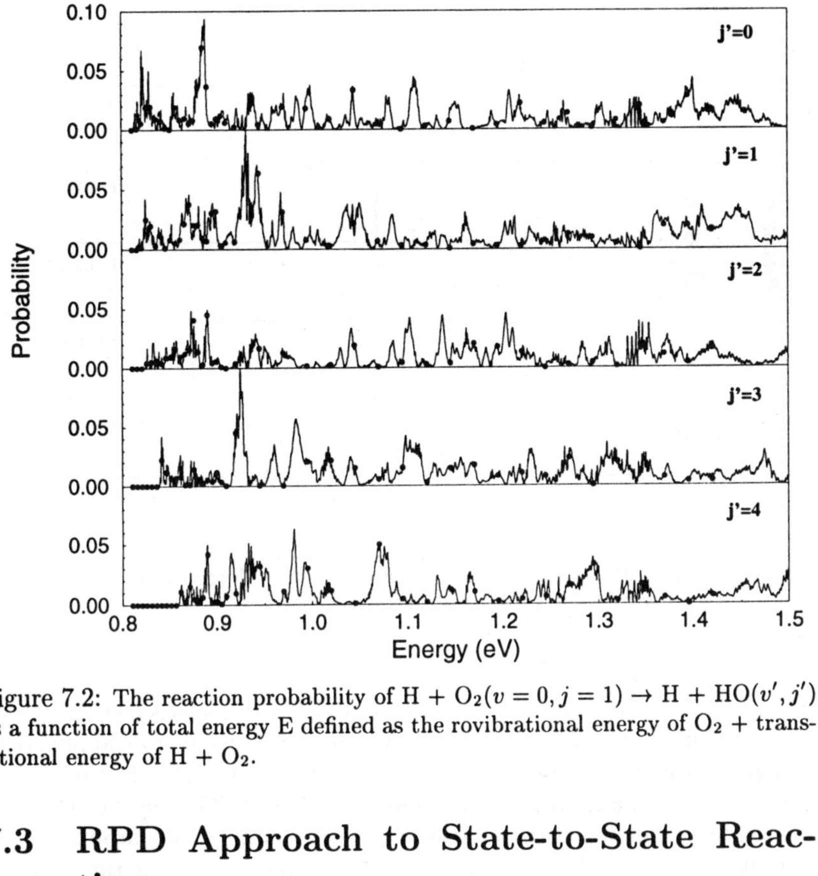

Figure 7.2: The reaction probability of H + $O_2(v = 0, j = 1) \rightarrow$ H + $HO(v', j')$ as a function of total energy E defined as the rovibrational energy of O_2 + translational energy of H + O_2.

7.3 RPD Approach to State-to-State Reactions

Motivations

It is well known that the most difficult problem in numerical computation of quantum reactive scattering is the choice of coordinates. If only total reaction probabilities (i.e., probabilities summed over final states of the product arrangement) are needed, it is quite reasonable (and often very efficient) to employ the Jacobi coordinates of the reactant arrangement in time-dependent wavepacket calculations, as has been discussed and shown in Secs. (6.4) and (6.6.1). However for state-to-state calculations, the use of a single set of Jacobi coordinates (corresponding to either reactant or product) is numerically inefficient. This is because the grid size or basis set

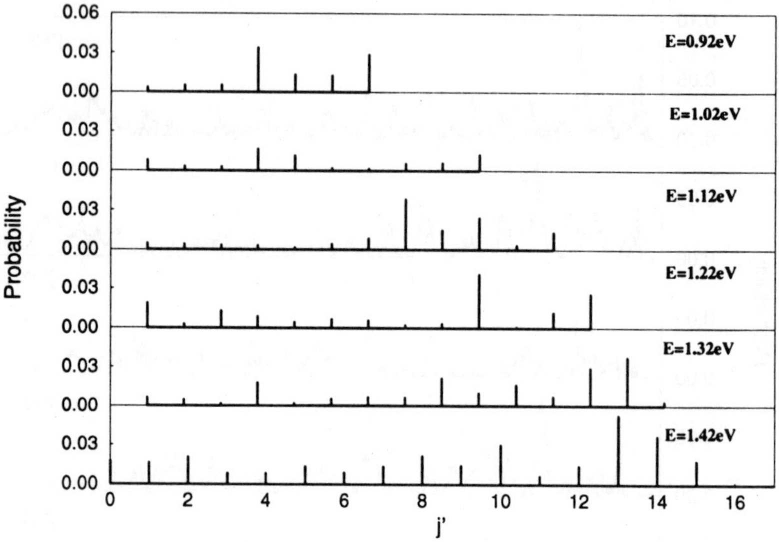

Figure 7.3: The OH rotational state distribution from the reaction of $H + O_2(v = 0, j = 1) \rightarrow H + HO(v' = 0, j')$ at six energies.

selected for one arrangement has to cover the entire space including both the reactant and product arrangements. For such state-to-state applications to large systems, the computational cost can be enormous in comparison with that for calculating total reaction probabilities. This has been demonstrated by a number of recent state-to-state dynamics calculations for the tetraatomic H_2 + OH reaction [131, 132] and its reverse reaction [134] by employing the reactant Jacobi coordinates for wavepacket propagation.

Because the scattering wavefunctions in different regions of arrangement are naturally described by their corresponding Jacobi coordinates, it is most desirable to carry out the quantum dynamics calculation in each arrangement separately using the corresponding Jacobi coordinates. A natural extension of the method of calculating total reaction probabilities described in Sec. 6.4 would be to extend the wavefunction propagation beyond the dividing surface into the product arrangement but the propagation is done only for the reactive component of the wavefunction. This requires one to partition the full scattering wavefunction into different components corresponding to different arrangements and to calculate them differently and separately.

7.3.1 The RPD Equations

The basic strategy of the RPD (Reactant-Product Decoupling) scheme is to partition the full TD wavefunction into a sum of reactant component Ψ_r and all product components $\Psi_p (p = 1, 2, 3, \cdots)$

$$\Psi = \Psi_r + \sum_p \Psi_p \tag{7.32}$$

The various component wavefunctions are required to satisfy the following *decoupled* equations [159]

$$\begin{cases} i\hbar\frac{\partial}{\partial t}|\Psi_r(t)\rangle = H|\Psi_r(t)\rangle - i\sum_p V_p|\Psi_r(t)\rangle \\ i\hbar\frac{\partial}{\partial t}|\Psi_p(t)\rangle = H|\Psi_p(t)\rangle + iV_p|\Psi_r(t)\rangle \end{cases} \tag{7.33}$$

where H is the full Hamiltonian and $-iV_p$ is the negative imaginary potential (absorbing potential) used to completely absorb the wavefunction $\Psi_r(t)$ in a narrow strip separating the reactant from the product as illustrated in Fig. 7.4. Equation (7.33) is *decoupled* in the sense that the solution for $\Psi_r(t)$ is independent of those for $\Psi_p(t)$ and the latter are independent of each other. If we sum the equations over all the component wavefunctions in Eq. (7.33), we recover the original Schrödinger equation for the full wavefunction Ψ. Thus solving the RPD Eq. (7.33) is formally equivalent to solving the original Schrödinger equation for any potentials $-iV_p$.

It is noted that solving for $\Psi_r(t)$ is completely independent of solving for $\Psi_p(t)$. If the absorbing potential V_p is chosen to be located past the transition state for reaction, then $\Psi_r(t)$ will be the correct representation of the full scattering wavefunction in the reactant and strong interaction regions where V_p is zero, provided that $\Psi_r(t)$ is perfectly absorbed. The second equation in Eq. (7.33) is an inhomogeneous equation with a time-dependent source term $iV_p\Psi_r(t)$ that provides the driving force towards the asymptotic region in the pth product arrangement space. Since the product component wavefunction $\Psi_p(t)$ needs to be nonzero only in the corresponding pth product space starting from where V_p becomes nonzero, its calculation involves only an inelastic propagation in that particular arrangement, completely independent of component wavefunctions of other product arrangements. Thus the RPD method naturally allows us to use different Jacobi coordinates to calculate different arrangement component wavefunctions. Because the source term $V_p\Psi_p(t)$ is confined to the reactant-product transition region only which is defined by the absorbing potential

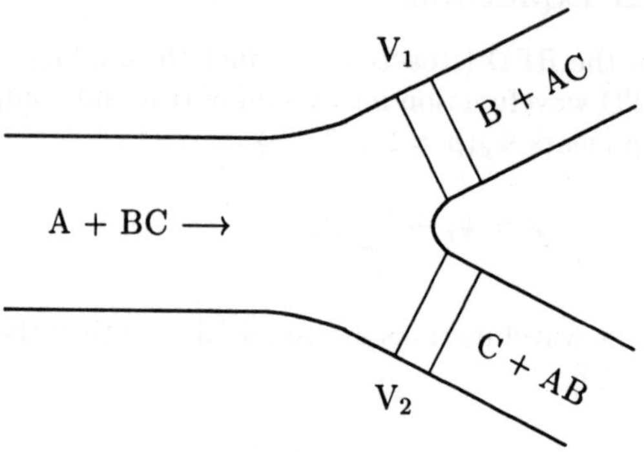

Figure 7.4: Illustrative drawing of absorbing potentials used in the RPD approach for state-to-state atom-diatom reactive scattering.

V_p, one does not need to calculate the full blown overlap matrix between basis functions of reactant and product arrangements for the source term.

7.3.2 Reactant Component Wavefunction

The numerical methods for solving $\Psi_r(t)$ in Eq. (7.33) have been developed for TD calculations of total reaction probabilities for atom-diatom reactions and for diatom-diatom reactions as in Secs. (6.4) and (6.6.1). The absorbing potentials V_p are placed just beyond the transition state to completely absorb the wavefunction in order to prevent reflection. Such TD calculations for Ψ_r in general can be efficiently carried out by using the Jacobi coordinates of the reactant arrangement, and the basic procedures for wavepacket propagation are identical to those described in Sec. 6.4. The main difference in the present calculation for Ψ_r, however, is that we do not discard the absorbed pieces of wavefunction $V_p\Psi_r(t)$ but instead store them in a proper representation on computer disk for later calculation of the product component $\Psi_p(t)$ [159]. For that purpose, an efficient method to transform the source term $V_p\Psi_r(t)$ to the product Jacobi coordinates is needed and is discussed in Sec. 7.3.5. Since the method for time-dependent calculation of $\Psi_r(t)$ has already been described in Sec. 6.4, it is therefore not repeated here.

7.3.3 Product Component Wavefunction

In the RPD method, one calculates the Ψ_p component wavefunction *independently* of other product components. The calculation of Ψ_p involves an inelastic propagation of the wavefunction $\Psi_p(t)$ for the desired product channel p using the pth product Jacobi coordinates. The formal solution for $\Psi_p(t)$ can be written as

$$|\Psi_p(t+\Delta)\rangle = e^{-\frac{i}{\hbar}H\Delta}|\Psi_p(t)\rangle + \frac{1}{\hbar}e^{-\frac{i}{\hbar}H\Delta}\int_t^{t+\Delta} e^{-\frac{i}{\hbar}H(t-t')}V_p|\Psi_r(t')\rangle\, dt' \quad (7.34)$$

If we use the trapezoidal rule to evaluate the integral for a short time step Δ, we obtain a propagation equation for $\Psi_p(t)$

$$|\Psi_p(t+\Delta)\rangle = e^{-\frac{i}{\hbar}H\Delta}\left[|\Psi_p(t)\rangle + \frac{\Delta}{2\hbar}V_p\Psi_r(t)\rangle\right] + \frac{\Delta}{2\hbar}V_p|\Psi_r(t+\Delta)\rangle \quad (7.35)$$

If we define a new product wavefunction $\widetilde{\Psi}_p(t) = \Psi_p(t) + \frac{\Delta}{2\hbar}V_p\Psi_r(t)$, which is everywhere the same as $\Psi_p(t)$ except in the absorption region, we obtain the simplified propagation equation for $\widetilde{\Psi}_p(t)$

$$\boxed{|\widetilde{\Psi}_p(t+\Delta)\rangle = e^{-\frac{i}{\hbar}H\Delta}|\widetilde{\Psi}_p(t)\rangle + \frac{\Delta}{\hbar}V_p|\Psi_r(t+\Delta)\rangle} \quad (7.36)$$

which is the final working formula for the calculation of $\widetilde{\Psi}_p(t)$.

Using Eq. (7.36), one carries out the standard split-operator propagation for $\widetilde{\Psi}_p(t)$ at each time step and simply adds in the second term in Eq. (7.36) afterwards. Since $\widetilde{\Psi}_p(t)$ is only nonzero in the product arrangement space starting from where V_p is turned on, the calculation of $\widetilde{\Psi}_p(t)$ is a much simpler inelastic scattering problem. Although a coordinate transformation is still required to generate $V_p\Psi_r(t)$ in the proper representation, this is much easier than the calculation of the full-blown overlap matrix because $V_p\Psi_r(t)$ is only nonzero in a very small region of space where the absorption potential operates. Also since $\widetilde{\Psi}_p(t)$ is everywhere the same as $\Psi_p(t)$ except in the absorbing region, we could directly use $\widetilde{\Psi}_p(t)$ to extract the final state dynamics information such as state-to-state S matrix elements or reaction probabilities. In this TD formalism, we need to store the calculated source term $\xi_p(t) = V_p\Psi_r(t)$ on computer disk at every time step which requires one to represent $\xi_p(t)$ in Jacobi coordinates of the product arrangement at each time step via coordinate transformation.

An alternative approach to calculating ψ_p is to use the time-independent version of Eq. (7.33)

$$\begin{cases} E|\psi_r(E)\rangle = H|\psi_r(E)\rangle -i\sum_p V_p|\psi_r(E)\rangle \\ E|\psi_p(E)\rangle = H|\psi_p(E)\rangle +iV_p|\psi_r(E)\rangle \end{cases} \tag{7.37}$$

and solve the time-independent product wavefunction $\psi_p(E)$ by wavepacket propagation

$$\begin{aligned} |\psi_p(E)\rangle &= iG^+(E)V_p|\psi_r(E)\rangle \\ &= \frac{1}{\hbar}\int_0^\infty dt e^{\frac{i}{\hbar}Et}e^{-\frac{i}{\hbar}Ht}V_p|\psi_r(E)\rangle \end{aligned} \tag{7.38}$$

where the homogeneous term is zero. In Eq. (7.38), the wavefunction $\psi_p(E)$ is obtained by propagating the energy-dependent wavepacket $\xi_p(E) = V_p\psi_r(E)$ and performing the Fourier transform afterwards for each desired energy. The wavefunction $\psi_r(E)$ is obtained by Fourier transforming the time-dependent wavefunction $\Psi_r(t)$ which is obtained by wavepacket propagation as described before. This approach is attractive if dynamics at a limited number of energies are desired because we only need to store the source term $\xi_p(E) = V_p\psi_r(E)$ and perform the coordinate transformation for the number of energies needed.

Essentially all the detailed formulas of basis functions described previously for propagating $\Psi_r(t)$ can be used for calculating $\Psi_p(t)$ in Eq. (7.36) or $\psi_p(E)$ in Eq. (7.38), except that the definitions of the basis functions are for the pth product arrangement instead of the reactant arrangement. Since the product wavefunction $\Psi_p(t)$ is zero in the strong interaction region, the propagation of Ψ_p only involves an inelastic process and therefore the basis set used to represent $\Psi_p(t)$ is considerably smaller than that of $\Psi_r(t)$ described in Sec. 6.4. Consequently, the computational cost for calculating $\Psi_p(t)$ is generally insignificant compared to that for $\Psi_r(t)$. An important practical matter in applying the RPD approach to state-to-state reactive scattering is the use of good absorbing potentials. It is found that state-to-state S matrix elements are generally more sensitive to parameters of absorbing potentials than are total reaction probabiities, This is expected because state-to-state dynamics is more sensitive to details of the potential energy surface. Therefore more care needs to be exercised in choosing the absorbing parameters in order to obtain accurate state-to-state S matrix elements or reaction probabilities.

7.3.4 Extraction of State-to-State S Matrix Elements

After the wavepacket is fully developed in the inelastic region of the specific product arrangement with proper absorbing boundary conditions, one can straightforwardly perform the final state analysis in the asymptotic region (R' large) to obtain the energy-dependent S matrix elements or reaction probabilities by Fourier transforming the TD wavefunction $\Psi_p(t)$ to $\psi_p(E)$ at a large asymptotic distance

$$\psi_p(E) \overset{R'\to\infty}{\longrightarrow} \sqrt{\frac{\mu_p}{2\pi\hbar^2}} \left(\sum_m S_{pm,ri} \frac{e^{ik_m R'}}{\sqrt{k_m}} |\varphi_{pm}\rangle \right) \tag{7.39}$$

where μ_p is the reduced translational mass and φ_{pm} the internal channel function in the pth product arrangement. If any long range elastic potential is present such as the centrifugal potential, one needs to replace the plane wave function $\exp(ik_m R')$ by the outgoing Hankel or other appropriate radial function. The calculation of S matrix elements can be done using the amplitude formalism described in Sec. 7.2.2. Alternatively, if only the absolute square of the S matrix element or reaction probability is required, one can avoid specifying the specific form of the radial function and instead evaluate the flux to obtain converged reaction probabilities at relatively shorter radial distances (see Sec. 7.2.2 for details).

7.3.5 A Collocation Quadrature Scheme

In order to carry out the TD propagation for Ψ_p, we need to re-express the source term $\xi_p(t) = V_p\Psi_r(t)$ in terms of the product basis set. This involves the numerical calculation for the expansion coefficients

$$\xi_{pn}(t) = \langle\phi_n|\xi_p(t)\rangle \tag{7.40}$$

where ϕ_n are the N basis functions of the product arrangements. Because $\xi_p(t)$ and ϕ_n are defined with respect to basis functions of different arrangements, the numerical evaluation of Eq. (7.40) involves a coordinate transformation between reactant and product arrangements as described in the previous subsection. This can be computationally expensive since the numerical integrations are inherently multidimensional, and the transformation has to be done at each time step. Therefore efficient methods have to be used to minimize the computational cost for this step. For this purpose, we can use a collocation quadrature scheme to efficiently calculate the integral in Eq. (7.40). For clarity, we drop the subscript p and the variable t in the following discussion.

The integral in Eq. (7.40) can be evaluated by an N term summation

$$\xi_n = \sum_i W_{ni}\xi(\bar{q}_i) \tag{7.41}$$

where $\bar{q}_i$ are N prefixed multidimensional points of the Jacobi coordinates defined in the product arrangement and W_{ni} is a undetermined weighting matrix. For an atom-diatom system, ϕ_n is the product of translation, vibration and rotation functions and $\bar{q}_i$ denotes (R_i, r_i, θ_i). The matrix W_{ni} can be obtained by a simple matrix inversion

$$\mathbf{W} = \mathbf{\Phi}^{-1} \tag{7.42}$$

where the matrix element $\mathbf{\Phi}_{ni}$ is just the value of ϕ_n at the quadrature point $(\bar{q}_i)$. This collocation choice of the weighting matrix guarantees that the orthogonality of the overlap integral is strictly preserved

$$<\phi_n|\phi_m> = \sum_i \mathbf{W}_{ni}\mathbf{\Phi}_{im} = \delta_{nm}, \tag{7.43}$$

and the summation in Eq. (7.41) will be exact if the functions ξ span the N-dimensional vector space of ϕ_n. If the basis functions are not orthogonal, Eq. (7.42) is easily generalized to

$$\mathbf{W} = \mathbf{O}\mathbf{\Phi}^{-1} \tag{7.44}$$

where $\mathbf{O}$ is the basis overlap matrix $\mathbf{O}_{nm} = <\phi_n|\phi_m>$. Although the choice of N points can be rather arbitrary as long as the inverse $\mathbf{\Phi}^{-1}$ exists, it is best to use good quadrature points to minimize the numerical error. For direct product basis functions, a natural choice is DVR (discrete variable representation) points. For non-direct product basis functions, the choice of good points remains to be explored.

7.4 RPD Method for Triatomic Reactions

7.4.1 Transformation of Jacobi Coordinates

Since the reactant wavefunction $\Psi_r(t)$ is expressed in terms of basis functions defined in the reactant Jacobi coordinates while $\Psi_p(t)$ is defined in terms of the product basis functions, we need to perform a coordinate transformation between reactant and product arrangements. The transformation of Jacobi coordinates from the reactant arrangement A + BC $(\mathbf{R}, \mathbf{r})$

to the product arrangement C + AB ($\mathbf{R}', \mathbf{r}'$) is given by (cf. Fig. 4.2)

$$\begin{pmatrix} \mathbf{R}' \\ \mathbf{r}' \end{pmatrix} = \begin{pmatrix} a_2 & a_3 \\ 1 & a_1 \end{pmatrix} \begin{pmatrix} \mathbf{R} \\ \mathbf{r} \end{pmatrix}, \tag{7.45}$$

where $a_1 = -m_C/(m_C + m_B)$, $a_2 = -m_A/(m_A + m_B)$, and $a_3 = a_1 a_2 - 1$. By inverting the matrix relation, one can obtain the transformation formula from AB + C to A + BC as well. In scalar form, the above transformation relation can be explicitly written as

$$r' = \sqrt{R^2 + (a_1 r)^2 + 2a_1 Rr\cos\theta} \tag{7.46}$$

$$R' = \sqrt{(a_2 R)^2 + (a_3 r)^2 + 2a_2 a_3 Rr\cos\theta} \tag{7.47}$$

$$\cos\theta' = \frac{1}{R'r'}(a_2 R^2 + a_1 a_3 r^2 + (a_1 a_2 + a_3)Rr\cos\theta). \tag{7.48}$$

Similarly, one can obtain the transformation relation from A + BC to B + AC,

$$r'' = \sqrt{R^2 + (b_1 r)^2 + 2b_1 Rr\cos\theta} \tag{7.49}$$

$$R'' = \sqrt{(b_2 R)^2 + (b_3 r)^2 + 2b_2 b_3 Rr\cos\theta} \tag{7.50}$$

$$\cos\theta'' = \frac{1}{R''r''}(b_2 R^2 + b_1 b_3 r^2 + (b_1 b_2 + b_3)Rr\cos\theta) \tag{7.51}$$

with $b_1 = m_B/(m_C + m_B)$, $b_2 = m_A/(m_A + m_C)$, and $b_3 = b_1 b_2 - 1$.

7.4.2 Numerical Test for H + H_2 Reaction

In this section, we demonstrate the applicability of the RPD method to the test case of the 3D H + H_2 reaction for zero total angular momentum (J=0) on the LSTH potential energy surface [74]. Although this is an atom-diatom application, the computational issues discussed here are generally applicable to reactive scattering involving more than three atoms.

As indicated previously, the most sensitive parameters in the RPD approach to state-to-state scattering are those of absorbing potentials defined in terms of the radial coordinate of the product B + AC.

$$V_{abs} = -i\alpha\left(\frac{R' - R'_1}{R'_2 - R'_1}\right)^{\beta}, \qquad R'_1 < R < R'_2 \tag{7.52}$$

for the shaded regions as schematically illustrated in Fig. 7.4. The values of R'_1 and $R'_2 - R'_1$ determine the starting position and the width of the absorbing potential, respectively. The absorbing potentials used to block the product arrangements have been tested extensively for H + H_2. We found that excellent results can be obtained for $R_1' \geq 3.25\,a_0$ and $R_2' - R_1' \geq 1.75a_0$. The value of α is generally in the range of [0.05, 0.15] a.u.. and β between 1.0 and 3.0.

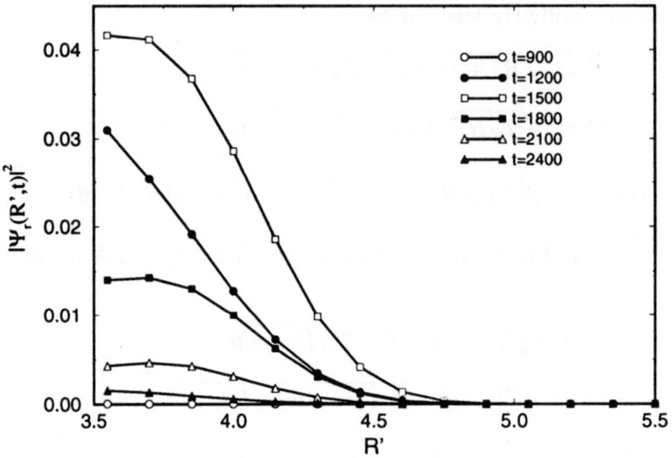

Figure 7.5: Absolute square of the reactant wavefunction $|\Psi_r(t)|^2$ plotted as a one-dimensional function of the radial coordinates of the product arrangement at various propagation times. The other two internal degrees of freedom have been integrated over.

Figure 7.5 shows one-dimensional plots of the reactant wavefunction $\Psi_r(t)$ at various propagation times as a function of the translation coordinate R' of the product arrangement where the absorbing potential V_p is defined in the range of $3.5a_0$ to $5.5a_0$. No observable reflections are present in Fig. 7.5 and the wavefunctions are fully absorbed before the end of absorption at $R' = 5.5a_0$. Similar plots are shown in Fig. 7.6 for the time-dependent source term $\xi_p(t) = V_p\Psi_r(t)$. Here we see that the R'-dependence of $\xi_p(t)$ behaves exactly as expected: it has a maximum in the middle of the absorbing region and decays to zero toward both ends of the absorbing region.

The propagation of $\Psi_p(t)$ in the product arrangement is quite straightforward and relatively trivial in comparison to that of $\Psi_r(t)$ since it only involves an inelastic propagation. For example, only four H_2 vibrational

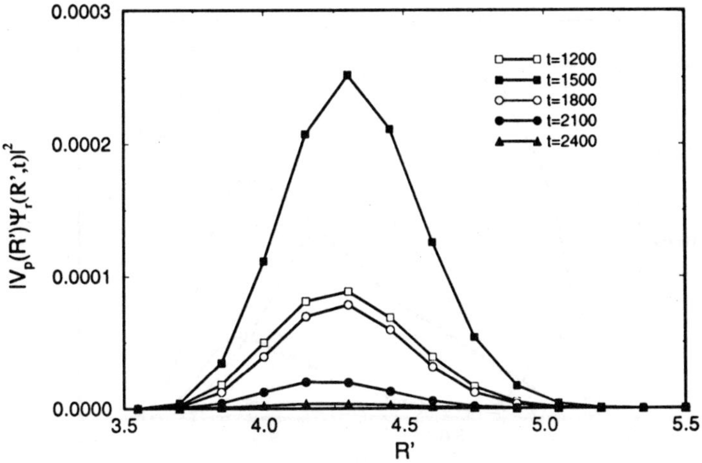

Figure 7.6: The absolute square of the source term $|V_p\Psi_r(t)|^2$ plotted as a one-dimensional function of the radial coordinate of the product arrangement at various propagation times.

functions are used in the basis expansion of $\Psi_p(t)$ compared to about 30 H_2 vibrational functions used in the expansion of $\Psi_r(t)$. Thus only four collocation-quadrature points (DVR points) are used in Eq. (7.41). The number of rotational functions is also reduced by about a factor of two in the calculation of $\Psi_p(t)$. Figure 7.7 shows one-dimensional plots of the product wavefunction $\Psi_p(t)$ as a function of the radial coordinate R' in which vibrational and rotational degrees of freedom are integrated out. It is clear from Fig. 7.7 that $\Psi_p(t)$ is negligible in the strong interaction region and picks up amplitude when getting close to $R' = 3.5a_0$ where the absorbing potential is turned on. This is exactly what we expect $\Psi_p(t)$ to be, because $\Psi_r(t)$ is a good representation of the full wavefunction already except near and in the absorbing region. The calculated reaction probabilities from the RPD approach can be compared to those from the time-independent S matrix Kohn variational calculation. Figure 7.8 shows a plot of $H + H_2$ reaction probabilities to final vibrational states (summed over final rotational states). These results are in excellent agreement with the time-independent calculation at all energies shown in the figure. More complete state-to-state reaction probabilities and their comparisons with time-independent variational calculations can be found in [160] and [161].

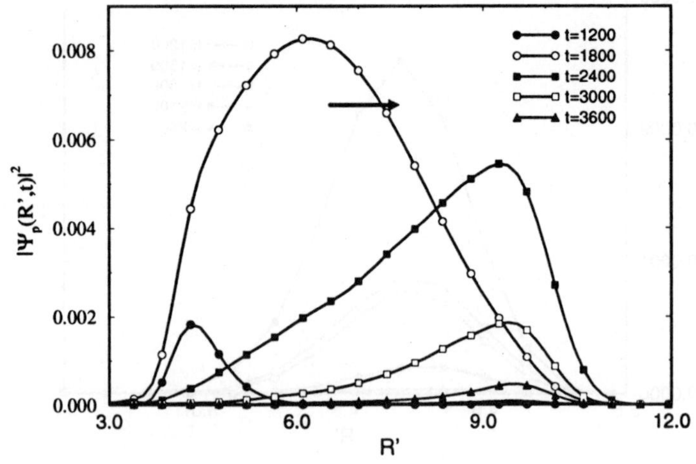

Figure 7.7: Same as Fig. 7.5 except for the product component wavefunction $|\Psi_p(t)|^2$.

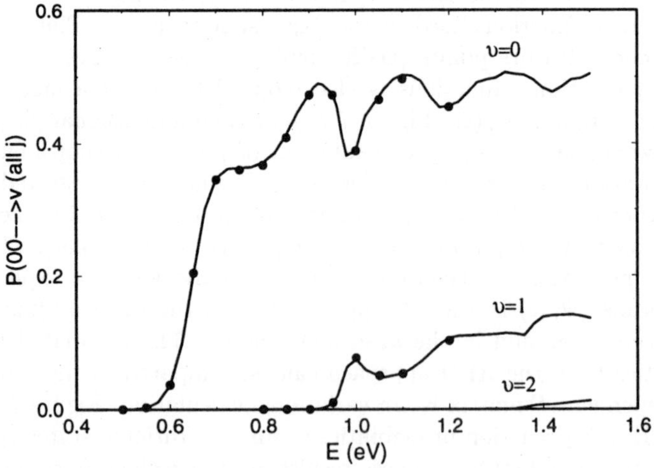

Figure 7.8: Comparison of final vibration specific reaction probabilities from the RPD calculation (solid lines) with the time-independent variational calculation (filled circles) for the H + H_2 reaction.

Chapter 8

Dynamics of Chemical Reactions

8.1 Scattering Resonance

8.1.1 Introduction

One of the most important results of quantum scattering theory is the phenomenon of resonance. Resonance plays an important role in bimolecular collisions and unimolecular reactions. For example, the product branching ratio and product state distribution in a bimolecular reaction or unimolecular dissociation are often controlled by resonance states. The resonance states are also called metastable species in chemistry and their existence can be detected and lifetimes can be measured experimentally by spectroscopic and other methods. The formation of resonance states or metastable species can arise from a variety of energy transfer processes such as bimolecular collisions, photoabsorption, or other energetic processes. The theoretical treatment of resonance is a difficult task partly because resonance states are not so well defined as bound states. Rigorously speaking, resonance states are continuum states that possess some bound state characteristics such as locality. The properties of resonance states can vary widely from those that are essentially continuum scattering states to those that behave almost like a bound state depending on their relative lifetimes.

The scattering resonance is generally classified as a shape or Feshbach resonance. In a shape resonance, the particle is temporarily trapped in a one-dimensional potential well by its own centrifugal barrier through which the particle will eventually escape by tunneling as illustrated in Fig. 8.1. The Feshbach resonance refers to multidimensional scattering in which the energy associated with the scattering coordinate is temporarily depleted through coupling to other degrees of freedom. As a result, the system is temporarily trapped and forms a quasibound or metastable complex. A Feshbach resonance is also called a dynamical resonance and can exist even for a scattering system on a purely repulsive potential surface. Resonances are inherently quantum mechanical and occur at some discrete energies called resonance energies that are usually associated with the formation of quasi bound states. The two most important measures of a resonance are resonance energy E_R and resonance lifetime τ (or resonance width Γ).

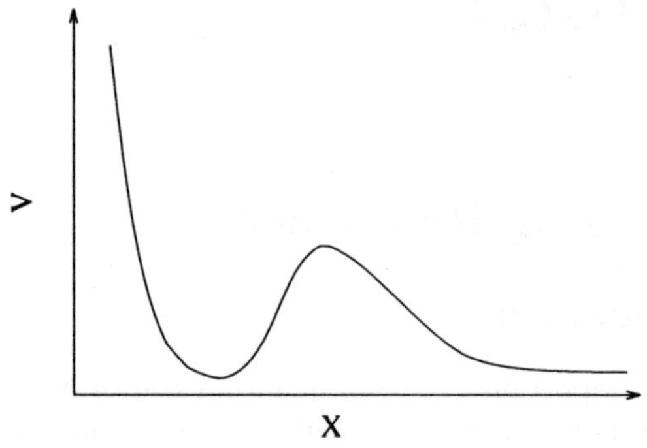

Figure 8.1: One dimensional potential for shape resonance.

8.1.2 Phase Shift and Time Delay

Let us first discuss resonance in elastic scattering in order to illustrate the main features of the resonance phenomenon. For a fixed partial wave l, elastic scattering is characterized by the phase shift δ_l which is usually a smooth function of energy. Near a resonance energy, however, δ_l is a rapidly changing function of energy. It is then convenient to separate the phase shift δ_l into a slowly varying function of energy (background phase shift) δ_l^0 and a resonance part δ_l^R

$$\delta_l = \delta_l^0 + \delta_l^R \tag{8.1}$$

Near the resonance energy E_l^R, the resonance phase shift has the standard energy-dependence [48]

$$\boxed{\tan(\delta_l^R) = \frac{\Gamma_l}{E_l^R - E}} \tag{8.2}$$

where Γ_l is called half-width of the resonance. Equation (8.2) shows that as the scattering energy passes through the resonance energy E_l^R, the phase shift δ_l^R changes from 0 to π, and therefore the S matrix $S = \exp(2i\delta_l)$ changes its phase by 2π. Figure 8.2 illustrates the energy-dependence of the resonance phase shift δ_l^R.

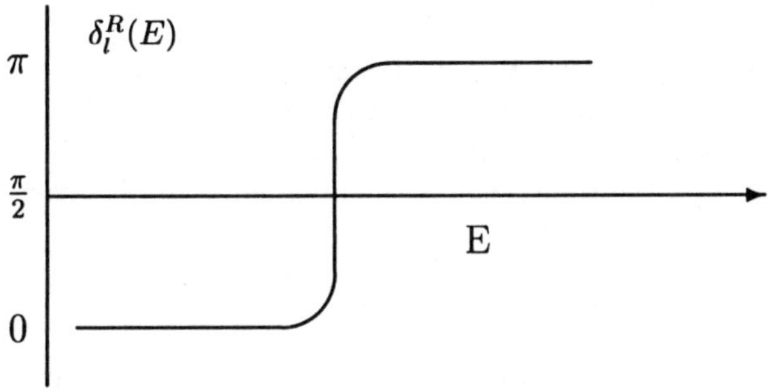

Figure 8.2: Energy-dependence of the scattering resonance phase shift $\delta_l^R(E)$.

In order to give a more direct physical interpretation of the phase change, we examine the asymptotic form of the outgoing scattering wave-function by including the dynamical phase factor $\exp(-iEt)$

$$\psi(R,t) \overset{R\to\infty}{\longrightarrow} e^{-\frac{i}{\hbar}Et+ikR+i\delta_l} \tag{8.3}$$

By differentiating with respect to energy the constant phase equation

$$\varphi_l(E) = -\frac{1}{\hbar}Et + kR + \delta_l = \text{const} \tag{8.4}$$

we obtain an equation for the traversal time of the outgoing wave

$$\begin{aligned} T_l &= \hbar\left(\frac{dk}{dE}R + \frac{d\delta_l}{dE}\right) \\ &= \frac{R}{v} + \hbar\frac{d\delta_l}{dE} \\ &= T^0 + T_l^d \end{aligned} \tag{8.5}$$

The first term T^0 in Eq. (8.5) is the traversal time (or direct time-delay) through a distance R with a constant velocity $v = \frac{\hbar k}{m}$. The second term

$$T_l^d = \hbar \frac{d\delta_l}{dE} \tag{8.6}$$

is the "extra" or collision time delay due to interaction with the scattering potential.

At the resonance energy E_l^R, the resonance phase shift is given by Eq. (8.2) and the corresponding time delay takes the form

$$\boxed{T_l^R = \frac{\hbar \Gamma_l}{(E - E_l^R)^2 + \Gamma_l^2}} \tag{8.7}$$

which is the well known Breit-Wigner formula [48] and has a Lorenzian shape. If the time delay due to direct scattering T^0 is negligible, the scattering time-delay is given by the resonance time delay $T_l^d \simeq T_l^R$. At the resonance energy E_l^R, the time delay is the maximum and is given by

$$T_l^R \simeq \frac{\hbar}{\Gamma_l} \tag{8.8}$$

This result has a physically intuitive interpretation: the resonance state has an energy uncertainty of Γ_l and thus has an associated lifetime given by the uncertainty relation (8.8). At off resonance energies $|E - E_l^R| \gg \Gamma_l$, the time delay T_l^R is negligible

$$T_l^R \simeq \frac{\hbar \Gamma_l}{(E - E_l^R)^2} \approx 0 \tag{8.9}$$

Thus, the time delay is a direct measurement of the resonance lifetime.

8.1.3 Lifetime Matrix

Since the S matrix in elastic scattering is given by $S_l = \exp(2i\delta_l)$, the time delay can also be expressed in terms of the energy derivative of the S matrix directly

$$T_l^R = \frac{d\delta_l}{dE} = -i\frac{\hbar}{2} S_l^\dagger \frac{dS_l}{dE} \tag{8.10}$$

In multichannel scattering, such as inelastic or reactive scattering, S is a square matrix whose dimension is equal to the number of open asymptotic channels. It is then plausible to generalize the time delay defined in Eq. (8.10) to multichannel scattering by defining a hermitian lifetime matrix

$$\boxed{\mathbf{T}_{life} = -i\frac{\hbar}{2}\mathbf{S}^\dagger \frac{d\mathbf{S}}{dE}} \tag{8.11}$$

whose hermiticity can be easily proved by employing the unitary property of the S matrix ($\mathbf{S}^\dagger\mathbf{S} = \mathbf{I}$)

$$\mathbf{T}_{life}^\dagger = i\frac{\hbar}{2}\frac{d\mathbf{S}^\dagger}{dE}\mathbf{S} = -i\frac{\hbar}{2}\mathbf{S}^\dagger\frac{d\mathbf{S}}{dE} = \mathbf{T}_{life} \tag{8.12}$$

The lifetime matrix can thus be diagonalized to yield eigenvalues that are real functions of energy. These eigenvalues are interpreted as lifetimes associated with the resonance states of the system (if any). This procedure, called lifetime analysis, was proposed by Smith [162] to detect and analyze resonance states of collision systems. The method has proven to be quite useful and effective in analyzing scattering resonances.

8.1.4 Implication of Resonance in Cross Sections

Although resonances are quite abundant and often show up in the energy-dependence of the calculated S matrix or reaction probabilities for fixed partial waves or total angular momenta, the experimental verification of these resonances for bimolecular collisions is not very easy. This is because in collision experiments, one can only measure reaction cross sections which are quantities summed over all accessible total angular momenta or partial waves. In molecular collisions, the number of contributing partial waves is usually large ($10 \sim 100$). For example the cross section for elastic scattering given by Eq. (4.165) takes the form

$$\begin{aligned}\sigma(E) &= \frac{4\pi}{k^2}\sum_l (2l+1)\sin^2\delta_l \\ &\simeq \frac{4\pi}{k^2}\sum_l (2l+1)\frac{\Gamma_l^2}{(E_l^R - E)^2 + \Gamma_l^2}\end{aligned} \tag{8.13}$$

Since resonance energies are dependent on the total angular momentum l and usually shift to higher energy as l increases due to the centrifugal potential, the summation over l in the cross section often quenches resonance peaks that are present in the energy-dependence of the S matrix element at fixed partial waves. Thus, measuring the energy dependence of the cross section does not usually reveal the existence of resonance in collisions. One often has to utilize or combine the results of other experimental methods to help detect the existence of resonance, if any. One successful experimental approach is to study the photodetachment spectrum by forming stable negative ions from the collision complex in order to eliminate the need of summing over all partial waves [163].

However, if the resonance is sufficiently narrow such that the resonance width Γ is narrower than the shift of resonance energy with l, i.e., $\Gamma < \Delta E_l^R$, then it is possible for resonance structure to show up in the energy-dependence of integral cross sections. Collisional resonance could also be studied by measuring the differential cross section in a molecular beam experiment. Since the differential cross section measures the angular distribution of the reaction product which is often dominated by the contribution of a small fraction of partial waves, it is often a more sensitive probe of resonance. Of course, the most thorough and detailed investigation of resonance is possible by theoretical studies and calculations. In order to have a rigorous theoretical understanding of the resonance phenomenon and provide some mathematical support, we present in the following section a formal theory on resonance (Feshbach theory) [164].

8.2 Formal Theory of Resonance

8.2.1 Complex Symmetric Operator

If an operator A satisfies the condition

$$A^\dagger = A^* \tag{8.14}$$

it is a complex symmetric operator. In analyzing resonance properties and in other applications, one often encounters complex symmetric operators. For example, the Green's function operators $G^\pm$ are complex symmetric operators, as are real hermitian operators. It is thus useful to discuss some general properties of complex symmetric operators.

If ϵ_λ and ϕ_λ are eigenfunctions of a complex symmetric operator A

$$A\phi_\lambda = \epsilon_\lambda \phi_\lambda \tag{8.15}$$

the eigenvalues ϵ_λ are in general complex. The orthogonality condition for the eigenfunctions can be derived by the relation

$$\begin{aligned} <\phi_{\lambda'}^*|A|\phi_\lambda> - <\phi_{\lambda'}^*|A^{\dagger *}|\phi_\lambda> &= \epsilon_\lambda <\phi_{\lambda'}^*|\phi_\lambda> - <A^*\phi_{\lambda'}^*|\phi_\lambda> \\ &= (\epsilon_\lambda - \epsilon_{\lambda'}) <\phi_{\lambda'}^*|\phi_\lambda> = 0 \end{aligned} \tag{8.16}$$

Thus the eigenfunctions can be normalized to

$$\boxed{<\phi^*_{\lambda'}|\phi_\lambda>= \int \phi_{\lambda'}\phi_\lambda d\tau = \delta_{\lambda'\lambda}} \tag{8.17}$$

This normalization is different from that for eigenfunctions of hermitian operators, because it completely fixes the phase of the eigenfunctions. If the eigenfunction is multiplied by a phase factor $e^{i\alpha}$, the normalization becomes

$$<\phi^*_\lambda|\phi_\lambda>= e^{i2\alpha} \neq 1 \tag{8.18}$$

In addition, the normalization imposed in Eq. (8.17) also implies that the real part of the eigenfunction ϕ_λ is orthogonal to its imaginary part

$$<\phi^R_\lambda|\phi^I_\lambda>= 0 \tag{8.19}$$

For the special case of real hermitian operators, the eigenfunctions can always be chosen real, and the orthogonality condition in Eq. (8.17) becomes equivalent to the standard orthogonality condition for real eigenfunctions of hermitian operators.

If we define matrix elements of a complex symmetric operator A in any basis set φ_m by the prescription

$$A_{mn} =<\varphi^*_m|A|\varphi_n>, \tag{8.20}$$

it is straightforward to prove that the matrix A_{mn} is complex symmetric

$$A_{mn} = A_{nm} \tag{8.21}$$

8.2.2 Projection Operators and Feshbach Partitioning

Since resonance states by their nature are scattering wavefunctions that belong to continuum space, the rigorous treatment of resonances is in the realm of scattering theory. For ordinary scattering wavefunctions, the amplitude of the wavefunction in the interaction region is relatively small compared to that in the continuum region. At a resonance energy, however, the scattering wavefunction is characterized by a large amplitude in the interaction region relative to that in the asymptotic region. Thus, the resonance states possess some bound state characteristics in the interaction region, and it is often convenient to treat them like bound states in the interaction region. The following describes the formal theory of resonance utilizing projection operators [164].

Since the most prominent feature of the resonance wavefunction is the dominance of the wavefunction in the interaction region, it is ideal to separate the scattering wavefunction Ψ into a resonance dominated bound piece Ψ_Q and a regular scattering piece Ψ_P

$$\Psi = \Psi_P + \Psi_Q \tag{8.22}$$

where Ψ_Q is nonzero only in a finite inner region. If we use R_0 to define a finite range in the scattering coordinate, then we require

$$\Psi_Q(R > R_0) = 0 \tag{8.23}$$

and Ψ_P satisfies the correct asymptotic scattering boundary conditions such as given in Sec. 4.4.3. The partition of the total wavefunction Ψ into Ψ_P and Ψ_Q can be done formally by introducing projection operators P and Q such that

$$\Psi_P = P\Psi \tag{8.24}$$

$$\Psi_Q = Q\Psi \tag{8.25}$$

where the projection operators satisfy the standard relations

$$\begin{cases} P^2 = P \\ Q^2 = Q \\ P^\dagger = P \\ Q^\dagger = Q \\ PQ = 0 \end{cases} \tag{8.26}$$

and $P + Q = 1$. If we multiply the Schrödinger equation

$$(E - H)(\Psi_P + \Psi_Q) = 0 \tag{8.27}$$

by P and Q operators from the left, respectively, we obtain the coupled equations

$$(E - H_{PP})\Psi_P = H_{PQ}\Psi_Q \tag{8.28}$$

$$(E - H_{QQ})\Psi_Q = H_{QP}\Psi_P \tag{8.29}$$

where various operators are defined as

$$\begin{cases} H_{PP} = PHP \\ H_{QQ} = QHQ \\ H_{PQ} = PHQ \\ H_{QP} = QHP = H_{PQ}^\dagger \end{cases} \tag{8.30}$$

The coupled Eq. (8.29) is convenient for analyzing resonance properties of the wavefunction because it explicitly separates the bound component of the wavefunction Ψ_Q from the continuum part Ψ_Q . Integrating the first equation in Eq. (8.29) gives rise to a formal solution for Ψ_P

$$\Psi_P = \Psi_0 + G^+_{PP}(E) H_{PQ} \Psi_Q \tag{8.31}$$

where Ψ_0 is a solution of the equation

$$(E - H_{PP})\Psi_0 = 0 \tag{8.32}$$

and the Green's function is defined as

$$G^+_{PP}(E) = \lim_{\epsilon \to 0} \frac{1}{E - H_{PP} + i\epsilon} \tag{8.33}$$

The solution in Eq. (8.31) is then substituted into Eq. (8.29) to obtain an inhomogeneous equation for Ψ_Q

$$[E - H^C_{QQ}(E)]\Psi_Q = H_{QP}\Psi_0 \tag{8.34}$$

where the effective Hamiltonian is defined as

$$\boxed{H^C_{QQ}(E) = H_{QQ}(E) + H_{QP} G^+_{PP}(E) H_{PQ}} \tag{8.35}$$

Equation (8.34) is the desired formal equation that involves only the bound component Ψ_0.

Since the outgoing wave Green's function can be expressed in terms of the principal value Green's function, i.e.,

$$G^+_{PP}(E) = G^p_{PP}(E) - i\pi\delta(E - H_{PP}) \tag{8.36}$$

the effective Hamiltonian can be separated into a real part H^R_{QQ} and an imaginary part H^I_{QQ}

$$H^C_{QQ}(E) = H^R_{QQ}(E) + iH^I_{QQ}(E) \tag{8.37}$$

where the real and imaginary parts are defined, respectively, as

$$\begin{cases} H^R_{QQ}(E) = H_{QQ}(E) + H_{QP} G^p_{PP}(E) H_{PQ} \\ H^I_{QQ}(E) = -\pi H_{QP} \delta(E - H_{PP}) H_{QP} \end{cases} \tag{8.38}$$

It is important to note that both $H^R_{QQ}(E)$ and $H^I_{QQ}(E)$ are ***nonlocal*** and ***energy-dependent*** operators. One can also show further that the imaginary Hamiltonian $H^I_{QQ}(E)$ is ***negative*** definite, i.e., for any function ϕ

$$\begin{aligned} <\phi|H^I_{QQ}(E)|\phi> &= -\pi <\phi|H_{QP}\delta(E - H_{PP})H_{PQ}|\phi> \\ &= -\pi <H_{PQ}\phi|\delta(E - H_{PP})|H_{PQ}\phi> \\ &= -\pi <\xi|\delta(E - H_{PP})|\xi> \leq 0 \end{aligned} \tag{8.39}$$

The last inequality holds because the density operator $\rho(E) = \delta(E - H_{PP})$ is positive definite.

A little note on the projection operator is in order. So far we have not specified either of the projection operators P or Q. The basic criterion of choosing the projection operator is to separate the contribution of resonance scattering from that of direct scattering in the interaction region. Without more specifics, the projection operator cannot be uniquely defined. One has to rely on intuition to choose good projection operators. A convenient choice of projection operators is to include open channels only for operator P and closed channels only for operator Q, viz.,

$$\begin{aligned} P &= \sum_{i=1} i^{N_0} |\varphi_i><\varphi_i| \\ Q &= \sum_{i>N_0}^{\infty} |\varphi_i><\varphi_i| \end{aligned} \tag{8.40}$$

where N_0 is the number of asymptotically open channel functions φ_i of the internal Hamiltonian. This definition of the P and Q operators satisfies the standard definitions for projection operators in Eq. (8.30) and also guarantees that the Ψ_Q component vanishes outside the interaction region because it is composed of closed channels only.

8.2.3 Resonance States

Equation (8.34) is physically intuitive and mathematically attractive for analyzing resonance properties. To illuminate the salient feature of the Feshbach partitioning scheme, we solve Eq. (8.34) to yield the formal solution

$$\Psi_Q = [E - H^R_{QQ}(E) - iH^I_{QQ}(E)]^{-1} H_{QP}\Psi_0 \tag{8.41}$$

Since Ψ_Q is bound, we can expand it in any complete set of the $\mathcal{L}^2$ basis. In particular, we can choose ϕ_λ, the eigenfunctions of the complex operator $H^C_{QQ}(E)$, to expand Ψ_Q

$$\Psi_Q = \sum_\lambda \phi_\lambda C_\lambda \tag{8.42}$$

where ϕ_λ satisfies the eigenvalue equation

$$H^C_{QQ}|\phi_\lambda\rangle = E^c_\lambda(E)|\phi_\lambda\rangle \tag{8.43}$$

The *energy-dependent* complex eigenvalues can be written as

$$E^c_\lambda(E) = E^R_\lambda(E) - i\frac{\Gamma_\lambda(E)}{2} \tag{8.44}$$

Equation (8.39) proves that $\Gamma_\lambda \geq 0$.

Because the Hamiltonian $H^C_{QQ}(E)$ is *complex symmetric*

$$\begin{aligned} H^C_{QQ}(E)^{\dagger *} &= [H^R_{QQ}(E) + iH^I_{QQ}(E)]^{\dagger *} \\ &= [H^R_{QQ}(E) - iH^I_{QQ}(E)]^* \\ &= H^C_{QQ}(E) \end{aligned} \tag{8.45}$$

the orthogonality condition for the eigenfunction is defined in Eq. (8.17) for complex symmetric operators

$$\langle\phi^*_{\lambda'}|\phi_\lambda\rangle = \delta_{\lambda'\lambda} \tag{8.46}$$

Thus Eq. (8.41) can be solved to yield the solution for the bound component wavefunction

$$\Psi_Q = \sum_\lambda \frac{\langle\phi^*_\lambda|H_{QP}|\Psi_0\rangle}{E - E^R_\lambda + i\frac{\Gamma_\lambda}{2}}\phi_\lambda \tag{8.47}$$

Equation (8.47) gives the desired expression for the resonance wavefunction Ψ_Q. Here the complex eigenvalues $E^c_\lambda(E)$ and their corresponding eigenfunctions ϕ_λ are called resonance energies and resonance states. The real part of the resonance energy $E_\lambda(E)^R$ gives the ***position*** of the resonance and the imaginary part Γ_λ gives the ***full width*** of the resonance. Equation (8.47) explicitly shows that if the scattering energy E is near one of its resonance energies $E^R_\lambda(E)$, the dominant contribution in Eq. (8.47) comes from the resonance energy in question. For example, if the scattering energy $E \sim E_\lambda^R(E)$, then Ψ_Q can be approximated by a single term

$$\Psi_Q \simeq -2i \frac{<\phi_\lambda^*|H_{QP}|\Psi_0>}{\Gamma_\lambda} \phi_\lambda \tag{8.48}$$

For narrow resonance, Γ_λ is very small, and thus Ψ_Q can have a large amplitude in the interaction region. If the scattering energy is far away from the resonance energy, Ψ_Q will generally be small.

Perturbation calculation of resonance energy

In order to solve the complex eigenvalue problem $H_{QQ}^c \phi_\lambda = E_\lambda^c \phi_\lambda$, we partition H_{QQ}^c as

$$H_{QQ}^c = H_{QQ} + W_{QQ}^c \tag{8.49}$$

where W_{QQ}^c is defined as

$$W_{QQ}^c = H_{QP} G_{PP}^p(E) H_{PQ} - i\pi H_{QP} \delta(E - H_{PP}) H_{QP} \tag{8.50}$$

If we use eigenfunctions of the hermitian operator H_{QQ} to expand the complex eigenfunction ϕ_λ

$$\phi_\lambda = \sum_j \varphi_j^0 C_j \tag{8.51}$$

where φ_j^0 are solutions of the eigenvalue equation

$$H_{QQ} \varphi_j^0 = E_j^0 \varphi_j^0 \tag{8.52}$$

we substitute the above expansion for ϕ_λ in Eq. (8.43) to obtain an eigenvalue equation for the expansion coefficient

$$[\mathbf{E}^c - \mathbf{E}^0 - \mathbf{W}_{QQ}^c]\mathbf{C} = 0 \tag{8.53}$$

where

$$[\mathbf{W}_{QQ}^c]_{jj'} = <\varphi_j^0|W_{QQ}^c|\varphi_{j'}^0> \tag{8.54}$$

The linear algebraic Eq. (8.53) can be solved by straightforward matrix diagonalization of the complex symmetric matrix

$$\mathbf{H}_{QQ}^C = \mathbf{E}^0 + \mathbf{W}_{QQ}^c \tag{8.55}$$

to yield complex eigenvalues E^c_λ and eigenfunctions ϕ_λ.

If the resonance is isolated and narrow, we can treat W^c_{QQ} as a small perturbation in Eq. (8.53) and derive the complex resonance energy to first order

$$E^c_\lambda \approx E^0_\lambda + E^S_\lambda - i\frac{\Gamma_\lambda}{2} \tag{8.56}$$

where the real part of the perturbation energy given by

$$\boxed{E^S_\lambda = <\varphi^0_\lambda|H_{QP}G^p_{PP}(E)H_{PQ}|\varphi^0_\lambda>} \tag{8.57}$$

is called the energy shift while the imaginary part

$$\boxed{\Gamma_\lambda = 2\pi\ <\varphi^0_\lambda|H_{QP}\delta(E-H_{PP})H_{QP}|\varphi^0_\lambda>} \tag{8.58}$$

is the full width of resonance. The first order resonance wavefunction is given by standard perturbation theory

$$|\phi_\lambda> = |\varphi^0_\lambda> + \sum_{\lambda'}{}' \frac{<\varphi^0_{\lambda'}|W^c_{QQ}|\varphi^0_\lambda>}{E^0_\lambda - E^0_{\lambda'}}|\varphi^0_{\lambda'}> \tag{8.59}$$

It should be noted here that all these complex resonance energies and eigenfunctions are dependent on the scattering energy E as is obvious from the energy dependence of the complex operators.

8.2.4 Decay of Resonance

As is well known, a resonance state is characterized by its complex energy. The real part of the resonance energy E^R_λ specifies the position of the resonance to be found in the energy spectrum. The imaginary part Γ_λ, on the other hand, gives the lifetime of the resonance or its decay characteristic. The latter can be explicitly investigated by examining the time-dependence of the resonance component wavefunction Ψ_Q.

Suppose Ψ_Q is dominated by a single resonance; then from Eq. (8.47), we have

$$\Psi_Q(E) \simeq \frac{<\phi^*_\lambda|H_{QP}|\Psi_0>}{E - E^R_\lambda + i\frac{\Gamma_\lambda}{2}}\phi_\lambda \tag{8.60}$$

Since $\Psi_Q(E)$ is a component of the *stationary* scattering solution of the original Schrödinger equation $(E-H)(\Psi_P + \Psi_Q) = 0$, we can construct a time-dependent wavepacket corresponding to the Q component by

$$\Psi_Q(t) = \int_{-\infty}^{\infty} dE\ a(E)e^{-\frac{i}{\hbar}Et}\Psi_Q(E)$$
$$\simeq \int_{-\infty}^{\infty} dEa(E)e^{-\frac{i}{\hbar}Et}\frac{<\phi_\lambda^*|H_{QP}|\Psi 0>}{E - E_\lambda^R + i\frac{\Gamma_\lambda}{2}}\phi_\lambda \quad (8.61)$$

where $a(E)$ has a narrow energy distribution around the resonance energy $E_\lambda r^R$. The integral in Eq. (8.61) can be carried out by contour integration around the bottom half in complex energy space as shown in Fig. 8.3 to yield

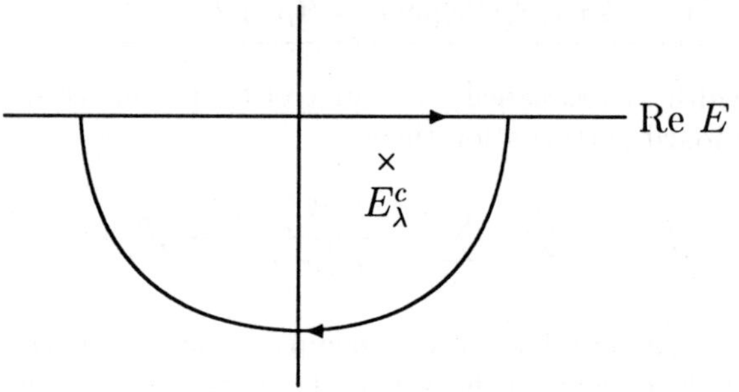

Figure 8.3: Contour used to integrate Eq. (8.61)

$$\Psi_Q(t) \simeq -2\pi i e^{-\frac{i}{\hbar}E_\lambda^C t}a(E_\lambda^c)\ <\phi_\lambda^*|H_{QP}|\Psi_0>\ \phi_\lambda$$
$$= e^{-\frac{i}{\hbar}E_\lambda^R t}e^{-\frac{\Gamma_\lambda}{2\hbar}t}\Psi_Q(0) \quad (8.62)$$

where

$$\Psi_Q(0) = a(E_\lambda^c)\ <\phi_\lambda^*|H_{QP}|\Psi_0>\ \phi_\lambda \quad (8.63)$$

Equation (8.62) shows an explicit time-dependence of $\Psi_Q(t)$ with a complex energy in the exponent. Thus the resonance wavefunction $\Psi_Q(t)$ decays exponentially

$$|\Psi_Q(t)|^2 \sim e^{-\frac{\Gamma_\lambda}{\hbar}t}|\Psi_Q(0)|^2 \quad (8.64)$$

with a characteristic lifetime of

$$\tau = \frac{\hbar}{\Gamma_\lambda} \tag{8.65}$$

8.2.5 Calculation of Resonance

The formal theory of resonance presented in the previous subsections provides a rigorous mathematical framework and justification for treating a scattering resonance as a quasibound state Ψ_Q. In practical applications, however, one often uses more direct and simple methods to calculate resonance energies and wavefunctions [166]. As mentioned before, resonance states are, rigorously speaking, continuum or scattering states that have bound state characteristics in the interaction region. Thus the most rigorous theoretical treatment of resonance is that of scattering theory as presented in Chapter 4. For example, we can examine the energy-dependence of the S matrix obtained from a scattering calculation to obtain both the position and width of the resonance, if any.

Most often, however, resonance information can be extracted from non-scattering calculations that are much easier than full-blown scattering calculations. As discussed in the preceding subsections, resonance wavefunctions are concentrated in the interaction region of the collision system with small but long tails extending to infinity. The narrower the resonance, the smaller the long tail of the wavefunction. In other words, narrow resonance wavefunctions are very much like bound state wavefunctions except for small and long tails. This makes it possible to use methods for bound states to calculate resonance wavefunctions. One of the widely used methods to calculate resonance energies is the stabilization method [165]. The stabilization method is based on the localization property of the resonance wavefunction in the interaction region. If the numerical grid employed in the bound state-like calculation for a resonance is large enough to enclose the localized part of the resonance wavefunction, the calculated eigenenergy should be almost independent of further increase of the grid size. In contrast, if the calculated wavefunction is not a highly localized resonance wavefunction but a continuum or scattering wavefunction spreading throughout the space, the calculated eigenenergy is a pseudo eigenenergy which will appreciably decrease as the grid size is increased. The situation is illustrated in Fig. 8.4.

Since the stabilization method involves simply bound state calculations, one can only obtain the real part of resonance energies or the positions of resonances. In principle, one can also estimate the resonance width from the slope of the resonance energy as a function of grid size from the stabilization

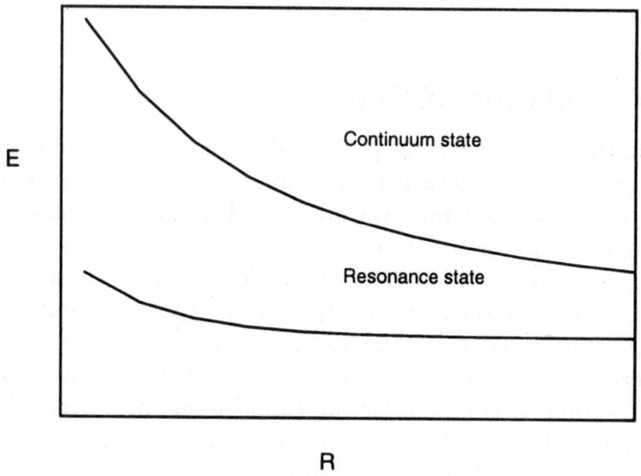

Figure 8.4: Stabilization plot of the calculated eigenenergy as a function of numerical grid to determine resonance energies. A resonance energy is identified by a flattering curve while an off-resonance energy is a monotonically decreasing function of the grid size R.

plot but the procedure is cumbersome and inaccurate. More sophisticated methods employ either complex basis functions or complex coordinates to construct a complex symmetric Hamiltonian matrix like what is described in the preceding subsection. A particularly convenient approach, however, is to simply add a local absorbing potential in the Hamiltonian but employ only real basis functions [167]. This is essentially an approximate numerical implementation of the formal Feshbach theory of resonance in which the nonlocal and energy-dependent complex Hamiltonian is replaced by a simple local (usually energy-independent) absorbing potential.

8.3 Reaction Rates and Transition State Theory

8.3.1 Cumulative Reaction Probability

At the heart of chemistry is the determination of reaction rate constants. As is known, the exact thermal rate constant for an elementary bimolecular reaction

$$A + B \longrightarrow P \tag{8.66}$$

can be rigorously calculated by Boltzmann averaging the reactive flux over the initial states and the collision energy

$$\begin{aligned} r &= <\sigma v> \\ &= \frac{4\pi}{Q_{int}} \sum_{fi} e^{-\epsilon_i/kT} \int_0^\infty \left(\frac{\mu}{2\pi kT}\right)^{3/2} v_i^3 \exp\left(-\frac{\mu v_i^2}{2kT}\right) \sigma_{fi}(v_i) dv_i \end{aligned} \tag{8.67}$$

where μ is the translational mass, ϵ_i is the eigenenergy of the internal state of the colliding partners, and v_i is the relative speed of the collision. The quantum partition function Q_{int} is defined as

$$Q_{int} = \sum_i e^{-\epsilon_i/kT} \tag{8.68}$$

where the summation is over all energetically accessible internal states of the reagents. The reaction cross section σ_{fi} in Eq. (8.67) is given by the formula (cf. Eq. (4.277))

$$\sigma_{fi} = \frac{\pi\hbar^2}{\mu^2 v_i^2} \sum_J (2J+1)|S_{fi}^J|^2 \tag{8.69}$$

where S_{fi}^J is the state-to-state reactive S matrix element.

Using the definitions in Eqs. (8.68) and (8.69), the rate Eq. (8.67) can be rearranged to give rise to the following result

$$r = \frac{kT}{2\pi\hbar} \frac{1}{Q_0} \int_0^\infty N(E) e^{-E/kT} / kT dE \tag{8.70}$$

where Q_0 is the total partition function for the reactants A + B

$$\begin{aligned} Q_0 &= Q_{int} \left(\frac{2\pi\mu kT}{h^2}\right)^{3/2} \\ &= Q_{int} Q_{trans} \\ &= Q_A Q_B \end{aligned} \tag{8.71}$$

and $N(E)$ is the "cumulative reaction probability " defined as the sum over both initial and final states of reaction probability

$$\boxed{N(E) = \sum_J (2J+1) \sum_{fi} |S_{fi}^J|^2} \tag{8.72}$$

Since the thermal rate constant is determined by the cumulative reaction probability $N(E)$ without any explicit reference to state-to-state quantities, it is desirable to directly calculate N(E) without having to calculate state-to-state reaction probabilities first and later sum over all states. Physically, the reaction rate is determined by the dynamics in a relatively small region near the transition state, so direct calculation of $N(E)$ should be computationally advantageous since it involves only short time dynamics in a small spatial region. Such approach is, at least, formally possible and there is an elegant formula for direct calculation of $N(E)$ by Miller [168]

$$\boxed{N(E) = \frac{1}{2}(2\pi\hbar)^2 Tr[\hat{F}\delta(E-H)\hat{F}\delta(E-H)]} \tag{8.73}$$

where $\hat{F}$ is the flux operator. Equation (8.73) has been widely applied to calculate cumulative reaction probabilities and rate constants [168].

8.3.2 Relation to Transition State Theory

If we perform a partial integration in Eq. (8.70), the rate constant can be rewritten in a more suggestive form

$$r = \frac{kT}{h}\frac{1}{Q_A Q_B}\int_0^\infty \rho(E)e^{-E/kT}dE \tag{8.74}$$

where $h = 2\pi\hbar$ and

$$\rho(E) = \frac{dN(E)}{dE} \tag{8.75}$$

The quantity $\rho(E)$ might be considered as a "density of states" from which we can define a "partition function"

$$\begin{aligned} Q_{ex}^{\ddagger} &= \int_0^\infty e^{-E/kT}\rho(E)dE \\ &= \frac{1}{kT}\int_0^\infty e^{-E/kT}N(E)dE \end{aligned} \tag{8.76}$$

Thus the rate equation (8.74) can be put in the form

$$\boxed{r = \frac{kT}{h}\frac{Q_{ex}^{\ddagger}}{Q_A Q_B}} \tag{8.77}$$

Equation (8.77) is in exactly the same form as the classical transition state theory (TST) expression for the rate constant

$$r_{tst} = \frac{kT}{h}\frac{Q_c^\ddagger}{Q_A Q_B} \tag{8.78}$$

where $Q_c^\ddagger$ is the true partition function at the transition state. However, it is important to point out that the rate Eq. (8.77) is the *exact* quantum mechanical result while the TST rate Eq. (8.78) is the classical transition state approximation. $Q_c^\ddagger$ is the true partition function at the transition state. Comparing the exact quantum rate expression Eq. (8.77) with the transition state expression Eq. (8.78), we can try to associate the quantum mechanical quantity $Q_{ex}^\ddagger$ with the quantum partition function at the transition state. Accordingly, the quantity $\rho(E) = N'(E)$ is given the meaning of "density of states".

The above interpretation should be meaningful if the transition state theory is valid. Thus the analogy to TST gives a physically intuitive meaning to the exact quantum cumulative reaction probability $N(E)$: it represents the total number of open channels (states) at total energy E at the "transition state". This interpretation of $N(E)$ is quite profound and significant because it provides a practical means to actually allow us to examine the validity of transition state theory by analyzing the energy dependence of the cumulative reaction probability $N(E)$. For the transition state interpretation to be valid, $N(E)$ should be a *nondecreasing* function of energy because it represents the total number of open states at the transition state at fixed energy E. In particular, the quantization of transition states should show up in the energy-dependence of the cumulative reaction probability as a step-like function of energy. For example, if we assume that the reaction can only proceed through quantized transition states, then the TST result for the cumulative reaction probability should behave like

$$N^\ddagger(E) \simeq \sum_J (2J+1) \sum_\nu h(E - E_\nu^\ddagger) \tag{8.79}$$

where $h(E)$ is a step function and $E_\nu^\ddagger$ are quantized transition state energies.

One could also examine the energy-dependence of the "density of states" $\rho(E)$ which should show spikes when a new transition state becomes open [169]. In fact, $\rho(E)$ is a more sensitive function of energy and thus should show more prominent structures of resonance, much like the energy spectrum. For example, Fig. 8.5a shows the calculated cumulative reaction probability as a function of energy for the H + H_2 reaction. The resonance structure is mostly buried in the energy dependence of Fig. 8.5. However, the energy-dependence of the corresponding "density of states" $\rho(E)$

is highly structured and shows very clearly individual resonance states as indicated by resonance state labels in Fig. 8.5b

Rate for unimolecular reaction

From the expression for the exact rate constant of a bimolecular reaction in Eq. (8.77), one can straightforwardly derive a similar rate expression for unimolecular reaction. If we consider the following forward and reverse reaction process

$$A + B \underset{k_{-1}}{\overset{k_2}{\rightleftharpoons}} C \tag{8.80}$$

when equilibrium is reached, the equilibrium constant K is given by the relation

$$K = \frac{k_2}{k_{-1}} = \frac{Q_C}{Q_A Q_B} \tag{8.81}$$

where Q_C is the partition function of molecule C. Thus k_{-1} can be written as

$$k_{-1} = k_2 \frac{Q_A Q_B}{Q_C} \tag{8.82}$$

Now using Eq. (8.77) for the bimolecular rate constant k_2, we arrive at the exact expression for k_{-1}

$$k_{-1}(T) = \frac{kT}{h} \frac{Q_{ex}^{\ddagger}}{Q_C} \tag{8.83}$$

By writing the partition function Q_C in terms of the density of states $\rho_c(E)$,

$$Q_C = \int \rho_c(E) e^{-E/kT} dE \tag{8.84}$$

and Eq. (8.76) for $Q_{ex}^{\ddagger}$, then we can obtain a corresponding micro canonical rate expression for the unimolecular reaction

$$\boxed{k_{-1}(E) = \frac{1}{h} \frac{N(E)}{\rho_c(E)}} \tag{8.85}$$

Equations (8.83) and (8.85) are exact expressions for the canonical and micro canonical rate of unimolecular reaction. If we replace the exact cumulative reaction probability $N(E)$ by the total number of open states $N^{\ddagger}(E)$ at the transition state, Eq. (8.85) becomes the standard RRKM theory for the unimolecular decay rate [170].

Further reading

Refs. [171–173] are good sources for further reading on the dynamical theory of chemical reactions.

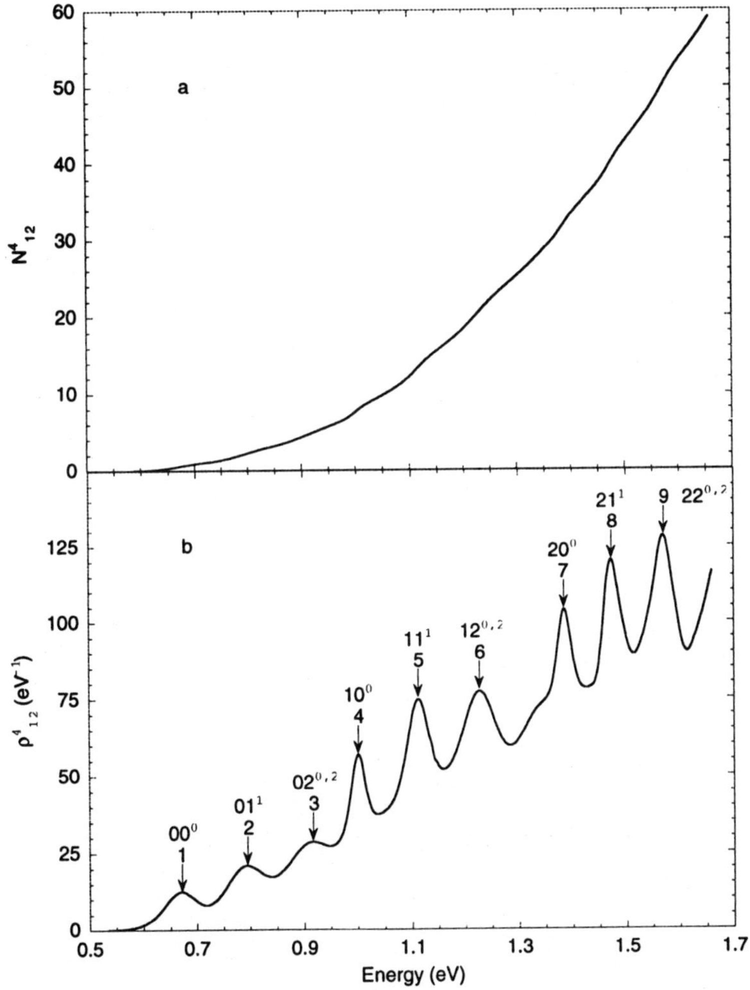

Figure 8.5: Reaction of H + H_2 for total angular momentum J=4. (a) Cumulative reaction probability. (b) Density of states. The numbers are attempted assignments for quantized transition states. (From Ref. [169] with permission, courtesy of Thomas C. Allison and Donald G. Truhlar)

Chapter 9

Photon Induced Molecular Processes

9.1 Introduction

This chapter discusses theories and computational methods for treating interactions of molecules with the radiation field. In the majority of practical situations, the interaction between a molecule and the radiation field can be treated as a small perturbation. This allows us to generally classify the dynamical process in terms of the order of interaction. In the first order process in which a single photon is either destroyed (absorption) or created (emission), the result is the well known golden rule formula. The second order process in which two photons are involved (one destroyed and another one created) corresponds to light scattering including resonance and non-resonance Raman scattering. An important class of molecular interactions with the radiation field is that of photofragmentation which is important in spectroscopy and photochemical reactions. Photon induced dissociation of molecules is of significant importance in photochemistry. For example, the dissociation of ozone (O_3) in the stratosphere plays a vital role in protecting us from ultraviolet radiation from the sun.

9.2 Time-Dependent Perturbation Theory

General formalism

The Hamiltonian of a molecular system can be written as the sum of an unperturbed Hamiltonian H_0 and an interaction H_{int}

$$H = H_0 + H_{int} \tag{9.1}$$

Using the interaction representation, the TD Schrödinger equation is written as

$$i\hbar \frac{\partial \Psi_I(t)}{\partial t} = H_I(t)\Psi_I(t) \tag{9.2}$$

where $H_I(t)$ is the interaction Hamiltonian in the interaction picture defined as

$$H_I(t) = e^{\frac{i}{\hbar}H_0 t} H_{int} e^{-\frac{i}{\hbar}H_0 t} \tag{9.3}$$

The TD wavefunction can be expanded in terms of eigenfunctions of H_0 as

$$|\Psi_I(t)\rangle = \sum_k C_k(t)|\Phi_k\rangle \tag{9.4}$$

where Φ_k are stationary eigenfunctions of H_0 satisfying

$$H_0|\Phi_k\rangle = E_k|\Phi_k\rangle \tag{9.5}$$

Equation (9.2) can be solved to yield equations for the expansion coefficients

$$\frac{dC_m(t)}{dt} = \frac{1}{\hbar} \sum_k \langle m|H_I(t)|k\rangle \, C_k(t) \tag{9.6}$$

where $|k\rangle$ stands for $|\Phi_k\rangle$.

Equation (9.6) can be integrated to give a formal solution for the expansion coefficient

$$C_m(t) = C_m(0) + \frac{1}{\hbar} \sum_k \int_0^t \langle m|H_I(t')|k\rangle \, C_k(t')dt' \tag{9.7}$$

If the interaction $H_I(t)$ can be treated as a small perturbation, we can use a parameter λ to characterize its strength and replace it by $\lambda H_I(t)$. This facilitates the expansion of the coefficients $C_k(t)$ in orders of λ

$$C_k(t) = \sum_{n=0} \lambda^n C_k^{(n)}(t) \tag{9.8}$$

By substituting Eq. (9.8) in Eq. (9.7) and equating terms with the same orders, we obtain an iterative equation for the expansion coefficient

$$\boxed{C_m^{(n)}(t) = C_m^{(n)}(0) + \frac{1}{i\hbar}\sum_k \int_0^t <m|H_I(t')|k> \, C_k^{(n-1)}(t')dt'} \tag{9.9}$$

Equation (9.9) gives a general perturbative solution for the expansion coefficient which is then used to generate wavefunctions to all orders of perturbation.

Of particular interest is the case when the system is in a particular eigenstate $|l>$ at $t = 0$. This can be represented by the initial condition for the wavefunction

$$C_k^{(n)}(0) = \delta_{kl}\delta_{n0} \tag{9.10}$$

By substituting Eq. (9.10) into Eq. (9.9), we arrive at the first order solution

$$C_m^{(1)}(t) = \frac{1}{i\hbar}\int_0^t dt \, <m|H_I(t')|l> \tag{9.11}$$

Similarly we can derive the second order solution

$$\begin{aligned} C_m^{(2)}(t) &= \frac{1}{i\hbar}\sum_k \int_0^t <m|H_I(t')|k> \, C_k^{(1)}(t')dt' \\ &= \frac{1}{(i\hbar)^2}\sum_k \int_0^t dt' \int_0^{t'} dt'' \, <m|H_I(t')|k><k|H_I(t'')|l> \end{aligned} \tag{9.12}$$

and so on.

Golden rule formula

In evaluating the time integral in the iterative solution to Eq. (9.9), it is important to include a time-dependent phase factor given by

$$<m|H_I(t')|k>=<m|H_{int}|k> \, e^{\frac{i}{\hbar}(E_m - E_k)t} \tag{9.13}$$

where $\omega_{mk} = (E_m - E_k)/\hbar$ and the definition of $H_I(t)$ in Eq. (9.3) has been used. Let's consider a common situation in which the interaction Hamiltonian H_{int} is a periodic function of time

$$H_{int} = V_{int}e^{\pm i\omega t} \tag{9.14}$$

where V_{int} is independent of time. The special case of a time-independent interaction is included in the above definition by simply setting $\omega = 0$.

For the interaction that acts in a time period $[0, T]$, the first order solution of Eq. (9.11) can be integrated explicitly to yield the result

$$\begin{aligned} C_m^{(1)}(T) &= \frac{1}{i\hbar}\int_0^T \langle m|H_I(t')|l\rangle \\ &= \frac{1}{i\hbar}\langle m|V_{int}|l\rangle \int_0^T e^{i(\omega_{ml}\pm\omega)t} \\ &= \frac{1}{i\hbar}\langle m|V_{int}|l\rangle \frac{1}{i\omega}\left[e^{i(\omega_{ml}\pm\omega)T}\right] \end{aligned} \tag{9.15}$$

The probability that the system will end up in state $|m\rangle$ from the initial state $|l\rangle$ *per unit time* is defined as

$$\begin{aligned} p_m^{(1)} &= |C_m^{(1)}(T)|^2/T \\ &= \frac{1}{\hbar^2(\omega_{ml}\pm\omega)^2}|\langle m|V_{int}|l\rangle|^2 \sin^2\left(\frac{\hbar}{2}(\omega_{ml}\pm\omega)T\right) \end{aligned} \tag{9.16}$$

In the limit $T \to \infty$, the above equation simplifies to

$$p_{ml}^{(1)} = \frac{2\pi}{\hbar}|\langle m|V_{int}|l\rangle|^2\delta(E_m - E_l \pm \hbar\omega) \tag{9.17}$$

where the limiting form of the δ-function

$$\delta(x) = \lim_{a\to\infty}\frac{1}{\pi}\frac{\sin^2(ax)}{ax^2} \tag{9.18}$$

has been used. For simplicity, the limit of the following integral can be directly employed to give rise to the δ-function in later discussions, i.e.

$$\lim_{T\to\infty}\frac{1}{T}\left|\frac{1}{i\hbar}\int_0^T e^{\frac{i}{\hbar}Et}dt\right|^2 = \frac{2\pi}{\hbar}\delta(E) \tag{9.19}$$

The δ-function in Eq. (9.17) enforces the energy conservation

$$E_m - E_l \pm \hbar\omega = 0 \tag{9.20}$$

The process with $E_m = E_l + \hbar\omega$ corresponds to absorption because the energy of the final state $|m\rangle$ is larger than that of the initial state $|l\rangle$ by absorbing a quantum of energy $\hbar\omega$. The reverse process with $E_m = E_l - \hbar\omega$ corresponds to emission in which the initial state $|l\rangle$ emits a quantum of energy $\hbar\omega$ to drop to the lower energy state $|m>$. If we sum over all degenerate final states, we obtain the well known *golden rule* formula

$$\begin{aligned} w_{fi} &= \int \frac{2\pi}{\hbar} | <f|V_{int}|i> |^2 \delta(E_m - E_l \pm \hbar\omega)\rho(E_f)dE_f \\ &= \boxed{\frac{2\pi}{\hbar} | <f|V_{int}|i> |^2 \rho(E_f)} \end{aligned} \tag{9.21}$$

where $\rho(E_f)$ is the density of final states. Equation (9.21) gives the transition probability per unit time from a given initial state $|i>$ to all degenerate final states $|f>$ with the same energy E_f.

9.3 Interaction of Molecules with the Radiation Field

9.3.1 Vector Potentials in Quantum Mechanics

The Hamiltonian of a nonrelativistic particle with mass m and electric charge q in the electromagnetic field is given by [174, 175]

$$H = \frac{1}{2m}\left(\mathbf{p} - \frac{q}{c}\mathbf{A}\right)^2 + V_0 \tag{9.22}$$

where V is the scalar potential and A is the vector potential . In quantum mechanics, the momentum $\mathbf{p}$ is replaced by the operator $-i\hbar\nabla$. If we denote the solution to the vector-free Hamiltonian as ψ^0 which satisfies the equation

$$\left[\frac{1}{2m}\mathbf{p}^2 + V_0\right]\psi^0 = 0 \tag{9.23}$$

then the solution to Eq. (9.22) can be written in the form

$$\psi = \exp\left(\frac{iq}{\hbar c}\int^{\mathbf{x}} \mathbf{A}(\mathbf{x}')\cdot d\mathbf{x}'\right)\psi^0 \tag{9.24}$$

This result can be proved by noting that

$$\begin{aligned} \left(-i\hbar\nabla - \frac{q}{c}\mathbf{A}\right)\psi &= \exp\left(\frac{iq}{\hbar c}\int^{\mathbf{x}} \mathbf{A}(\mathbf{x}')\cdot d\mathbf{x}'\right)\left(-i\hbar\nabla - \frac{q}{c}\mathbf{A}\right)\psi^0 \\ &+ \left(\frac{iq}{\hbar c}\right)\mathbf{A}(\mathbf{x})(-i\hbar)\psi^0 \\ &= \exp\left(\frac{iq}{\hbar c}\int^{\mathbf{x}} \mathbf{A}(\mathbf{x}')\cdot d\mathbf{x}'\right)(-i\hbar\nabla\psi^0) \end{aligned} \tag{9.25}$$

From the above relation, it is straightforward to obtain the following

$$\left(-i\hbar\nabla - \frac{q}{c}\mathbf{A}\right)^2 \psi = \exp\left(\frac{iq}{\hbar c}\int^{\mathbf{x}} \mathbf{A}(\mathbf{x}')\cdot d\mathbf{x}'\right)(-\hbar^2\nabla^2\psi^0) \qquad (9.26)$$

Thus using Eq. (9.23) one easily verifies that Ψ given by Eq. (9.24) is indeed a solution of the Hamiltonian equation.

In gauge transformation, the vector potential is changed to

$$\mathbf{A} \rightarrow \mathbf{A}' = \mathbf{A} + \nabla\varphi \qquad (9.27)$$

which keeps the magnetic field unchanged

$$\mathbf{B} = \nabla \times \mathbf{A} \qquad (9.28)$$

The corresponding wavefunction transforms as

$$\begin{aligned}\psi' &= \exp\left(\frac{iq}{\hbar c}\int^{\mathbf{x}} (\mathbf{A}(\mathbf{x}') + \nabla\varphi\cdot d\mathbf{x}'\right)\psi^0 \\ &= \psi \exp\left(\frac{iq}{\hbar c}\varphi(\mathbf{x})\right)\end{aligned} \qquad (9.29)$$

Thus the form of the wavefunction depends on the particular gauge we use for the vector potential.

Aharonov-Bohm effect

A remarkable effect of the vector potential on the wavefunction is the Aharonov-Bohm effect [176]. Consider the situation in which a coherent beam of charged particles such as electrons passes through a double slit and meets again at the detector as illustrated in Fig. 9.1. The particles traveling through the two separate paths 1 and 2 do not encounter the magnetic field ($\mathbf{B} = 0$) which is confined to the small region indicated in the figure. However, the wavefunction describing the particles reaching the detector is given by the superposition of two terms representing the two paths

$$\begin{aligned}\psi &= \psi_1 + \psi_2 \\ &= \exp\left(\frac{iq}{\hbar c}\int_{path1}^{\mathbf{x}} \mathbf{A}(\mathbf{x}')\cdot d\mathbf{x}'\right)\psi^0 + \exp\left(\frac{iq}{\hbar c}\int_{path2}^{\mathbf{x}} \mathbf{A}(\mathbf{x}')\cdot d\mathbf{x}'\right)\psi^0 \\ &= \left[1 + \exp\left(\frac{iq}{\hbar c}\oint \mathbf{A}(\mathbf{x}')\cdot d\mathbf{x}'\right)\right]\psi_1\end{aligned} \qquad (9.30)$$

where Eq. (9.24) for the wavefunction of each path has been used, and the closed path is completed along path 2 and then along path 1 in the opposite direction. By using Gauss's theorem, we can write

$$\begin{aligned}\oint \mathbf{A}(\mathbf{x}') \cdot d\mathbf{x}' &= \int \nabla \times \mathbf{A}(\mathbf{x}') \cdot d\mathbf{S} \\ &= \int \mathbf{B}(\mathbf{x}') \cdot d\mathbf{S} \\ &= \Phi_B \end{aligned} \tag{9.31}$$

where $\Phi_B = \int \mathbf{B}(\mathbf{x}') \cdot d\mathbf{S}$ is the total magnetic flux passing through the surface enclosed by the complete path. We thus obtain the expression for the interference pattern

$$|\psi|^2 = |\psi_0|^2 \left| \left[1 + \exp\left(\frac{iq}{\hbar c} \Phi_B \right) \right] \right|^2 \tag{9.32}$$

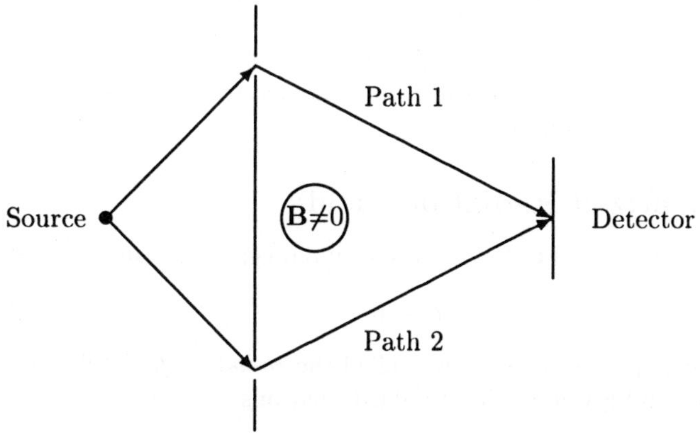

Figure 9.1: Illustration of Aharonov-Bohm effect.

This is a truly remarkable result because the interference pattern is determined by the total magnetic flux *even* though the particles experience no magnetic field or Lorentz force in the paths they traveled! This is purely a quantum effect with no classical explanation and has been experimentally verified.

Flux quantization

Another important consequence of Eq. (9.24) for wavefunctions in vector potentials is the quantization of magnetic flux trapped in a superconductor. Since the magnetic field is zero inside the superconductor due to the Meissner effect, the correlated electron pair inside the superconductor experiences no magnetic field. If the electron pair is transported through a closed path inside the superconductor, the wavefunction of the transported electron pair is given by

$$\psi = \psi^0 \exp\left(\frac{iq}{\hbar c}\oint \mathbf{A}(\mathbf{x}') \cdot d\mathbf{x}'\right) \tag{9.33}$$

where ψ^0 is the wavefunction of the electron pair before being transported. Since the wavefunction has to be single valued, we must have

$$\frac{iq}{\hbar c}\oint \mathbf{A}(\mathbf{x}') \cdot d\mathbf{x}' = 2n\pi \qquad (n = 0, \pm 1, \pm 2, ...) \tag{9.34}$$

or the flux must be quantized by the relation

$$\Phi_B = \oint \mathbf{A}(\mathbf{x}') \cdot d\mathbf{x}' = \frac{n\pi\hbar c}{e} \qquad n = 0, \pm 1, \pm 2, ... \tag{9.35}$$

where the charge of the electron pair is $q = 2e$. This important effect has been experimentally verified as well, and it helped to establish the electron pairing mechanism in superconductors.

9.3.2 Quantized Radiation Field

In the quantum theory of radiation, the Coulomb gauge is usually employed

$$\nabla \cdot \mathbf{A} = 0 \tag{9.36}$$

where the vector potential $\mathbf{A}$ is also called the *radiation field.* The electromagnetic fields are given by the standard relations

$$\begin{cases} \mathbf{B} = \Delta \times \mathbf{A} \\ \mathbf{E} = -\dfrac{1}{c}\dfrac{\partial \mathbf{A}}{\partial t} \end{cases} \tag{9.37}$$

The application of the standard quantization procedure leads to the quantized radiation field

$$\mathbf{A}(\mathbf{x}, t) = c\sum_{\mathbf{k},\alpha}\sqrt{\frac{\hbar}{2\omega V}}\left[a_{\mathbf{k},\alpha}(t)\epsilon_{\mathbf{k}}^{\alpha}e^{i\mathbf{k}\cdot\mathbf{x}} + a_{\mathbf{k},\alpha}^{\dagger}(t)\epsilon_{\mathbf{k}}^{\alpha}e^{-i\mathbf{k}\cdot\mathbf{x}}\right] \tag{9.38}$$

where V is the total volume in space, $\omega = |\mathbf{k}|c$ is the photon frequency, and $a_{\mathbf{k},\alpha}(t) = a_{\mathbf{k},\alpha}(0)\exp(-i\omega t)$. The operators $a_{\mathbf{k},\alpha}$ and $a^{\dagger}_{\mathbf{k},\alpha}$ are creation and annihilation operators satisfying the commutation relations for bosons

$$[a_{\mathbf{k},\alpha}, a^{\dagger}_{\mathbf{k}',\alpha'}] = a_{\mathbf{k},\alpha}a^{\dagger}_{\mathbf{k}',\alpha'} - a^{\dagger}_{\mathbf{k}',\alpha'}a_{\mathbf{k},\alpha} = \delta_{\mathbf{k},\mathbf{k}'}\delta_{\alpha,\alpha'} \tag{9.39}$$

$$[a_{\mathbf{k},\alpha}, a_{\mathbf{k}',\alpha'}] = [a^{\dagger}_{\mathbf{k},\alpha}, a^{\dagger}_{\mathbf{k}',\alpha'}] = 0 \tag{9.40}$$

The radiation gauge of Eq. (9.36) requires that the ***polarization*** vectors $\epsilon^{\alpha}_{\mathbf{k}}(\alpha = 1, 2)$ are ***perpendicular*** to the photon momentum (or propagation direction) $\mathbf{k}$, viz.,

$$\epsilon^{\alpha}_{\mathbf{k}} \cdot \mathbf{k} = 0 \qquad\qquad (\alpha = 1, 2) \tag{9.41}$$

Thus the propagation of the radiation field is perpendicular to the plane of polarization formed by $\epsilon^{1}_{\mathbf{k}}$ and $\epsilon^{2}_{\mathbf{k}}$ which are orthogonal unit vectors satisfying

$$\epsilon^{\alpha}_{\mathbf{k}} \cdot \epsilon^{\alpha'}_{\mathbf{k}} = \delta_{\alpha\alpha'} \tag{9.42}$$

The radiation field given by Eq. (9.38) involves the summation over all photons whose $\mathbf{k}$ and α values specify the momentum and polarization (or spin) of the photons. Although the photon is a boson with spin $S = 1$, only two of the three spin states exist due to the use of Coulomb gauge.

The Hamiltonian of the electromagnetic field defined by

$$H_{em} = \frac{1}{2}\int_V (\mathbf{B}\cdot\mathbf{B} + \mathbf{E}\cdot\mathbf{E})\, d\mathbf{x} \tag{9.43}$$

is thus quantized as well. By using Eqs. (9.37) and (9.38), one can derive the following result for the quantized Hamiltonian

$$\begin{aligned} H_{em} &= \frac{1}{2}\sum_{\mathbf{k},\alpha} \hbar\omega[a_{\mathbf{k},\alpha}a^{\dagger}_{\mathbf{k},\alpha} + a^{\dagger}_{\mathbf{k},\alpha}a_{\mathbf{k},\alpha}] \\ &= \sum_{\mathbf{k},\alpha}\left[N_{\mathbf{k},\alpha} + \frac{1}{2}\right]\hbar\omega \end{aligned} \tag{9.44}$$

where the number operator is defined by

$$N_{\mathbf{k},\alpha} = a^{\dagger}_{\mathbf{k},\alpha}a_{\mathbf{k},\alpha} \tag{9.45}$$

Thus we can use the occupation number for photons $N_{\mathbf{k},\alpha}$ to denote the quantum state of the radiation field $|N_{\mathbf{k},\alpha}>$.

9.3.3 One Photon Processes

The Hamiltonian describing the interaction of a molecule with the radiation field in Eq. (9.22) can be written as

$$H = H_0 + H_{int} \tag{9.46}$$

where the interaction Hamiltonian is

$$H_{int} = -\frac{q}{mc}\mathbf{A}\cdot\mathbf{p} + \frac{q^2}{2mc^2}\mathbf{A}\cdot\mathbf{A} \tag{9.47}$$

Here the Coulomb gauge $\nabla \cdot \mathbf{A} = 0$ has been used. The vector potential $\mathbf{A}(\mathbf{x}, t)$ is now a field operator defined in Eq. (9.38) which operates on photon states $|N_{\mathbf{k},\alpha}>$ at space and time of $(\mathbf{x}, t)$. The combined quantum state of matter and photon is given by the direct product of matter state $|l>$ and photon state $|N_{\mathbf{k},\alpha}>$

$$|l; N_{\mathbf{k},\alpha}> = |l>\ |N_{\mathbf{k},\alpha}> \tag{9.48}$$

By treating the interaction H_{int} as a small perturbation, we can calculate a first order process such as absorption discussed in Sec. 9.2 by evaluating the matrix element between initial and final states

$$\begin{aligned} <f; N_{\mathbf{k},\alpha} - 1|H_{int}|i; N_{\mathbf{k},\alpha}> &= -\frac{q}{mc} <f; N_{\mathbf{k},\alpha} - 1|\mathbf{A}\cdot\mathbf{p}|i; N_{\mathbf{k},\alpha}> \\ &= -\frac{q}{m}\sqrt{\frac{N_{\mathbf{k},\alpha}\hbar}{2\omega V}}\epsilon_{\mathbf{k}}^{\alpha}\cdot <f|e^{i\mathbf{k}\cdot\mathbf{x}}\mathbf{p}|i>\ e^{-i\omega t} \end{aligned} \tag{9.49}$$

where the second term in Eq. (9.47) does not contribute to the one photon process because the term $\mathbf{A}\cdot\mathbf{A}$ contains an even number of annihilation or creation operators. Similarly, we can obtain the interaction matrix element for the emission process

$$\begin{aligned} &<f; N_{\mathbf{k},\alpha} + 1|H_{int}|i; N_{\mathbf{k},\alpha}> = \\ &-\frac{q}{m}\sqrt{\frac{(N_{\mathbf{k},\alpha} + 1)\hbar}{2\omega V}}\epsilon_{\mathbf{k}}^{\alpha}\cdot <f|e^{-i\mathbf{k}\cdot\mathbf{x}}\mathbf{p}|i>\ e^{i\omega t} \end{aligned} \tag{9.50}$$

Absorption

The matrix element $<f|e^{-i\mathbf{k}\cdot\mathbf{x}}\mathbf{p}|i>$ is usually calculated by using the dipole approximation

$$e^{-i\mathbf{k}\cdot\mathbf{x}} \approx 1 \tag{9.51}$$

The dipole approximation is valid at relatively long wavelengths of the radiation field such that $\lambda \gg x$ or $x/\lambda \approx 0$. Within the dipole approximation, the golden rule formula of Eq. (9.21) for absorption becomes

$$w_{fi}^{abs} = \frac{2\pi}{\hbar}\frac{q^2}{m^2}\frac{N_{\mathbf{k},\alpha}\hbar}{2\omega_{fi}V}|\epsilon_{\mathbf{k}}^{\alpha}\cdot\mathbf{p}_{fi}|^2\rho(E_f) \tag{9.52}$$

where $\mathbf{p}_{fi} = <f|\mathbf{p}|i>$. The matrix element of the momentum operator can be rewritten as

$$\begin{aligned}\mathbf{p}_{fi} &= \frac{im}{\hbar} <f|[H_0, \mathbf{x}]|i> \\ &= -\frac{im(E_f - E_i)}{\hbar} <f|\mathbf{x}|i> \\ &= -\frac{im\omega_{fi}}{q}\mathbf{P}_{fi}\end{aligned} \tag{9.53}$$

where

$$\mathbf{P}_{fi} = <f|q\mathbf{x}|i> \tag{9.54}$$

is the matrix element of the dipole moment $\mathbf{P} = q\mathbf{x}$. For a system with more than one charged particle, the dipole moment $\mathbf{P}$ is given by the summation

$$\mathbf{P} = \sum_i q_i\mathbf{x}_i \tag{9.55}$$

Thus, Eq. (9.52) for absorption can be written in the form

$$\boxed{w_{fi}^{abs} = \pi n_{\mathbf{k},\alpha}\omega_{fi}|\epsilon_{\mathbf{k}}^{\alpha}\cdot\mathbf{P}_{fi}|^2\rho(E_f)} \tag{9.56}$$

where $n_{\mathbf{k},\alpha} = N_{\mathbf{k},\alpha}/V$ is the density of photons. Equation (9.56) gives the probability of absorbing a polarized photon per unit time.

Emission

A similar result can be obtained for the emission process

$$\boxed{w_{fi}^{em} = \pi(n_{\mathbf{k},\alpha} + \frac{1}{V})\omega_{fi}|\epsilon_{\mathbf{k}}^{\alpha}\cdot\mathbf{P}_{fi}|^2\rho(E_f)} \tag{9.57}$$

The quantum theory shows that even in the absence of the radiation field ($n_{\mathbf{k},\alpha} = 0$), there is still photon emission according to Eq. (9.57). Such emission is called spontaneous emission and the rate is given by

$$w_{fi}^{sem} = \frac{\pi\omega_{fi}}{V}|\epsilon_{\mathbf{k}}^{\alpha}\cdot\mathbf{P}_{fi}|^2\rho(E_f) \tag{9.58}$$

Thus, the total emission is the sum of *induced* emission and *spontaneous* emission

$$w_{fi}^{em} = w_{fi}^{iem} + w_{fi}^{sem} \tag{9.59}$$

where the induced emission w_{fi}^{iem} is given by exactly the same formula as absorption, i.e. equation (9.56). This is simply due to microscopic reversibility.

Since spontaneous emission emits photons in all directions and all polarizations, we need to sum over all degenerate photon states. Using the standard formula for the density of states for photons in a solid angle $d\Omega_{\mathbf{k}}$

$$\begin{aligned}\rho(E) &= \frac{L^3}{(2\pi)^3}\frac{d\mathbf{k}^3}{d(\hbar\omega_{fi})} \\ &= \frac{V\omega_{fi}^2 d\Omega_{\mathbf{k}}}{(2\pi)^3\hbar c^3}\end{aligned} \tag{9.60}$$

and the result of the following integral

$$\sum_{\alpha=1}^{2}\int d\Omega|\epsilon_{\mathbf{k}}^{\alpha}\cdot\mathbf{P}_{fi}|^2 = \frac{8\pi}{3}|\mathbf{P}_{fi}|^2 \tag{9.61}$$

we obtain the total rate of spontaneous emission

$$\boxed{W_{fi}^{spon} = \frac{\omega_{fi}^3|\mathbf{P}_{fi}|^2}{3\pi\hbar c^3}} \tag{9.62}$$

9.3.4 Two Photon Processes: Light Scattering

Kramers-Heisenberg formula

Besides first order absorption and emission, another important class of molecular interactions with light is light scattering, which includes Rayleigh scattering (long wavelength limit), Thomson scattering (short wavelength limit), and the Raman effect. In light scattering, the initial molecular system is in state i with an incident photon $(\mathbf{k}, \epsilon_{\mathbf{k}}^{\alpha})$, and the final molecular state is in state f with an outgoing photon $(\mathbf{k}', \epsilon_{\mathbf{k}'}^{\alpha'})$. Since light scattering is a two photon process and involves annihilation of one photon and creation of another photon, only terms containing $a^\dagger a$ or $aa^\dagger$ contribute to this process. In first order of H_{int}, the term contributing to this process is given by

$$C_{fi}^{(1)}(t) = \frac{1}{i\hbar}\left(\frac{q^2}{2mc^2}\right) <f; N_{\mathbf{k}',\alpha'}|\mathbf{A}\cdot\mathbf{A}|i; N_{\mathbf{k},\alpha}> e^{\frac{i}{\hbar}(E_f-E_i)t}$$

$$= \delta_{fi}\frac{-iq^2}{2mV\sqrt{\omega\omega'}}\epsilon_{\mathbf{k}'}^{\alpha'}\cdot\epsilon_{\mathbf{k}}^{\alpha}\int_0^t e^{i(\omega_{fi}+\omega'-\omega)t'}dt' \tag{9.63}$$

where photon frequencies are given by $\omega = |\mathbf{k}|c$ and $\omega' = |\mathbf{k}'|c$. Due to the presence of the Kronecker delta δ_{fi}, the molecular state remains the same, and therefore the term in Eq. (9.63) contributes to elastic scattering of photons only, since ω' must equal ω due to energy conservation.

In second order of H_{int} the term contributing to light scattering is given by the second order perturbation result in Eq. (9.12)

$$C_{fi}^{(2)}(t) = \frac{1}{(i\hbar)^2}\left(\frac{q}{mc}\right)^2\sum_l\int_0^t dt_1\int_0^{t_1}dt_2 <f|\mathbf{A}(t_1)\cdot\mathbf{p}|l><l|\mathbf{A}(t_2)\cdot\mathbf{p}|i>$$

$$= \frac{1}{(i\hbar)^2}\frac{c^2\hbar}{2V\sqrt{\omega\omega'}}\left(\frac{q}{mc}\right)^2\int_0^t dt_1\int_0^{t_1}dt_2\sum_l\Big[(\epsilon_{\mathbf{k}'}^{\alpha'}\cdot\mathbf{p}_{fl})(\epsilon_{\mathbf{k}}^{\alpha}\cdot\mathbf{p}_{li})$$

$$\times e^{i(\omega_{fl}+\omega')t_1}e^{i(\omega_{li}-\omega)t_2} + (\epsilon_{\mathbf{k}}^{\alpha}\cdot\mathbf{p}_{fl})(\epsilon_{\mathbf{k}'}^{\alpha'}\cdot\mathbf{p}_{li})$$

$$\times e^{i(\omega_{fl}-\omega)t_1}e^{i(\omega_{li}+\omega')t_2}\Big] \tag{9.64}$$

where the two terms in the brackets correspond to two intermediate photon states $|l;0>$ and $|l;\mathbf{k},\alpha,\mathbf{k}',\alpha'>$, respectively. After integrating over t_2, the above equation can be simplified to yield

$$C_{fi}^{(2)}(t) = \frac{ic^2}{2V\sqrt{\omega\omega'}}\left(\frac{q}{mc}\right)^2\sum_l\left[\frac{(\epsilon_{\mathbf{k}'}^{\alpha'}\cdot\mathbf{p}_{fl})(\epsilon_{\mathbf{k}}^{\alpha}\cdot\mathbf{p}_{li})}{E_l-E_i-\hbar\omega}\right.$$

$$\left.+\frac{(\epsilon_{\mathbf{k}}^{\alpha}\cdot\mathbf{p}_{fl})(\epsilon_{\mathbf{k}'}^{\alpha'}\cdot\mathbf{p}_{li})}{E_l-E_i+\hbar\omega'}\right]\int_0^t dt_1 e^{\frac{i}{\hbar}(E_f-E_i+\hbar\omega'-\hbar\omega)t_1} \tag{9.65}$$

In the limit $t\to\infty$, we can directly use the result of Eq. (9.19)

$$\left|\int_0^t dt_1 e^{\frac{i}{\hbar}(E_f-E_l+\hbar\omega'-\hbar\omega)t_1}\right|^2 = t2\pi\hbar\delta(E_f-E_i+\hbar\omega'-\hbar\omega) \tag{9.66}$$

to obtain the second order transition probability per unit time

$$w_{fi} = \frac{d}{dt}\left|C_{fi}^{(1)}(t) + C_{fi}^{(2)}(t)\right|^2$$

$$= \frac{2\pi}{\hbar}\frac{q^2\hbar}{2mV\sqrt{\omega\omega'}}\Bigg|\delta_{fi}\epsilon^{\alpha}_{\mathbf{k}}\cdot\epsilon^{\alpha'}_{\mathbf{k}'} - \frac{1}{m}\sum_l\left[\frac{(\epsilon^{\alpha'}_{\mathbf{k}'}\cdot\mathbf{p}_{fl})(\epsilon^{\alpha}_{\mathbf{k}}\cdot\mathbf{p}_{li})}{E_l - E_i - \hbar\omega}\right.$$
$$\left. + \frac{(\epsilon^{\alpha}_{\mathbf{k}}\cdot\mathbf{p}_{fl})(\epsilon^{\alpha'}_{\mathbf{k}'}\cdot\mathbf{p}_{li})}{E_l - E_i + \hbar\omega'}\right]\Bigg|^2 \delta(E_f - E_i + \hbar\omega' - \hbar\omega) \tag{9.67}$$

which is the well known Kramers-Heisenberg formula [177].

We can now proceed to calculate the transition probability for scattering of a photon ω' into a solid angle $d\Omega_{\mathbf{k}'}$

$$\begin{aligned} W_{fi}d\Omega_{\mathbf{k}'} &= w_{fi}\rho(E) \\ &= w_{fi}\frac{V\omega'^2}{(2\pi)^3\hbar c^3}d\Omega_{\mathbf{k}'} \\ &= \frac{q^4}{16\pi^2 V m^2 c^3}\left(\frac{\omega'}{\omega}\right)\Bigg|\delta_{fi}\epsilon^{\alpha}_{\mathbf{k}}\cdot\epsilon^{\alpha'}_{\mathbf{k}'} - \frac{1}{m}\sum_l\left[\frac{(\epsilon^{\alpha'}_{\mathbf{k}'}\cdot\mathbf{p}_{fl})(\epsilon^{\alpha}_{\mathbf{k}}\cdot\mathbf{p}_{li})}{E_l - E_i - \hbar\omega}\right. \\ &\quad \left. + \frac{(\epsilon^{\alpha}_{\mathbf{k}}\cdot\mathbf{p}_{fl})(\epsilon^{\alpha'}_{\mathbf{k}'}\cdot\mathbf{p}_{li})}{E_l - E_i + \hbar\omega'}\right]\Bigg|^2 d\Omega_{\mathbf{k}'} \end{aligned} \tag{9.68}$$

Dividing W_{fi} by the incoming photon flux c/V, we obtain the differential cross section for light scattering

$$\frac{d\sigma_{fi}}{d\Omega} = r_0^2\left(\frac{\omega'}{\omega}\right)\Bigg|\delta_{fi}\epsilon^{\alpha}_{\mathbf{k}}\cdot\epsilon^{\alpha'}_{\mathbf{k}'} - \frac{1}{m}\sum_l\left[\frac{(\epsilon^{\alpha'}_{\mathbf{k}'}\cdot\mathbf{p}_{fl})(\epsilon^{\alpha}_{\mathbf{k}}\cdot\mathbf{p}_{li})}{E_l - E_i - \hbar\omega}\right.$$
$$\left. + \frac{(\epsilon^{\alpha}_{\mathbf{k}}\cdot\mathbf{p}_{fl})(\epsilon^{\alpha'}_{\mathbf{k}'}\cdot\mathbf{p}_{li})}{E_l - E_i + \hbar\omega'}\right]\Bigg|^2 \tag{9.69}$$

where r_0 is defined as

$$r_0 = \frac{q^2}{4\pi mc^2}. \tag{9.70}$$

Raman scattering

An important technique in molecular spectroscopy is Raman scattering which is inelastic scattering of a photon off a target molecule. Figure 9.2 illustrates the Raman process. In Raman scattering, the first term in Eq. (9.69) does not contribute and is neglected to yield the Raman differential cross section

$$\boxed{\frac{d\sigma^{ram}_{fi}}{d\Omega} = \frac{r_0^2\omega'}{m^2\omega}\Bigg|\sum_l\left[\frac{(\epsilon^{\alpha'}_{\mathbf{k}'}\cdot\mathbf{p}_{fl})(\epsilon^{\alpha}_{\mathbf{k}}\cdot\mathbf{p}_{li})}{E_l - E_i - \hbar\omega} + \frac{(\epsilon^{\alpha}_{\mathbf{k}}\cdot\mathbf{p}_{fl})(\epsilon^{\alpha'}_{\mathbf{k}'}\cdot\mathbf{p}_{li})}{E_l - E_i + \hbar\omega'}\right]\Bigg|^2} \tag{9.71}$$

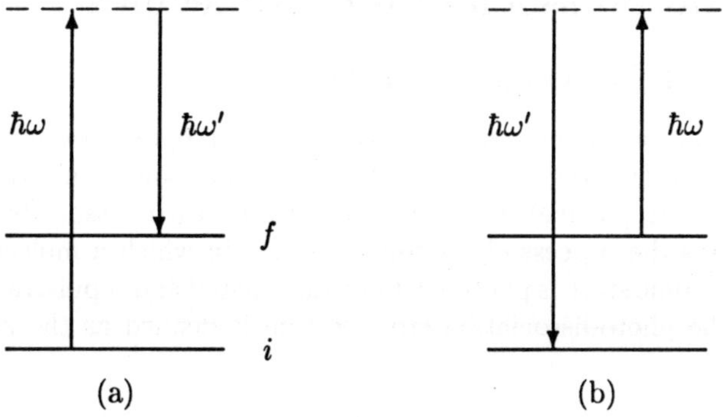

Figure 9.2: Illustration of (a) Stokes Raman Scattering and (b) Anti-Stokes Raman Scattering.

In resonance scattering, the scattering cross section is dominated by a resonance state I whose energy satisfies either the equation

$$E_I = E_i + \hbar\omega \tag{9.72}$$

or

$$E_I = E_i - \hbar\omega' \tag{9.73}$$

In such a resonance case, one needs to replace the energy of the virtual state E_l by the complex energy $E_I - i\frac{\Gamma_I}{2}$ in order to account for the damping effect. Thus Eq. (9.71) for resonance scattering becomes

$$\frac{d\sigma_{fi}^{raman}}{d\Omega} \simeq \frac{\omega\omega'^3}{16\pi^2 c^4}\left|\frac{(\epsilon_{\mathbf{k}'}^{\alpha'} \cdot \mathbf{P}_{fI})(\epsilon_{\mathbf{k}}^{\alpha} \cdot \mathbf{P}_{Ii})}{E_I - E_i - \hbar\omega - i\frac{\Gamma_I}{2}}\right|^2 \tag{9.74}$$

for the contribution of the first term and

$$\frac{d\sigma_{fi}^{raman}}{d\Omega} = \frac{\omega\omega'^3}{16\pi^2 c^4}\left|\frac{(\epsilon_{\mathbf{k}}^{\alpha} \cdot \mathbf{P}_{fI})(\epsilon_{\mathbf{k}'}^{\alpha'} \cdot \mathbf{P}_{Ii})}{E_I - E_i + \hbar\omega' - i\frac{\Gamma_I}{2}}\right|^2 \tag{9.75}$$

for the contribution of the second term. Here the matrix element of the momentum operator $\mathbf{p}$ has been replaced by that of the dipole moment via Eq. (9.53).

9.4 Photodissociation of Molecules

9.4.1 Half Collision Dynamics

Photodissociation of molecules involves a direct optical excitation (usually in the UV or VUV region) from the ground (bound) electronic state to an excited (repulsive) electronic state which then dissociates. Figure 9.3 illustrates the process of photodissociation in which a molecule in the ground electronic state is photoexcited to an excited and repulsive electronic surface. The photodissociation cross section is defined as the absorption

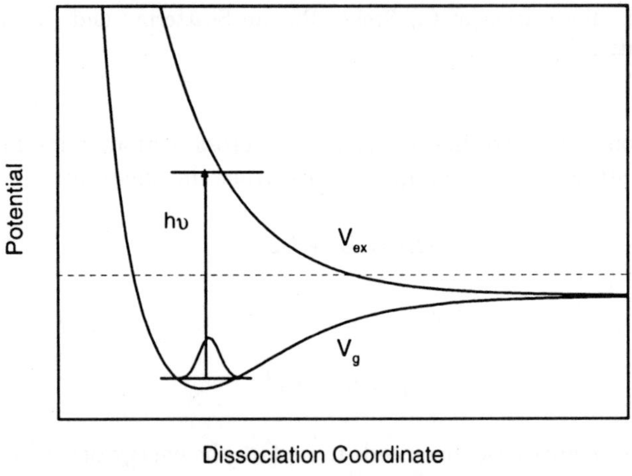

Figure 9.3: Illustration for direct photodissociation of molecules.

probability per unit time given in Eq. (9.56) divided by the photon flux $F_{photo} = n_{\mathbf{k},\alpha}c$

$$\boxed{\sigma_{fi} = \frac{\pi\omega}{c}|\epsilon_{\mathbf{k}}^{\alpha} \cdot \mathbf{P}_{fi}|^2 \rho(E_f)} \tag{9.76}$$

If we can write the molecular wavefunction as a direct product of the nuclear wavefunction ψ and the electron wavefunction ϕ^e for the initial and final states, as in the Born-Oppenheimer approximation, we can replace the matrix element of the dipole moment by the matrix element of the dipole function of the nuclear coordinates

$$\mathbf{P}_{fi} = <\psi_f^{(-)}|\mu_{fi}|\psi_i> \tag{9.77}$$

where μ_{fi} is the transition dipole function defined as

$$<\phi_f^e|\mathbf{P}|\phi_i^e> \tag{9.78}$$

and is a function of nuclear coordinates. Equation (9.76) can therefore be rewritten in the more familiar form in the photodissociation literature

$$\sigma_{fi} = \frac{\pi\omega}{c}| <\psi_f^{(-)}|\epsilon_{\mathbf{k}}^{\alpha} \cdot \mu_{fi}|\psi_i> |^2 \rho(E_f) \tag{9.79}$$

The fundamental quantity in Eq. (9.79) is the generalized Franck-Condon (FC) factor

$$A_f = | <\psi_f^{(-)}|\chi> |^2 \tag{9.80}$$

where the final state $\psi_f^{(-)}$ is an outgoing scattering state and χ is defined as

$$|\chi>= \epsilon_{\mathbf{k}}^{\alpha} \cdot \mu_{fi}|\psi_i> \tag{9.81}$$

where ψ_i is the initial wavefunction. In order to calculate this FC factor, one needs to calculate both the bound wavefunction $|\psi_i>$ and the scattering wavefunction $\psi_f^{(-)}$. The calculation of the latter is the main source of difficulty. The dissociation process starts from the interaction region on the repulsive potential surface where the scattering wavefunction on the repulsive surface $\psi_f^{(-)}$ has overlap with the bound state wavefunction ψ_i on the ground potential surface. Since in photodissociation, the dynamical process is initialized from the interaction region, it is also called half-collision dynamics to distinguish it from full-collision dynamics in bimolecular collisions in which the process starts from the asymptotic region and ends in the asymptotic region.

9.4.2 Theoretical Treatment

The standard TD approach to photodissociation is to simply convert the FC factor into a TD expression [178]

$$\begin{aligned} A_f &= <\psi_f^{(-)}|\chi> \\ &= <\lim_{t\to\infty} e^{\frac{i}{\hbar}Ht}e^{-\frac{i}{\hbar}H_0t}\Phi_f|\chi> \\ &= \lim_{t\to\infty} <\Phi_f|\Psi_I(t)> \end{aligned} \tag{9.82}$$

where the IR (interaction representation) wavefunction $\Psi_I(t)$, as defined in Eq. (6.4), is given by

$$\Psi_I(t) = e^{\frac{i}{\hbar}H_0 t} e^{-\frac{i}{\hbar}Ht} |\chi> \tag{9.83}$$

The total cross section for photodissociation is obtained by summing over all final states

$$\begin{aligned}\sigma_{tot} &= \frac{\pi\omega}{c}\sum_f |A_f|^2 \rho(E) \\ &= \frac{\pi\omega}{c} <\chi|\delta(E-H)|\chi> \\ &= \boxed{\frac{\omega}{2\hbar c}\int_{-\infty}^{\infty} e^{\frac{i}{\hbar}Et} <\chi|e^{-\frac{i}{\hbar}Ht}|\chi>\, dt}\end{aligned} \tag{9.84}$$

where the completeness relation

$$\sum_f |\psi_{fE}^{(-)}> \rho(E) <\psi_{fE}^{(-)}| = \delta(E-H) \tag{9.85}$$

has been used and bound states, if any, have been projected out from χ. Thus the total absorption cross section for photodissociation is given by the Fourier transform of the autocorrelation function $A(t) = <\chi|\chi(t)>$.

In the TD calculation for direct photodissociation, the wavepacket propagation is carried out on a repulsive potential energy surface, which does not generally support any bound states and the propagation usually takes only a short time for direct photodissociation. Therefore numerical calculation of cross sections for direct photodissociation is generally quite straightforward, at least in principle. However, complication arises when more than one fragment can result from the photodissociation. In such cases, the scattering wavefunction $\psi_{fE}^{(-1)}$ is a reactive scattering wavefunction and its calculation requires the theoretical machinery for reactive scattering. However, if we are only interested in the absorption spectrum or the total dissociation cross section without regard to final state distribution, the theoretical treatment is simplified. According to Eq. (9.84) the absorption spectrum is simply given by a Fourier transform of the autocorrelation function $A(t) = <\chi|\chi(t)>$ which has no explicit reference to specific asymptotic states of the fragments. Perhaps more important is the fact that the absorption spectrum is usually determined by a short range interaction on the repulsive potential energy surface. This is because the autocorrelation function $<\chi|\chi(t)>$ vanishes if $\chi(t)$ travels outside the interaction region where $\chi(0)$ is nonzero, provided no recurrence is present.

If the product state distribution from photodissociation is needed, one can use Eq. (9.82) to compute partial cross sections. In such cases, one can propagate the time-dependent wavefunction $\Psi_I(t)$ until the interaction is over and then project $\Psi_I(t)$ onto specific asymptotic states of the fragment Φ_{fE}. Since only a finite grid is used in the numerical TD calculation of the wavefunction, there will be an artificial boundary reflection of the Schrödinger wavefunction from the end of the numerical grid. Thus one has to use a large enough grid to accommodate the wavefunction. It was shown that by using the interaction representation, it is possible to alleviate the boundary reflection problem or even eliminate it completely in numerical calculations [97, 179]. This is possible because the IR wavefunction is stationary in the limit of $t = \infty$ while the Schrödinger wavefunction is not.

For many direct photodissociation problems, the dynamics is determined by relatively short time interactions in which the molecules in question quickly fall apart due to highly repulsive forces on the excited potential surface. For such direct dissociation problems, the TD wavepacket calculation is ideal because one only needs to propagate the wavefunction for a very short time in a small region of configuration space. A detailed book has been written by Schinke [180] on the topic of photodissociation dynamics of small molecules which is a good source for more information on photodissociation of molecules.

9.4.3 Dissociation of H_2O

Dissociation of water following photoexcitation to its electronically excited states has been well studied both experimentally and theoretically. In particular, the photodissociation of water from the first excited $\widetilde{A}$ state

$$H_2O(\widetilde{X} + \hbar\nu \longrightarrow H_2O(\widetilde{A}) \longrightarrow H(^2S) + OH(X^2\Pi) \tag{9.86}$$

provides a prototypical model for complete theoretical analysis of photofragmentation dynamics of triatomic molecules. The dissociation dynamics of water on the $\widetilde{A}$ surface is relatively simple and involves short time dynamics on a repulsive potential energy surface. Thus it is ideal to carry out the dissociation dynamics calculation using the time-dependent wavepacket approach because it involves only a short time propagation of the wavepacket.

To illustrate the main feature of photodissociation dynamics for water, we show some results of a TD wavepacket calculation for the photodissociation of H_2O in the first absorption band of $\widetilde{A}$. In the dynamics calculation presented here, the ground potential energy surface of Jensen [181] is used for the bound state calculation of water, and the potential surface of Engel

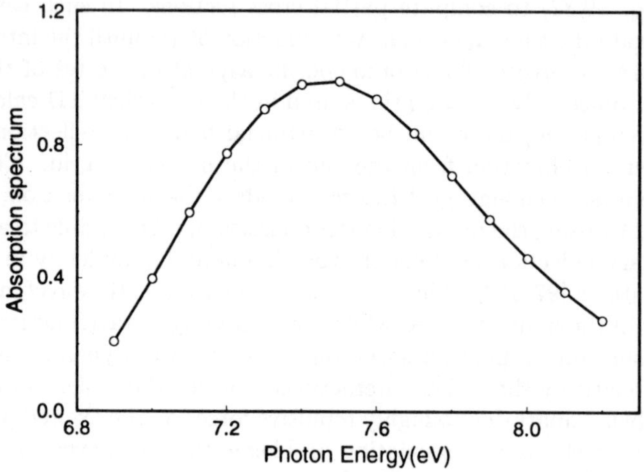

Figure 9.4: Total absorption spectrum of photodissociation of water in the first excited $\tilde{A}$ state.

et al. [182] is used for the dissociation dynamics calculation on the excited $\tilde{A}$ state. For simplicity, the transition dipole function is approximated by a constant. Since the photodissociation dynamics of water on the $\tilde{A}$ state is determined by a short time repulsive interaction, the three-dimensional dynamics calculation is relatively straightforward. Figure 9.4 shows the calculated total absorption spectrum of water in the ground rovibrational state. As shown in the figure, the absorption spectrum of water shows a broad and smooth feature, reflecting a typical short repulsive dynamics on the excited $\tilde{A}$ state.

Figure 9.5 shows the rotational state distribution of OH from the dissociation of H_2O. The OH rotational state distribution is essentially independent of the stretching excitation of H_2O as shown in the figure. This is because the stretching motion is essentially decoupled from the bending or rotation of OH, as is discussed very thoroughly in Ref. [182].

9.4.4 Dissociation of H_2O_2

Background Information

The photodissociation of polyatomic molecules beyond triatomics presents new dynamical features that are absent in the photodissociation of triatomic

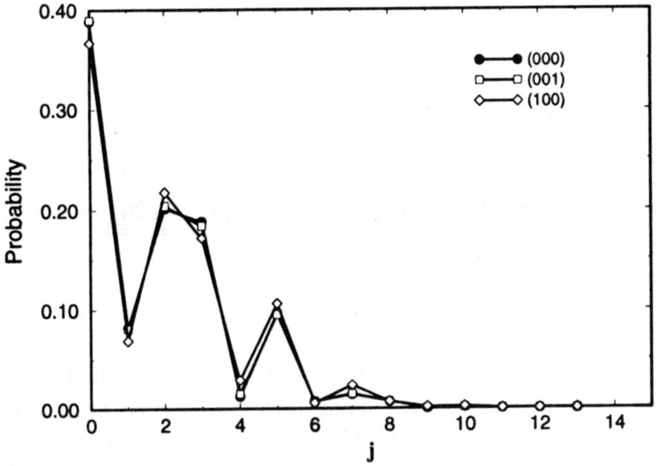

Figure 9.5: Rotational state distributions of OH from photodissociation of H_2O in the ground state (000) and two stretching excited states (100) and (001).

molecules. For example, a polyatomic molecule can fragment into more than one molecular species and there are correlations among the internal motions of the fragments. The photodissociation dynamics of H_2O_2 has been extensively studied by several experimental groups in recent years. The measurements in the UV photodissociation of H_2O_2 include scalar properties, e.g. the rotational state distribution of OH, as well as vector properties, e.g. the correlation between $\mathbf{j}_{oh}$ and recoil velocity $\mathbf{v}$ at several photolysis wavelengths (193, 248, and 266 nm in both beam and bulk). Experiments have shown that the photodissociation of H_2O_2 produces OH radicals almost entirely in the ground vibrational state, although about 3 eV excess energy is available for partitioning among all the degrees of freedom. This indicates that the coupling between the high-frequency OH vibration and the recoil translation of the fragments is very weak, which provides an ideal condition for freezing the OH bonds in the dynamics calculation. In addition, it is found that most of the dissociation energy is released as the relative translational energy between the OH fragments, while only a small fraction goes into OH rotations. These experimental evidences suggest that the UV photodissociation of H_2O_2 is characterized by a short-time direct process resulting primarily from the strong repulsion between the OH fragments.

Table 9.1: Potential parameters for the excited Ã and B̃ surfaces of H_2O_2 (atomic units).

Surface	α	β	γ	η
Ã	13.05	1.57	0.42	0.109
B̃	13.05	1.57	-0.24	0.026

Results of theoretical calculation

The time-dependent method described in Sec. 9.4 is applied to the photodissociation problem of H_2O_2. There are two excited states $\tilde{A}$ and $\tilde{B}$ that are directly accessible through photoexcitation at 248 nm. In the theoretical dynamics calculations, dissociation on each surface is calculated separately and no coupling between the two surfaces is considered. The repulsive surfaces are modeled by a simple 2D function that depends only on the inter-atomic distance R_{OO} and the torsional angle ϕ

$$V_{ex} = \alpha \exp(-\beta R_{OO})(1 + \gamma \cos\phi + \eta \cos^2\phi) \tag{9.87}$$

and the parameters for both surfaces are given in Table 9.1. The details of the numerical calculation can be found in [137, 183] and are omitted here.

Because there are two molecular fragments resulting from the photodissociation of H_2O_2, we can investigate the correlation relations between the rotational angular momenta of the two diatomic fragments. Let j_1 and j_2 denote, respectively the rotational quantum number of the two OH fragments. Fig. 9.6 shows the distribution $P_{|j_1-j_2|}$ which is defined as the probability distribution of the variable $|j_1 - j_2|$. The $P_{|j_1-j_2|}$ can be regarded as some sort of "coarse grained" final state distribution P_{j_1,j_2}, although containing less detailed information but showing a clear picture of the scalar correlation between j_1 and j_2. As shown in Fig. 9.6, the highest probability is for the two OH fragments to have rotation quantum numbers differing by one quantum (30% probability) upon 248 nm photolysis. The probability decreases as the rotation quantum number difference between the two OH diatoms increases, with states of $|j_1 - j_2| \leq 2$ making up about 65% of the products. This clearly reflects the fact that the initial geometric positions of the two OH diatoms are more or less dynamically equivalent.

It is also interesting to examine the vector correlation relation between rotational angular momenta $\mathbf{j}_1$ and $\mathbf{j}_2$ of the OH fragments. Since the total angular momentum J is zero, the projection of the angular momentum $\mathbf{j}_{12}$ on the body-fixed axis must also be zero; therefore the $\mathbf{j}_{12}$ vector must be perpendicular to the recoil velocity of the OH fragments as shown in Fig. 9.7. If we use the most probable values to represent the vector lengths of $\mathbf{j}_1$, $\mathbf{j}_2$, and $\mathbf{j}_{12}$ in Fig. 9.7, we can calculate the angle α between $\mathbf{j}_1$ and $\mathbf{v}$ (relative recoil velocity). Here j_{12} is the angular momentum of $\mathbf{j}_{12} = \mathbf{j}_1 + \mathbf{j}_2$. Since the angle α gives a measurement of the correlation between $\mathbf{j}_1$ and $\mathbf{v}$ and reflects the strength of the torsional torque, the measurement of j_{12} also gives similar information regarding the vector correlation of $\mathbf{j}_1$ and $\mathbf{v}$. A small α (small j_{12}) reflects a strong torsional torque and a weak bending torque, and vice versa. Using the most probable values to represent the lengths of vectors in Fig. 9.7, we find $\alpha \sim 40^0$ on the $\tilde{A}$ surface and $\alpha \sim 20^0$ on the $\tilde{B}$ surface (assuming $|\mathbf{j}_1| = |\mathbf{j}_2|$). This argument is only qualitative, however, because $\mathbf{j}_1$, $\mathbf{j}_2$, and $\mathbf{p}$ do not generally lie in the same plane. The vectors $\mathbf{j}_1$ or $\mathbf{j}_2$ are only confined to the surface of a cone with fixed angle α when both j_1 and m_1 are determined.

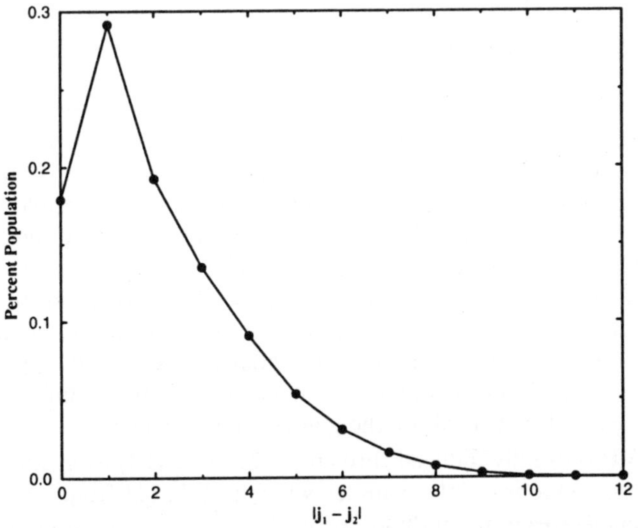

Figure 9.6: Distribution of $|j_1 - j_2|$ at 248 nm from photodissociation of H_2O_2. Here j_1 and j_2 are rotational quantum numbers of two OH fragments. The result is insensitive to the $\widetilde{A}$ or $\widetilde{B}$ surface.

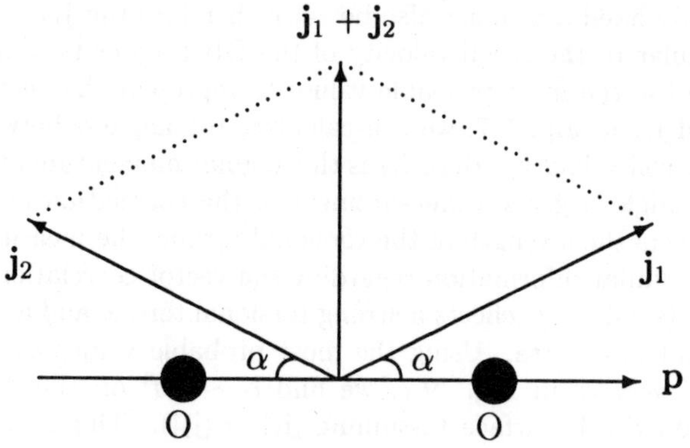

Figure 9.7: The vector correlation relation of $\mathbf{j}_1$ and $\mathbf{j}_2$ of the two HO fragments following the photodissociation of H_2O_2 at fixed total angular momentum J=0.

9.5 Vibrational Predissociation of Molecules

9.5.1 Introduction

For relatively small molecules, experimental evidence suggests that vibrational predissociation (VP) is primarily responsible for the observed linewidths, as opposed to intramolecular vibrational relaxation (IVR). Recent advances in infrared laser techniques have made it now possible to experimentally prepare a well-defined excited rovibrational state and to probe individual quantum states of the fragments resulting from laser excitation. This kind of state-to-state experiment demands accurate quantum mechanical calculations for comparison in order to elucidate the underlying dynamics and the intra- and inter-molecular forces. The long predissociation lifetime signals weak coupling between the initial and final vibrational states which makes it ideal for the golden rule treatment.

The VP is usually induced through an infrared (IR) excitation of one of the vibrational modes of the complex which is different from UV excitation of electronic states in photodissociation. In addition, the photodissociation in the latter case occurs on the excited PES which is often purely repulsive and involves only short time dynamics. However, VP occurs on the ground electronic PES which is attractive and often supports many bound and quasi-bound states. Figure 9.8 illustrates a VP process. Most VP dynamics calculations are treated by TI scattering methods. Recently, a TD golden rule treatment has been reported [184, 185]. Since the TD golden rule method is an efficient method for weakly bound systems, we present in the following the theoretical derivation and treatment for vibrational predissociation within the golden rule approximation.

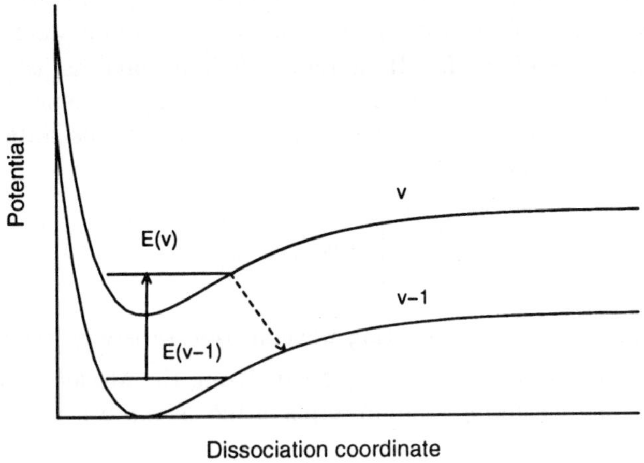

Figure 9.8: Process for vibrational predissociation. The system is photo promoted to the vibrational state v, which then dissociates via vibrational coupling to the v-1 state.

There are a number of well known factors that are very important to the VP lifetime. The first and the most important factor is the vibrational coupling potential, V_{fi}, between the initial (excited) and the final (de-excited) vibrational states. Obviously, the stronger the vibrational coupling, the shorter the lifetime. The second factor is the amount of total excess energy released during the dissociation. This is given by $E_R = E_{ex} - E_b$, where E_{ex} is the optical excitation energy and E_b is the initial bound state energy. Generally speaking, the larger the E_R, the longer the lifetime. The third factor is E_T, the amount of kinetic energy released which is equal to the difference between the total excess energy E_R and the internal (rovibrational) energy E_{int} to be shared by the fragments. If the total excess energy is the same, the process that releases less kinetic energy will generally give a shorter lifetime. This is known as the "momentum gap rule" [186], which, although not rigorous, is quite useful in a qualitative way.

9.5.2 Time-dependent Golden Rule Treatment

In this section, we give a theoretical derivation of the TD golden rule formula for treating vibrational predissociation problems. Before infrared or vibrational excitation, the molecule is in a given vibrational state. After excitation, the molecule is promoted to an excited vibrational state which serves as the initial state $|\chi_i>$ in our theoretical treatment. Rigorously speaking, this excited state is a resonance or quasi-bound state. But from a dynamical point of view, it can be treated like a bound state. The continuum eigenstate of the full Hamiltonian H is denoted as $|\psi_{fE}^{(-)}>$ whose quantum numbers are given by the composite index f. The probability per unit energy that the system will end up in the continuous state f is given by the overlap integral

$$\frac{dP_{fi}}{dE} = \left| <\psi_{fE}^{(-)}|\chi_i> \right|^2 \tag{9.88}$$

where $|\psi_{fE}^{(-)}>$ is the full scattering wavefunction satisfying the stationary Hamiltonian equation $(E-H)|\psi_{fE}^{(-)}>=0$. Using the Mϕller operator, one can express the scattering state in terms of the asymptotic state,

$$\begin{aligned} |\psi_{fE}^{(-)}> &= \Omega_{-}|\phi_{fE}^{(-)}> \\ &= \lim_{t\to\infty} e^{\frac{i}{\hbar}Ht} e^{-\frac{i}{\hbar}H_f t}|\phi_{fE}^{(-)}> \end{aligned} \tag{9.89}$$

where H_f is the vibrational adiabatic Hamiltonian for the vibrational state v_f, and $|\phi_{fE}^{(-)}>$ is the continuum eigenstate of H_f satisfying the Schrödinger equation, $(E-H_f)|\phi_{fE}^{(-)}>=0$. Utilizing Eq. (9.89), one can rewrite Eq. (9.88) as

$$\frac{dP_{fi}}{dE} = \lim_{t\to\infty} \left| <\phi_{fE}^{(-)}|\Psi_I(t)> \right|^2 \tag{9.90}$$

where the interaction representation wavefunction is defined as

$$|\Psi_I(t)>= e^{\frac{i}{\hbar}H_f t} e^{-\frac{i}{\hbar}Ht}|\chi_i> \tag{9.91}$$

In the rigorous full quantum treatment [32], one solves for the full time-dependent wavefunction in Eq. (9.91) which can be computationally expensive. In the golden rule treatment, $\Psi_I(t=\infty)$ is obtained by a first order approximation as follows,

$$\begin{aligned}|\Psi_I(\infty)\rangle &= |\chi_i\rangle - i\int_0^\infty dt' e^{\frac{i}{\hbar}H_f t'} V e^{-\frac{i}{\hbar}Ht'}|\chi_i\rangle \\ &\simeq |\chi_i\rangle - i\int_0^\infty dt' e^{\frac{i}{\hbar}H_f t'} V e^{-\frac{i}{\hbar}(E_R - i\Gamma/2)t'}|\chi_i\rangle \\ &= |\chi_i\rangle + (H_f - E_R + i\Gamma/2)^{-1} V|\chi_i\rangle, \end{aligned} \tag{9.92}$$

where $V = H - H_f$ is the coupling potential that induces the dissociation and E_R and Γ are, respectively, the resonance energy and the decay width. The projection of $|\Psi_I(\infty)\rangle$ onto the scattering state $|\phi_{fE}^{(-)}\rangle$ gives

$$\langle\phi_{fE}^{(-)}|\Psi_I(\infty)\rangle \simeq (E_f - E_R + i\Gamma/2)^{-1} \langle\phi_{fE}^{(-)}|V|\chi_i\rangle\,. \tag{9.93}$$

By utilizing the unitary condition

$$\sum_f P_{fi} = \sum_f \int dE_f \left| \langle\phi_{fE}^{(-)}|\Psi_I(\infty)\rangle \right|^2 = 1, \tag{9.94}$$

we arrive at the relation

$$\begin{aligned} 1 &= \sum_f \int dE_f \frac{|\langle\phi_{fE}^{(-)}|V|\chi_i\rangle|^2}{|E_f - E_R + i\Gamma/2|^2} \\ &\simeq \sum_f |\langle\phi_{fE}^{(-)}|V|\chi_i\rangle|^2 \int dE_f \frac{1}{|E_f - E_R + i\Gamma/2|^2} \\ &= \frac{\pi}{\Gamma/2}\sum_f \left| \langle\phi_{fE_R}^{(-)}|V|\chi\rangle \right|^2 . \end{aligned} \tag{9.95}$$

where the result of contour integration has been used to evaluate the integral. Equation (9.95) is the desired time-independent golden rule result for the partial decay width Γ_f

$$\boxed{\Gamma_f = 2\pi\left| \langle\phi_{fE}^{(-)}|V|\chi_i\rangle \right|^2} \tag{9.96}$$

The total decay width Γ or FWHM (full width at half maximum) is simply the sum of all partial widths $\Gamma = \sum_f \Gamma_f$ and the lifetime is defined as the inverse of the total decay width

$$\tau = \frac{1}{\Gamma}. \tag{9.97}$$

If we invoke the Mϕller operator again as in Eq. (9.89) for the scattering state $\phi_{fE}^{(-)}$, Eq. (9.96) becomes

$$\Gamma_f = 2\pi \left| <\phi_{fE}^0|\Phi_I(t=\infty)> \right|^2, \tag{9.98}$$

where the TD wavefunction in the interaction representation is defined as

$$|\Phi_I(t)>= e^{\frac{i}{\hbar}H_0 t} e^{-\frac{i}{\hbar}H_f t}|\phi_c> \tag{9.99}$$

Here the initial wavefunction is defined by the continuum portion of $|\phi_c>= \hat{P}_c|\chi_i>$ where $\hat{P}_c$ is the projection operator. The operator $\hat{P}_c$ can be constructed by

$$\hat{P}_c = I - \sum_b |E_b><E_b| \tag{9.100}$$

where the $|E_b>$ are bound states of H_f. The state $|\phi_{fE}^0>$ is the "free" scattering state satisfying the asymptotic equation

$$(E - H_0)|\phi_{fE}^0>= 0, \tag{9.101}$$

and the "residual" potential is given by

$$V_f = H_f - H_0. \tag{9.102}$$

If we make vibrational adiabatic approximation, i.e., we write $|\chi_i>= |v_i>|\bar{\chi}_i>$ and $|\phi_{fE}^0>= |v_f>|\bar{\phi}_{fE}^0>$ where "barred" states have one less vibrational degree of freedom. we can factor out the vibrational state and obtain a simpler equation with one less degree of freedom

$$\boxed{\Gamma_f = 2\pi \left| <\bar{\phi}_{fE}^0|\bar{\Phi}_I(t=\infty)> \right|^2,} \tag{9.103}$$

where $\bar{\Phi}_I(t=0) = \hat{P}V_{fi}\chi_i$ and $V_{fi} =<v_f|V|v_i>$. Thus the vibrational degree of freedom is eliminated from the dynamics calculation.

9.5.3 $\mathrm{HeCl_2}(v) \rightarrow \mathrm{He} + \mathrm{Cl_2}(v-1)$

Equations (9.99) and (9.103) have been applied to the calculation of VP dynamics for $HeCl_2$ [185]. For $HeCl_2$ with total angular momentum J=0, the

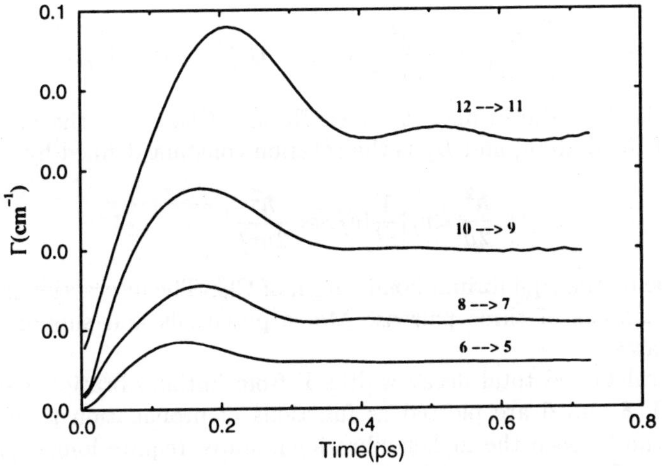

Figure 9.9: Decay width Γ for vibrational predissociation of $HeCl_2(v, n = 0)$ as a function of propagation time for initial states v=6, 8, 10 and 12.

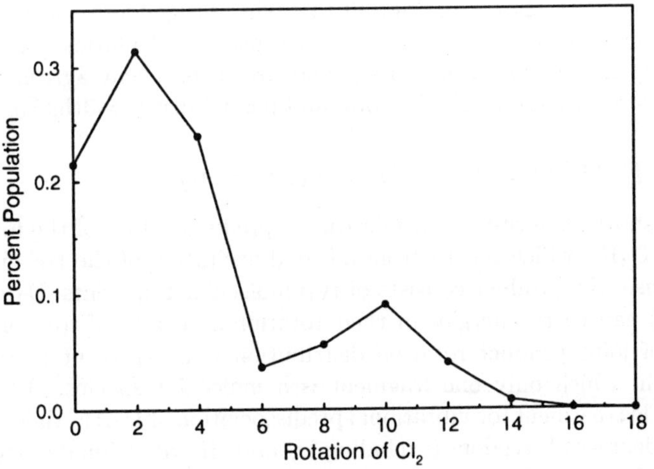

Figure 9.10: Rotational state distribution of Cl_2 following dissociation of $HeCl_2(v = 8, n = 0) \rightarrow He + Cl(v = 7)$.

vibrationally adiabatic Hamiltonian can be written in Jacobi coordinates (cf. Fig. 3.1) as

$$H_f = -\frac{\hbar^2}{2\mu}\frac{\partial^2}{\partial R^2} + + \frac{\mathbf{j}^2}{2\mu R^2} + \epsilon_{v_f} + B_f \mathbf{j}^2 + <v_f|V|v_f> \tag{9.104}$$

where μ is the reduced mass between He and Cl_2, ϵ_{v_f} is the vibrational energy of the diatom, and B_f is the rotation constant defined by

$$\frac{\hbar^2}{2\mu} <v_f|\frac{1}{r^2}|v_f> \approx \frac{\hbar^2}{2\mu r_e^2} \tag{9.105}$$

with r_e being the equilibrium bond length of Cl_2. The interaction potential $V(R,\gamma)$ consists of three pairwise Morse potentials and supports a few bound states.

The calculated total decay widths Γ from initial vibrational states of v=12, 10, 8 and 6 are plotted as functions of propagation time in Fig. 9.9. As can be seen the higher vibrational states require longer propagation times. This can be explained by the different amount of excess energy released from different initial vibrational states due to anharmonicity of vibrations. The smaller energy release (high initial v) requires longer propagation time than larger energy release (low initial v). The decay widths all converge with less than 1 ps propagation time. The rotational state distribution of Cl_2 from the dissociation of $HeCl_2(v = 8, n = 0)$ to He + $Cl_2(v = 7)$ is shown in Fig. 9.10. The TD golden rule results are in excellent agreement with the full wave packet calculation [32] and the time-independent calculation [187], and are in excellent agreement with experimental measurements for total and partial decay widths [187].

9.5.4 $D_2HF(v = 1) \rightarrow D_2 + HF(v = 0)$

We now show some results for vibrational predissociation of the four-atom complex D_2HF which results from infrared excitation of the HF vibration. In this case, the product consists of two molecular fragments, D_2 and HF, and both can carry energies in their rotational degrees of freedom. This feature of joint product rotation distribution is not present in triatomic systems in which only one fragment is a molecular species. In the TD golden rule treatment for vibrational predissociation of D_2HF, there are four internal degrees of freedom (excluding D_2 and HF vibrational coordinates) shown in Fig. 3.2. The Hamiltonian H_f can be written as

$$H_f = -\frac{\hbar^2}{2\mu}\frac{\partial^2}{\partial R^2} + \frac{\mathbf{L}^2}{2\mu R^2} + B_1\mathbf{j}_1^2 + B_2\mathbf{j}_2^2 \tag{9.106}$$

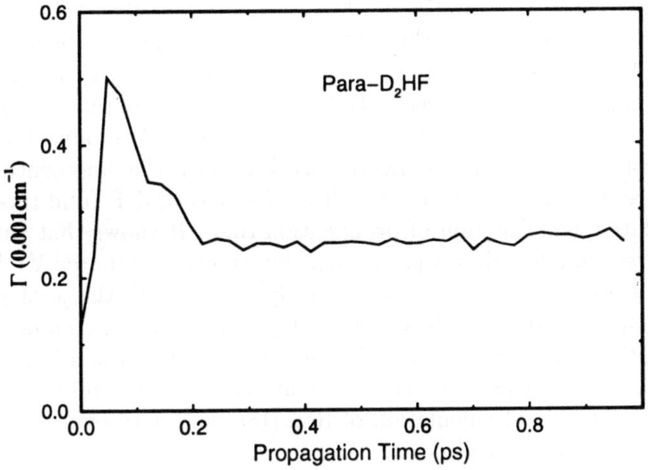

Figure 9.11: Decay width Γ for vibrational predissociation in para-$D_2(v = 0)HF(v = 1) \rightarrow D_2(v = 1) + HF(v = 0)$ (odd rotation of D_2) as a function of propagation time.

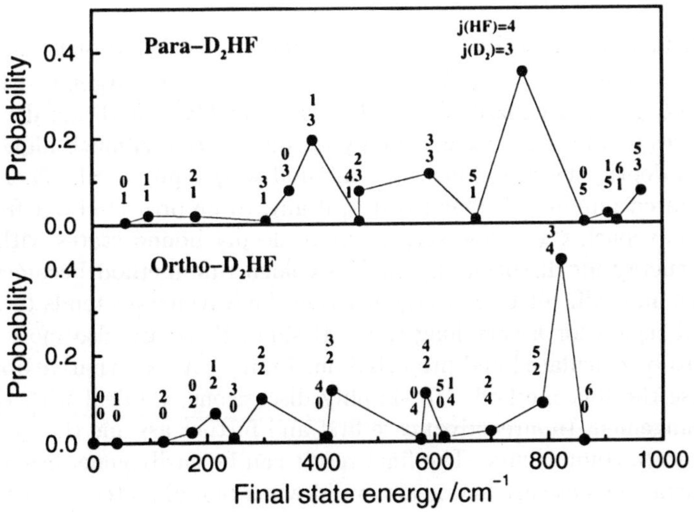

Figure 9.12: Final rotational state distributions for para- and ortho-$D_2HF(v = 1) \rightarrow D_2(v = 0, j_1) + HF(v = 0, j_2)$. The upper and lower numbers refer to the rotational quantum numbers of HF and D_2, respectively.

where μ is the reduced mass between D_2 and HF, **L** is the orbital angular momentum operator, $\mathbf{j}_1$ and $\mathbf{j}_2$ are the rotational angular momenta of D_2 and HF, and B_1 and B_2 are their corresponding rotational constants.

The TD golden-rule calculation is carried out using the ***ab initio*** PES of Ref. [188]. The VP process in D_2HF(v=1) is dominated by the V→V energy transfer from the HF(v=1) mode to the D_2(v=1) mode which is the most favorable decay channel and is consistent with the "momentum gap" rule. Figure 9.11 shows the decay width Γ for para-D_2HF (odd rotation of D_2) calculated as a function of propagation time. It shows that the decay width converges in less than 1 ps propagation and the computed VP lifetime is about 1.5 ns. In addition to the total decay width Γ, the joint product rotation distributions of both fragments D_2 and HF are also shown in Fig. 9.12 for both para- and ortho-D_2HF where the rotation quantum numbers of both fragments are determined simultaneously. These results are in good agreement with the TI calculation of Ref. [188] and with experiment [189] as discussed in more detail in Ref. [190].

9.6 Flux Method for Photofragmentation

9.6.1 Single Arrangement

The fundamental quantity in a photofragmentation calculation is the Franck-Condon factor $F_n = | <\psi_{nE}^{(-)}|\chi> |^2$ where $\psi_{nE}^{(-)}$ is an outgoing scattering state and χ is a localized wavepacket. For weakly bound van der Waals complexes in which the excess energy for dissociation is much higher than the well depth, the TD golden rule method works quite well. There are a very limited number of bound states, if any, to be projected out from the initial wavepacket χ. However, for more deeply bound states with small excess energy for dissociation, the TD golden rule method becomes cumbersome and difficult to converge because the wavepacket tends to stay in the well region for a very long time. Besides, there are also more bound states to be calculated and projected out from χ. An alternative approach is to use the flux method. To simplify discussions, we deal with the single arrangement (nonreactive) case first and further assume that χ has no bound state components. The final result can be easily generalized to the multi-arrangement case and is also valid when bound states are present as long as the surface for flux evaluation is located outside the bound state region.

Under the above assumptions, $\psi_{nE}^{(-)}$ forms a complete basis set and we can expand $|\chi>$ in this basis,

$$
\begin{aligned}
|\chi> &= \sum_n \int_0^\infty dE |\psi_{nE}^{(-)}><\psi_{nE}^{(-)}|\chi> \\
&= \sum_n \int_0^\infty dE |\psi_{nE}^{(-)}> a_n(E)
\end{aligned} \tag{9.107}
$$

where $|\psi_{nE}^{(-)}>$ is energy-normalized, i.e.,

$$
<\psi_{mE'}^{(-)}|\psi_{nE}^{(-)}> = \delta_{mn}\delta(E'-E) \tag{9.108}
$$

and satisfies the asymptotic boundary condition,

$$
\psi_{nE}^{(-)} \stackrel{R\to\infty}{\longrightarrow} \sqrt{\frac{\mu}{2\pi\hbar^2}} \left[-\frac{e^{ik_n R}}{\sqrt{k_n}}|\phi_n> + \sum_m S_{mn}^* \frac{e^{-ik_m R}}{\sqrt{k_m}}|\phi_m> \right], \tag{9.109}
$$

where $|\phi_n>$ are channel basis functions and S_{mn} are S matrix elements. It is useful to point out that if χ is real, one can use either ψ_{nE}^- or ψ_{nE}^+ to calculate the F-C factor.

If we define a stationary state

$$
\begin{aligned}
|A> &= 2\pi\delta(E-H)|\chi> \\
&= 2\pi \sum_n |\psi_{nE}^{(-)}> a_n(E)
\end{aligned} \tag{9.110}
$$

and use the definition of the Green's function,

$$
G^\pm = G^P \mp i\pi\delta(E-H), \tag{9.111}
$$

we can write

$$
|A> = |A^+> + |A^-> \tag{9.112}
$$

where

$$
\boxed{|A^\pm> = \pm i G^\pm |\chi> .} \tag{9.113}
$$

From the above definitions for the state $|A>$, the asymptotic boundary conditions for $|A^\pm>$ are seen to be

$$
A^+(R) \stackrel{R\to\infty}{\longrightarrow} -i\sqrt{\frac{2\pi\mu}{\hbar^2}} \sum_n \frac{e^{ik_n R}}{\sqrt{k_n}} |\phi_n> a_n(E) \tag{9.114}
$$

and

$$A^-(R) \stackrel{R\to\infty}{\longrightarrow} i\sqrt{\frac{2\pi\mu}{\hbar^2}}\sum_m \frac{e^{-ik_m R}}{\sqrt{k_m}}|\phi_m> b_m(E) \tag{9.115}$$

where the coefficient $b_m(E)$ is defined as

$$b_m(E) = -\sum_n S^*_{mn} a_n(E). \tag{9.116}$$

If χ is real, as it is in most cases, A will also be real and the relation $A^+(R) = A^{-*}(R)$ must hold. Thus, the coefficients $a_n(E)$ and $b_n(E)$ must be the complex conjugates of each other, i.e.,

$$a_n(E) = b^*_n(E) = -\sum_m S_{nm} a^*_m(E). \tag{9.117}$$

From Eq. (9.114), one can easily obtain the needed coefficient $a_n(E)$ at any fixed large distance R_∞ as

$$a_n(E) = i\sqrt{\frac{\hbar^2 k_n}{2\pi\mu}} e^{-ik_n R_\infty} A^+_n(R_\infty) \tag{9.118}$$

where $A^+_n(R) =<R| <\phi_n|A^+>$. Equation (9.118) provides a convenient means for calculating the partial decay width in photofragmentation dynamics. The calculation of $A^+_n(R)$ has to be carried out at a large distance R_∞ in order for the plane wave representation to be valid for A^+_n. Of course, one could also replace the plane wave by a distorted wave such as a Ricatti-Bessel function when the centrifugal potential is present. Since the partial decay width or product state distribution is given by the absolute square of $a_n(E)$, a more efficient approach is to evaluate the flux of $|A^+>$ at a distance beyond which only elastic scattering is present. From the definition for $|A^+>$ in Eq. (9.113), it is easy to see that $|A^+>$ satisfies the inhomogeneous Schrödinger equation,

$$(E - H)|A^+>= i|\chi> . \tag{9.119}$$

Since $\chi(R)$ is localized in the interaction region in photofragmentation processes, the wavefunction $A^+(R)$ satisfies the regular Schrödinger equation outside the region where χ is nonzero (outer region). Thus the total flux is a constant in the outer region where χ is zero. The solution in the outer region is analogous to that of the homogeneous Schrödinger equation. This allows the calculation of the decay width to be carried out at a much shorter distance and with a much smaller numerical grid.

For calculation of the partial decay width using the flux method, one needs to select the surface for flux evaluation to lie outside the region beyond which inelastic scattering effects are negligible. The partial decay width can be calculated using the following flux formula

$$|a_n(E)|^2 = \frac{\hbar}{2\pi}\mathrm{Re}\left[A_n^{+*}(R)\hat{v}A_n^{+}(R)\right]\Bigg|_{R=R_L}$$
$$= \boxed{\frac{\hbar^2}{2\pi\mu}\mathrm{Im}\left(A_n^{+*}(R)\frac{d}{dR}A_n^{+}(R)\right)\Bigg|_{R=R_L}} \tag{9.120}$$

where R_L is chosen to lie outside the region of inelastic scattering. Equation (9.120) is valid as long as no inelastic transitions are present at distance $R=R_L$ or larger. The main advantage of Eq. (9.120) is that there is no need to know the specific form of the asymptotic radial function as long as no inelastic scattering exists beyond R_L. Thus Eq. (9.120) is preferable in time-dependent computations.

9.6.2 Multi-arrangement

The main result—Eq. (9.120) from the previous section—is also valid in the case of rearrangement when properly generalized. Here we give brief derivations to generalize the results from the previous section to the multi-arrangement process. First, Eq. (9.107) is generalized to include multi-arrangement

$$|\chi\rangle = \sum_{\gamma n}\int_0^\infty dE|\psi_{\gamma nE}^{(-)}\rangle\, a_{\gamma n}(E), \tag{9.121}$$

where the summation now includes also an arrangement index γ. The asymptotic condition for the reactive scattering wavefunction now reads

$$\psi_{\gamma nE}^{(-)} \overset{R_\alpha\to\infty}{\longrightarrow} \sqrt{\frac{\mu_\alpha}{2\pi\hbar^2}}\Big[-\frac{e^{ik_{\alpha n}R_\gamma}}{\sqrt{k_{\alpha n}}}|\phi_{\alpha n}\rangle\,\delta_{\alpha\gamma}$$
$$+\sum_m S_{\alpha m,\gamma n}^{*}\frac{e^{-ik_{\alpha m}R_\alpha}}{\sqrt{k_{\alpha m}}}|\phi_{\alpha m}\rangle\Big] \tag{9.122}$$

Using the above condition for the wavefunction, the asymptotic boundary condition for A^+ now takes the form

$$A^+ \overset{R_\alpha \to \infty}{\longrightarrow} -i\sqrt{\frac{2\pi\mu_\alpha}{\hbar^2}} \sum_n \frac{e^{ik_{\alpha n}R}}{\sqrt{k_{\alpha n}}} |\phi_{\alpha n}> a_{\alpha n}(E) \tag{9.123}$$

and

$$A^- \overset{R_\alpha \to \infty}{\longrightarrow} i\sqrt{\frac{2\pi\mu_\alpha}{\hbar^2}} \sum_m \frac{e^{-ik_{\alpha m}R}}{\sqrt{k_{\alpha m}}} |\phi_{\alpha m}> b_{\alpha m}(E) \tag{9.124}$$

where

$$b_{\alpha m}(E) = -\sum_{\gamma n} S^*_{\alpha m,\gamma n} a_{\gamma n}(E). \tag{9.125}$$

As in the single arrangement case, if χ is real, one must have the relation

$$a_{\alpha m}(E) = b^*_{\alpha m}(E) = -\sum_{\gamma n} S_{\alpha m,\gamma n} a^*_{\gamma n}(E). \tag{9.126}$$

Equation (9.120) is now generalized to the multi-arrangement case as

$$|a_{\alpha n}(E)|^2 = \frac{\hbar^2}{2\pi\mu_\alpha} \mathrm{Im}\left[A^{+*}_{\alpha n}(R)\frac{d}{dR}A^+_{\alpha n}(R)\right]\Bigg|_{R=R_{\alpha L}} . \tag{9.127}$$

where μ_α is the reduced mass associated with the translational motion in the α arrangement.

9.6.3 Time-Dependent Flux Calculation

The calculation of Eq. (9.120) for partial decay width requires the value of the stationary wavefunction $A^+(R)$ and its derivative $A^{+\prime}(R)$ at a given large distance R. One could, of course, directly solve the stationary Eq. (9.119) for $A^+(R)$ using various numerical methods [191]. However, the computational time scales as N^3 with the number of channels N when using such stationary methods. We can use a time-dependent approach to calculate $A^{+\prime}$ by the standard $E \to t$ Fourier transform

$$\begin{aligned} |A^+> &= iG^+|\chi> \\ &= \frac{1}{\hbar}\lim_{\epsilon \to 0}\int_0^\infty dt e^{\frac{i}{\hbar}(E-H+i\epsilon)t}|\chi> \\ &= \frac{1}{\hbar}\int_0^\infty dt e^{Et}|\psi(t)> . \end{aligned} \tag{9.128}$$

where the time-dependent wavefunction is defined as

$$|\psi(t)\rangle = \lim_{\epsilon\to 0} e^{-\frac{i}{\hbar}(H-i\epsilon)t}|\chi\rangle \tag{9.129}$$

The small quantity ϵ can be viewed as a legitimate means for introducing an optical potential that is used to avoid artificial boundary reflections [192]. Thus $A^{+}_{\alpha n}(R_L)$ can be calculated by a half Fourier transform of the time-dependent wavefunction evaluated at a large distance R_L,

$$\boxed{A^{+}_{\alpha n}(R_L) = \frac{1}{\hbar}\int_0^{\infty} dt e^{\frac{i}{\hbar}Et} \langle\phi_{\alpha n}|\psi(t, R_L)\rangle} \tag{9.130}$$

and similarly for the derivative $A^{+\prime}_{n}(R_L)$

$$\boxed{A^{+\prime}_{\alpha n}(R_L) = \frac{1}{\hbar}\int_0^{\infty} dt e^{\frac{i}{\hbar}Et} \langle\phi_{\alpha n}|\psi'(t, R_L)\rangle\,.} \tag{9.131}$$

Equations (9.120) and (9.130) are the final working formulae for state-to-state dynamics calculations of photofragmentation. They involve a standard wavepacket propagation in Eq. (9.129) and a half Fourier transform in Eq. (9.130) at a fixed large distance R_L.

So far we have assumed that χ has no bound state components or they have been projected out from χ if any exist. This is actually not necessary in the flux approach as long as the distance R_L is chosen to be large enough such that the bound state components of χ vanish at $R \geq R_L$. Thus one does not need to worry about bound state components because they do not contribute to the flux.

VP lifetime of HF-DF

The isotopically substituted mixed dimer HFDF exists as two isomers, HF–DF and DF–HF as shown in Fig. 9.13, which provides an interesting case for studying mode sensitive vibrational predissociation. Thus, vibrational predissociation of the complex can occur through infrared excitation of either the HF stretching mode or the DF stretching mode. For each excited monomer (HF or DF), there are two isomers with quite different vibrational coupling strengths depending on whether the H (or D) is "free" or "bound" (cf. Fig. 9.13). Therefore there are four different types of vibrational excitations denoted by HF–*DF ("bound" D excitation), *DF–HF ("free" D excitation), DF–*HF ("bound" H excitation), and *HF–DF ("free" H excitation). For the symmetric HF dimer it is well known that the lifetime depends sensitively on whether the excitation is ν_1 ("free" H vibration) or ν_2 ("bound" H vibration). Experimental study of HFDF is desirable because the two product fragments HF and DF can be experimentally distinguished. Therefore the correlation of product rotational excitations to the initial mode of excitation can be determined unambiguously. For the symmetric HFHF, however, one can not distinguish one product rotation from another. Recent experiments have measured lifetimes for various combinations of infrared excitation for HFDF.

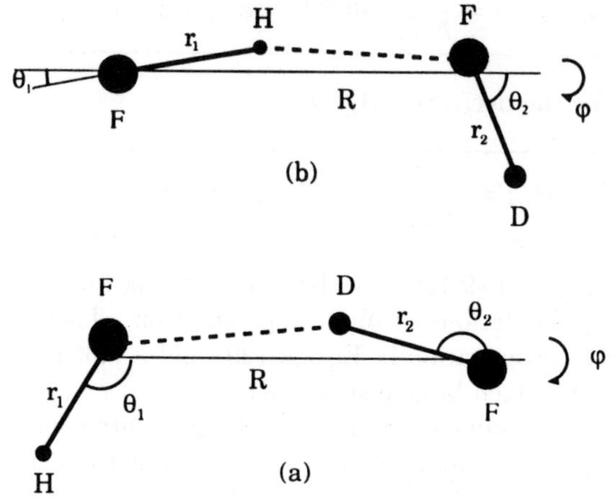

Figure 9.13: Equilibrium configurations of (a) HF–DF and (b) DF–HF.

In the theoretical treatment of predissociation dynamics for HFDF, we explicitly included four intermolecular degrees of freedom and treated the remaining two high frequency intramolecular HF bonds by adiabatic approximation. The flux method described in the previous section has been applied to the VP dynamics study of HFDF to obtain the VP lifetime from IR excitation of both HF and DF vibrations. For the $(HF)_2$ system, it is well known that the proton donor ("bound" H) excitation ν_2 gives stronger vibrational coupling than does the proton acceptor ("free" H) excitation ν_1, and the ν_2 lifetime is much shorter than the ν_1 lifetime. The excess energy in DF-excited dissociation for HFDF is about 1000 cm^{-1} less than that in HF excitation.

Table 9.2 lists the lifetimes for four different combinations: HF(v=1)-DF ("free" H excitation), HF-DF(v=1) ("bound" D excitation), DF-HF(v=1) ("bound" H excitation), and DF(v=1)-HF ("free" D excitation) together

Table 9.2: Vibrational predissociation lifetime for HFDF and HFHF in nanosecond (ns). E_{ex} is the excitation energy and E_R is the excess energy for dissociation, all in units of cm^{-1}.

Species	E_{ex}	E_R	τ_{exp}	τ_{theo}
HF–*DF	2838^a	1695	1.3^a	7.6
**DF–HF	2882^a	1804	5.5^a	19
DF–*HF	3868	2789	0.57^a	4.0
**HF–DF	3927	2785	$19.4^{a,d}$	160
HF–*HF$^{\pm}$	3868	2811	0.48^b	3.0^c
**HF–HF^{+}	3931	2874	24.8^b	36.6^c
**HF–HF^{-}	3931	2874	16.7^b	42.0^c

[a]Ref. [193]; [b]Ref. [196]; [c]Ref. [138]; [d]A strong J-dependence is observed for this transition and the lifetime ranges from 4.8 to 80 ns.

with experimental results. For the HF excited complex, the theoretical lifetime [194] for "free" H excitation is 40 times longer than that for "bound" H excitation. For the DF excited complex, the theoretical lifetime for "free" D excitation is about 2.5 times longer than that for "bound" D excitation. The agreement between theory and experiment is reasonably good, but the SBSQDE PES of Quack and Suhm [195] used in the quantum calculation needs further improvement in order to obtain better agreement with experiment.

9.7 The RPD Approach to Photofragmentation Dynamics

9.7.1 Theoretical Formulation

Since the fundamental basis of the RPD (reactant-product decoupling) approach for reactive scattering discussed in Sec. 7.3 is the separation of different components of the full wavefunction, the RPD approach can also be generalized to treat other dynamical processes such as inelastic scattering to separate different internal excitations. In particular, the RPD approach can be easily applied to treat photofragmentation (or half-scattering) dynamics [197, 198]. In a photodissociation process, for example, the dynamical quantity needed is the generalized Franck-Condon factor $| <\psi_f^{(-1)}|\chi> |^2$ which can be efficiently calculated by using time-dependent methods described in previous sections. The application of the RPD approach to such TD calculations of final state distributions in photodissociation is ideal because the interaction component wavefunction ψ_{int} is now nonzero only in the strong interaction region. As a result, the calculation of ψ_{int} is restricted to a relatively small strong interaction region while the calculations of the product components ψ_p are to be carried out in their respective arrangement spaces, exactly as in bimolecular reactive scattering.

In a general photofragmentation process, the initial molecule can simultaneously dissociate into more than one arrangement (fragment) channel. Consequently, the calculation of the TD wavefunction is a multi-arrangement reactive scattering problem. The main difference in the TD dynamics calculation for photofragmentation is that the initial wavepacket is localized in the "strong" interaction region on the dissociative PES while in bimolecular collision, the initial wavepacket is located in the asymptotic region of the initial arrangement space. Therefore the RPD treatment for bimolecular reactive scattering in Sec. 7.3 can be easily generalized to the photodissociation process. In order to obtain the specific product state distribution in photodissociation, one needs to project out the product states from the wavefunction $\psi(t)$ in the asymptotic region of the desired arrangement channel space.

Let us denote the initial wavepacket by $\psi(t=0)$, which in the present case, is given by the initial bound state wavefunction χ, i.e., $\psi(t=0)=\chi$ (assuming a constant transition dipole moment). The time propagation of this wavepacket is given by $\psi(t) = \exp(-iHt/\hbar)\chi$. Using the RPD approach, the wavefunction $\psi(t)$ can be split into the interaction component $\psi_{int}(t)$ and the sum of product components $\psi_\lambda(t)$ (λ=1, 2, ... , N)

$$\psi(t) = \psi_{int}(t) + \sum_\lambda \psi_\lambda(t) \tag{9.132}$$

where λ is the arrangement (fragment) channel label. The RPD equation for ψ_{int} is therefore defined as,

$$i\hbar\frac{\partial\psi_{int}(t)}{\partial t} = H\psi_{int}(t) - i\sum_\lambda V_\lambda\psi_{int}(t), \tag{9.133}$$

where the summation is over *all* arrangement channels, i.e., $\lambda = 1, ..., N$, and again, each absorbing potential V_λ is located just beyond the point-of-no-return for arrangement λ. The wavefunction $\psi_{int}(t)$ now describes the wavefunction of the molecule in the strong interaction region only, and is absorbed outside this region when it enters the point-of-no-return spaces. The various arrangement components of the full wavefunction that describe the corresponding fragmentation channels satisfy the RPD equation

$$i\hbar\frac{\partial\psi_\lambda(t)}{\partial t} = H\psi_\lambda(t) + iV_\lambda\psi_{int}(t) \qquad (\lambda = 1, ..., N) \tag{9.134}$$

If we add Eqs. (9.133) and (9.134) (for *all* arrangements), we obtain the Schrödinger equation for the full wavefunction

$$i\hbar\frac{\partial}{\partial t}\psi(t) = H\psi(t) \tag{9.135}$$

which proves the equivalence of the RPD equations to the original Schrödinger equation satisfied by the full wavefunction $\psi(t)$.

A major attractive feature in the current RPD approach to photodissociation is that the interaction wavefunction ψ_{int} is restricted to the strong interaction region only. This significantly facilitates the calculation of ψ_{int} because a much smaller numerical grid (interaction region only) is needed to calculate it. In comparison, the calculation of the reactant component wavefunction ψ_r in bimolecular reactive scattering requires a numerical grid that covers both the strong interaction region and the reactant arrangement space. Since the calculation of ψ_{int} constitutes the major computational cost, the RPD approach to photofragmentation dynamics is extremely attractive.

The calculation of the arrangement channel component, ψ_λ, involves only inelastic scattering and it can be carried out exactly the same way as in the RPD approach to bimolecular reactive scattering described in Sec. 7.3. For example, the wavefunction $\psi_\lambda(t)$ can be calculated by using the split operator scheme via Eq. (7.36)

$$|\tilde{\psi}_\lambda(t+\Delta)> = e^{-\frac{i}{\hbar}H\Delta}|\tilde{\psi}_\lambda(t)> + \frac{\Delta}{\hbar}V_\lambda|\psi_{int}(t+\Delta)> \tag{9.136}$$

where $\tilde{\psi}_\lambda(t) = \psi_\lambda(t) + \frac{\Delta}{2\hbar}V_\lambda\psi_{int}(t)$. Since $\tilde{\psi}_\lambda(t)$ is everywhere the same as $\psi_\lambda(t)$ except in the absorbing region, we could directly use $\tilde{\psi}_\lambda(t)$ to extract the final state-specific dynamics such as the product state distribution. Alternatively, one could use Eq. (7.38) to directly calculate $\psi_\lambda(E)$. Once the desired component wavefunction $\psi_\lambda(t)$ or $\psi_\lambda(E)$ has been calculated, specific final states can be projected out at a large radial distance using the methods discussed in Sec. 7.2.2.

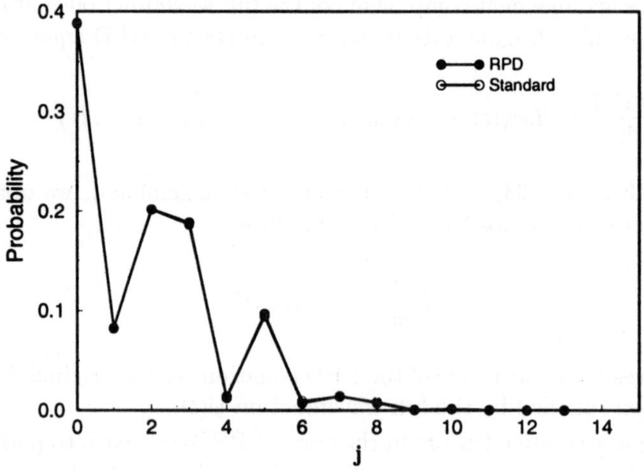

Figure 9.14: Comparison of the rotation distribution (normalized) of OH from the RPD calculation to that of the standard TD calculation using a large basis set and a large numerical grid. The H_2O was in the ground state (000).

Numerical test for photodissociation of H_2O

The RPD method has been applied to calculating the product state distribution of OH in the photodissociation of water in the first excited $\widetilde{A}$ state [198] using the same potential energy surfaces as in Sec. 9.4.3.

Figure 9.14 gives the calculated rotational state distribution of OH from the dissociation of H_2O in the ground vibrational state. As shown in the figure, the result from the present RPD calculation is indistinguishable from the standard TD calculation obtained by using a large basis set and a large numerical grid to cover all arrangement spaces. This demonstrates the numerical accuracy of the RPD approach to state-to-state photodissociation dynamics calculations.

Chapter 10

Molecular Reactions on Surfaces

10.1 Introduction

Heterogeneous catalysis of chemical reactions on solid surfaces is of great practical importance in many areas of science and technology. A major challenge to chemists today is to understand how the structure and properties of the surface of a catalyst affect the catalytic activity for a specific chemical reaction, and how the internal state of the reactant molecule influences the catalytic reactions. Although numerous studies on the kinetics of catalytic processes have been carried out in the past several decades, many detailed dynamical aspects of catalysis are far from clearly understood. A thorough understanding of the dynamics of reactions on surfaces is essential in achieving the ultimate goal of designing and controlling catalytic reactions in pollution control and other important industrial processes.

Gas-surface reactions present another challenge to quantum dynamics theory. Current dynamics studies of gas-surface processes can generally be divided into two categories based on different levels of complexity in the theoretical treatment. The first level of theory treats the solid surface as a rigid surface in which surface atoms are simply fixed at their equilibrium positions, or in other words, the surface is treated as a "cold" surface. The next level of theory treats the motions of surface atoms explicitly but its practical implementation is significantly more complicated, if not impossible. Therefore most theoretical studies, in which surface motion is explicitly included, are classical. Although one can include surface motions either explicitly such as in molecular dynamics, or implicitly such as in Langevin dynamics, the outcome of such a classical treatment is not always clear. On the other hand, the rigid-surface theory, although approximate, is clearly defined and physically transparent, and is susceptible to the treatment of high level dynamical theories such as quantum mechanics. Thus it is possible to develop rigorous theoretical methodologies to treat molecular reactions on rigid surfaces and to provide accurate predictions for surface reaction processes in which the effect of surface motion, often manifested by surface temperature effects, is negligible. In the following, we develop practical theoretical models for treating dynamics of dissociative adsorption of diatomic molecules on rigid surfaces.

10.2 Theoretical Models

10.2.1 Dissociative Adsorption on Metal Surfaces

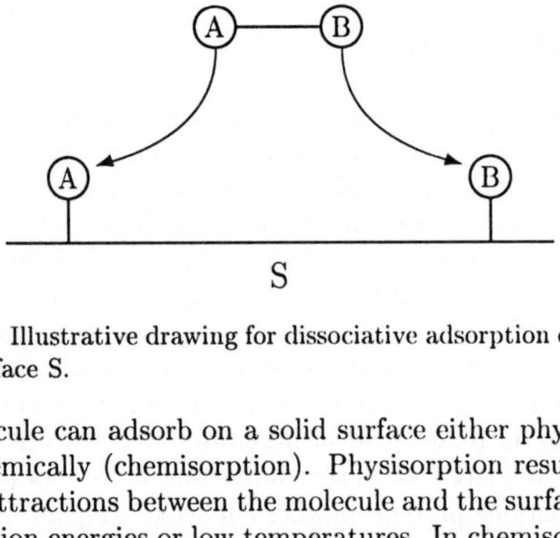

Figure 10.1: Illustrative drawing for dissociative adsorption of diatomic molecule AB on a surface S.

A molecule can adsorb on a solid surface either physically (physisorption) or chemically (chemisorption). Physisorption results from weak van der Waals attractions between the molecule and the surface and thus occurs at low collision energies or low temperatures. In chemisorption, the attraction between the molecule and the surface is due to the much stronger chemical force and the molecule forms chemical bonds with the surface. Many molecules readily dissociate upon adsorption and the phenomenon is called dissociative adsorption (or dissociative chemisorption). In dissociative adsorption, the breaking of molecular bonds is accompanied by the formation of new chemical bonds between the adsorbate and surface, much like molecular reaction in the gas phase. For example, the process of dissociative adsorption of a diatomic molecule on a surface S can be represented by the equation

$$AB + S \longrightarrow AS + BS \tag{10.1}$$

which is quite similar to that of an atom-diatom reaction. Figure 10.1 illustrates the process of dissociative adsorption for a diatomic molecule AB on a surface. Dissociative adsorption on a metal is often an activated process much like that in a gas phase reaction. Thus many theoretical methods developed for gas phase reaction dynamics studies can be readily applied to studying molecule-surface reactions with slight modifications. The following sections discuss theoretical models that have been currently employed in quantum dynamics studies of surface reactions.

10.2.2 Flat Surface Model

In the 3D flat-surface model for the interaction of a diatomic molecule with a solid surface, a molecule-surface interaction potential energy surface (PES) is constructed for a chosen dissociation site, and is assumed to be dependent only on three degrees of freedom: the diatomic bond length r, the perpendicular distance from the center of the diatom to the surface Z, and the polar angle θ as shown in Fig. 10.2. In this model, the potential is independent of the azimuthal angle ϕ and the lateral coordinates of the center of mass of the diatom (X,Y). It is worth mentioning, however, that the name "flat-surface" approximation is somewhat misleading because although the dynamics calculation does not include any surface corrugation explicitly, the effect of surface corrugation is somewhat incorporated in the PES which is constructed for a specific surface site. Thus for molecular interactions with the same solid surface but at different surface sites, one needs to construct a different PES in this model.

The Hamiltonian for a diatomic molecule in the flat surface model can be written in terms of the coordinates defined in Fig. 10.2

$$H = H_0 + U \tag{10.2}$$

with H_0 given by

$$H_0 = -\frac{\hbar^2}{2M}\frac{\partial^2}{\partial Z^2} - \frac{\hbar^2}{2\mu}\frac{\partial^2}{\partial r^2} + V_d(r) \tag{10.3}$$

$$U = \frac{\mathbf{j}^2}{2\mu r^2} + V(Z, r, \theta) \tag{10.4}$$

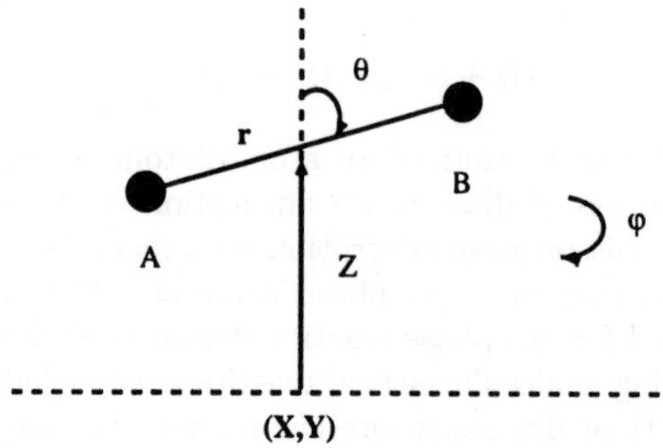

Figure 10.2: Molecular coordinates for diatom-surface collision where (X,Y,Z) are the center-of-mass coordinates of the diatom, r is the diatomic distance, θ is the polar angle, and ϕ is the out-of-plane azimuthal angle. The lateral coordinates (X,Y) are fixed in the fixed site model.

where M and μ are, respectively, the center-of-mass and reduced mass of the diatomic, and $\mathbf{j}$ is the angular momentum operator. The potential $V_d(r)$ is a Morse-like diatomic potential and $V(Z, r, \theta)$ is the molecule-surface interaction potential. As usual, the standard substitution $\Psi \to \Psi/r$ for the wavefunction has been assumed. The quantum mechanical treatment for diatomic dissociation on a flat-surface is essentially similar to that of atom-diatom reaction. In the coupled-channel wavepacket approach, the time-dependent wavefunction satisfying the Schrödinger equation $i\hbar\frac{\partial}{\partial t}\Psi(t) = H\Psi(t)$ can be expanded in diatomic rovibrational eigenfunctions as

$$\Psi^m_{v_0 j_0}(Z, r, \theta, t) = \sum_{t,n,j} F^m_{tnj,v_0 j_0}(t) u^n_t(Z) P_{jm}(\theta) \chi_n(r) \tag{10.5}$$

where $(v_0 j_0 m)$ denotes the initial rovibrational state of the diatomic molecule, $u^n_t(Z)$ is the translational basis function, $P_{jm}(\theta)$ is the normalized associated-Legendre polynomial, and $\chi_n(r)$ is the vibrational basis function. We note here that the rotation projection quantum number m is a conserved quantum number in the flat-surface model because the PES is independent of the azimuthal angle ϕ.

A non-direct product basis expansion method described in Sec. 6.6 for gas-phase reaction is also used in gas-surface reactions. This method is actually ideal for gas-surface reactions because the skewing angle of the PES is strictly 90^0 (see Fig. 10.3). A collinear model study showed explicitly that the required number of quasiadiabatic diatomic vibrational functions is only large near the potential saddle point region [199]. By using the nondirect product basis described in Sec. 6.6, the translational basis function is defined by

$$u_t^n(Z) = \begin{cases} \sqrt{\frac{2}{Z_4 - Z_1}} \sin \frac{n\pi(Z - Z_1)}{Z_4 - Z_1} & n \le n_{asy} \\ \sqrt{\frac{2}{Z_2 - Z_1}} \sin \frac{n\pi(Z - Z_1)}{Z_2 - Z_1} & n > n_{asy} \, , \end{cases} \tag{10.6}$$

where n_{asy} is chosen to be the number of energetically open vibrational channels plus a few closed vibrational channels of the diatom AB. The vibration eigenfunction $\chi_n(r)$ of the diatom AB is obtained by solving the eigenvalue equation

$$\left[-\frac{\hbar^2}{2\mu}\frac{\partial^2}{\partial r^2} + V_d(r)\right] \chi_n(r) = \varepsilon_n \chi_n(r), \tag{10.7}$$

where the diatomic reference potential $V_d(r)$ is defined to allow the vibrational function $\chi_n(r)$ to span a large interatomic distance r in order to allow diatomic dissociation to occur. After the basis functions have been selected, we can use the split-operator method to carry out wavepacket propagation,

$$\Psi(Z, r, \theta, t + \Delta) = e^{-iH_0\Delta/2} e^{-iU\Delta} e^{-iH_0\Delta/2} \Psi(Z, r, \theta, t), \tag{10.8}$$

where H_0 and U are defined in Eqs. (10.2), (10.3), and (10.4). Since the total number of vibration basis functions n_{max} is much larger than n_{asy}, the computationally intensive part of the calculation is limited to the interaction region $Z \in [Z_1, Z_2]$ where reaction occurs. In the asymptotic region, the amount of computation is more or less similar to that for inelastic scattering. The procedure for wavepacket propagation is essentially identical to that of gas-phase reaction described in the previous section and will not be detailed here. It should be mentioned here that the time-dependent wavefunction is absorbed at the edges of the grid to avoid boundary reflections.

The initial state-selected total dissociation probability of the diatom is obtained by projecting out the energy-dependent reactive flux. If ψ_{iE}^+ denotes the time-independent (TI) full scattering wavefunction, where the labels i and E denote the initial state and energy, the total dissociation probability from an initial state i can be obtained by evaluating the reactive flux. If the full stationary scattering wavefunction is normalized as

$$<\psi_{iE}^+|\psi_{iE'}^+> = 2\pi\hbar\delta(E - E') \tag{10.9}$$

the total dissociation probability can be calculated by the flux formula

$$P_i^R(E) = \frac{\hbar}{m_r} \mathrm{Im}[<\psi_{iE}^+|\delta(r - r_0)\frac{\partial}{\partial r}|\psi_{iE}^+>]. \tag{10.10}$$

Since the TD wavefunction can be expanded in terms of the TI wavefunctions which form a complete set

$$\begin{aligned} |\psi_i(t)\rangle &= e^{-\frac{i}{\hbar}Ht}\psi_i(0) \\ &= \frac{1}{2\pi\hbar}\int dE e^{-\frac{i}{\hbar}Et}|\psi_{iE}^{+}\rangle\, a_i(E) \end{aligned} \tag{10.11}$$

the TI scattering wavefunction can be obtained by Fourier transform

$$|\psi_{iE}^{+}\rangle = \frac{1}{a_i(E)}\int_{-\infty}^{\infty} e^{\frac{i}{\hbar}(E-H)t}|\psi_i(0)\rangle\, dt. \tag{10.12}$$

The coefficient $a_i(E)$ can be easily evaluated from the free asymptotic function ϕ_{iE} by standard manipulation

$$\begin{aligned} a_i(E) &= \langle\psi_{iE}^{+}|\psi_i(0)\rangle \\ &= \lim_{t\to-\infty} \langle\phi_{iE}|e^{\frac{i}{\hbar}H_0 t}e^{-\frac{i}{\hbar}Ht}|\psi_i(0)\rangle \\ &= \langle\phi_{iE}|\psi_i(0)\rangle, \end{aligned} \tag{10.13}$$

where the last equation holds because the initial wavepacket $\psi_i(0)$ is located in the asymptotic region with the incoming wave only.

10.2.3 Fixed Site Corrugated Surface Model

So far the 3D flat-surface model has been quite successful in providing qualitative and even some quantitative dynamics information for hydrogen dissociation on metals such as the role of hydrogen vibration and rotation in dissociative adsorption on Cu(111) [200–204]. However, the inherent limitation of the flat-surface model dictates that it cannot provide information on surface corrugation and its effect on molecular adsorption. One would like to investigate dynamical effects of dissociative adsorption in the presence of surface corrugation. In order to obtain this information, one needs to go beyond the flat-surface model and to include surface corrugation explicitly. It is relatively straightforward to extend the 3D flat-surface model to the 4D fixed-site model which includes local surface corrugation explicitly [203, 204]. In the 4D fixed-site model, the azimuthal angle ϕ is explicitly included while the lateral coordinates of the center of mass of the diatom are still fixed at a given site, usually at a symmetric site. The inclusion of the azimuthal angle ϕ in the dynamics calculation is a natural extension of the 3D flat-surface model and enables us to treat both local surface corrugation and surface site-specificity explicitly. This 4D model should provide a reasonable description for normal incidence of hydrogen over the symmetric sites of metal surfaces.

In the 4D model, the Hamiltonian remains the same as was given in the preceding section for the 3D model except that the potential now depends on four coordinates. The expansion of the wavefunction in Eq. (10.5) is now generalized to include the ϕ-dependence explicitly

$$\Psi_{v_0 j_0 m_0}(Z, r, \theta, \phi, t) = \sum_{t,n,j,m} e^{im\phi} F^{m}_{tnjm, v_0 j_0 m_0}(t) u^{n}_{t}(Z) \times Y_{jm}(\theta, \varphi) \chi_n(r) \qquad (10.14)$$

The remaining treatment is essentially identical to that of the flat-surface model described in the previous subsection. At this stage, important differences from the flat-surface treatment emerge. First, the m quantum number of the diatomic rotation is no longer conserved and the allowed Δm transitions are determined by the symmetry of the chosen site. As shown in the appendix of Ref. [202], the allowed Δm transitions are given by $\Delta m = \pm$Nk (k=0,1,2,...) where N is the rotation symmetry of the impact site with the possible values of N=1,2,3,4, and 6 for a crystal surface. Second, the 4D model describes the migration of the diatoms through the potential valleys correctly, which shows explicitly the dissociation path and adsorption sites. Neither of these two features can be represented in the flat-surface model. In fact, in the flat-surface model one has to artificially cut off the potential after the molecule passes through the reaction barrier in order to avoid the artificial boundary reflection from diffusion barriers. Thus the 4D fixed-site model provides a much more realistic description of the chemisorption process. We note here that we use the word "local" surface corrugation because the center of mass of the diatom is not allowed to move on the crystal surface in the fixed site model.

10.2.4 A Selection Rule for Homonuclear Diatoms

A selection rule has been uncovered in the flat-surface studies [201,203,205]. This rule relates the dissociation properties of a homonuclear diatom on a surface to its nuclear symmetry. Specifically, if the magnetic quantum number m of a homonuclear diatom can change by even quanta only ($\Delta m = \pm 2, 4, ...$) and the initial rotation state of the diatom satisfies $j + m$=odd, then the reaction probability is essentially forbidden at low energies for thermally neutral or endothermic dissociation on the surface. The selection rule immediately applies in two cases discussed in this chapter. First, in the flat-surface model for which the magnetic quantum number m is conserved, this selection rule holds rigorously. Secondly in the fixed site dissociation model, if the fixed site has even rotation symmetry for which the magnetic quantum number m can change by even quanta only, the selection rule also holds exactly. For a high symmetry fixed site with odd rotation symmetry, the $\Delta m \neq 0$ transition probabilities are expected to be small in general and the selection rule should be a very good approximation.

The derivation of this selection rule is quite straightforward and is given below. For a homonuclear molecule such as H_2, the Hamiltonian is invariant with respect to the exchange of the two hydrogen atoms. Thus the nuclear wavefunction must be properly symmetrized with respect to the exchange of two identical atoms. Before dissociating on the surface, the rotational wavefunction of a homonuclear diatom in the gas phase is described by spherical harmonics

$$\Psi_{gas} \sim Y_{jm}(\theta, \phi) = P_{jm}(\theta)e^{im\phi} \tag{10.15}$$

and the gas phase wavefunction Ψ_{gas} has exchange symmetry

$$P^{ex}_{gas} = (-)^j \tag{10.16}$$

After dissociating on the surface, the wavefunction for two adsorbed identical atoms contains

$$\Psi_{ad} \sim \exp(im'\phi) \begin{cases} [\chi_{v_1}(1)\chi_{v_2}(2) \pm [\chi_{v_2}(1)\chi_{v_1}(2)] & (v_1 \neq v_2) \\ \chi_{v_1}(1)\chi_{v_2}(2) & (v_1 = v_2) \end{cases} \tag{10.17}$$

where χ_v is the vibrational eigenfunction of the adatom.

If both adatoms are in the same vibrational state $v_1 = v_2$, the wavefunction for the adsorbates will have the exchange symmetry

$$P^{ex}_{ad} = (-)^{m'} \tag{10.18}$$

The conservation of exchange symmetry requires that $P^{ex}_{gas} = P^{ex}_{ad}$ or

$$(-)^j = (-)^{m'} \tag{10.19}$$

which can be written in terms of the rotational symmetry of the initial wavefunction of the diatom

$$F = (-)^{j+m} = (-)^{j-m} = (-)^{m'-m} = (-)^{\Delta m} \tag{10.20}$$

The situations can be classified as follows.

1). F is even: the symmetry condition (10.20) can always be satisfied since the elastic channel $\Delta m = 0$ is always allowed. A corollary is that if only odd $\Delta m \neq 0$ transitions are allowed in this case (such as at a fixed site with odd rotation symmetry), then the $v_1 = v_2$ states of the adatoms must be associated with the quantum number $m' = m$.

2). F is odd: the symmetry condition (10.20) can only be satisfied when Δm is odd. A corollary is that if only even $\Delta m \neq 0$ transitions are allowed in this case (such as at a fixed site with even rotation symmetry), then the vibrational states $v_1 = v_2$ of the adatoms must have different vibrational quantum numbers or $v_1 \neq v_2$.

It is thus clear that the selection rule applies in the second case when F is odd and only *even* Δm transitions are allowed (such as at a fixed site with even rotation symmeter). This symmetry relation dictates that for any homonuclear diatom whose rotation state satisfies $j + m$=odd and only even Δm transitions are allowed on the surface, the vibrational state of the two dissociated adatoms cannot be the same due to exchange symmetry.

When these conditions are satisfied, this selection rule prevents the population of $v_1 = v_2$ states of the adatoms including the *lowest* vibrational state $v_1 = v_2 = 0$. For endothermic or near thermally neutral dissociation of homonuclear diatoms on surfaces at very low energies, this symmetry restriction could exclude the *only* energetically accessible ground vibrational state of the adatoms from being populated, and thus render the dissociation of homonuclear diatoms symmetry forbidden. In addition, the exclusion of the other $v_1 = v_2$ vibrational states of the adatoms could severely limit the number of product channels for population at relatively low energies. This could significantly decrease the dissociation probability. However, if the molecule-surface reaction is a highly exothermic process, or when the total energy is very high, the closing of the $(v_1 = v_2)$ product states due to this selection rule will probably not have a significant effect on the dissociation probability because there are a sufficient number of allowed product channels with $v_1 \neq v_2$ for energy to flow into.

10.2.5 Fully Corrugated Surface Model

Ultimately, the exact dynamics study of dissociative adsorption of a diatomic molecule on a corrugated static surface should include the lateral motion of the center of mass of the diatom, i.e., include six degrees of freedom. The 6D Hamiltonian could be written in terms of the gas-phase

Jacobi coordinates (molecular coordinates) as in Fig. 10.2

$$H = -\frac{\hbar^2}{2M}\left[\frac{\partial^2}{\partial X^2} + \frac{\partial^2}{\partial Y^2} + \frac{\partial^2}{\partial Z^2}\right] - \frac{\hbar^2}{2\mu}\frac{\partial^2}{\partial r^2} + \frac{\mathbf{j}^2}{2\mu r^2} + V(X, Y, Z, r, \theta, \phi) \tag{10.21}$$

where (X,Y) are the lateral Cartesian coordinates of the center-of-mass (COM) of the diatom. This representation of the Hamiltonian is natural for describing inelastic scattering of a diatom from the surface, and is also suitable for describing the process of dissociative adsorption. However, this set of molecular coordinates is not very suitable when the diatomic bond length r is very large because the wavefunction is highly peaked around θ=90^0. One alternative is to simply use the atomic coordinates that are ideal for describing the diffusion process on the solid surface, and the Hamiltonian is simply expressed in atomic Cartesian coordinates. However, atomic coordinates are not very efficient for describing gas-phase processes.

One compromise is to use a set of coordinates intermediate between the molecular and atomic coordinates described below. We denote this set by (X, Y, Z, z, ρ, ϕ) where (z, ρ, ϕ) are the cylindrical coordinates of the interatomic vector $\mathbf{r}$ defined by the equation

$$\begin{cases} z = r\cos\theta \\ \rho = r\sin\theta \end{cases} \tag{10.22}$$

Using the following substitution for the wavefunction

$$\Psi = \rho^{-\frac{1}{2}}\psi \tag{10.23}$$

one can derive the Schrödinger equation for ψ

$$i\hbar\frac{\partial}{\partial t}\psi(t) = H'\psi(t) \tag{10.24}$$

where

$$H' = -\frac{\hbar^2}{2M}\left[\frac{\partial^2}{\partial X^2} + \frac{\partial^2}{\partial Y^2} + \frac{\partial^2}{\partial Z^2}\right] - \frac{\hbar^2}{2\mu}\left[\frac{\partial^2}{\partial \rho^2} + \frac{1}{\rho^2}(\frac{1}{4} + \frac{\partial^2}{\partial \phi^2}) + \frac{\partial^2}{\partial z^2}\right] + V(X, Y, Z, \rho, \phi, z) \tag{10.25}$$

The wavefunction ψ can be expanded in basis functions of the coordinates (X, Y, Z, z, ρ, ϕ) and the DVR representation can be used for all the coordinates.

10.3 The H_2/Cu(111) System

10.3.1 Background

Study of the H_2/Cu system has developed into an important testing ground for understanding the dynamics of activated dissociative adsorption of molecules at surfaces as well as the reverse process—associative desorption. It is well established that molecular hydrogen dissociates upon adsorption on Cu. However, the detailed dynamics of dissociative adsorption of H_2 on copper is still not fully understood despite extensive research efforts devoted to the subject. Since the pioneering molecular beam experiments of Balooch et al. [206] in 1974 for adsorption of hydrogen on several low-index faces of copper, many new experiments investigating the adsorption and desorption of H_2 on Cu have been reported. Molecular beam experiments [206–208] showed that the sticking probability of hydrogen on Cu rapidly increases as the incident kinetic energy of hydrogen is increased. It is also found that the sticking probability is a function only of the component of the incident kinetic energy that is normal to the surface. Because of limited energy resolution and lack of state-selection in the molecular beam experiments, it was not clear whether the increase of the sticking probability was mainly the effect of the increasing kinetic energy of hydrogen or the enhanced population of vibrationally excited hydrogen when the beam temperature was raised.

In addition to measurements of sticking probability, angular and kinetic energy distributions of H_2 desorbing from copper surfaces have also been measured. It is observed that the mean kinetic energy of the H_2 and D_2 desorbing from Cu(100) and Cu(111) surfaces are far in excess of the average energy anticipated from thermal equilibrium at surface temperatures. Also desorption experiments [210] showed that the product branching ratio P(v=1)/P(v=0) of desorbed H_2 and D_2 from Cu exceeds the Boltzmann ratio by more than a factor of 50 while the rotational state distribution of H_2 is found to be Boltzmann-like [210]. Based on the arguments of detailed balance or microreversibility, the hot vibrational population of desorbed hydrogen strongly suggests that hydrogen adsorption should be enhanced by vibrational excitation which indicates a later barrier for adsorption.

The H_2/Cu system has been extensively studied theoretically as well. Because hydrogen dissociation on Cu is an activated process, the tunneling of hydrogen can play an important role in dissociative adsorption. Therefore it is desirable to employ quantum mechanical methods in theoretical studies. Recently, three dimensional quantum dynamics calculations for dissociation of hydrogen on Cu using empirical potential energy surfaces predicted a strong dependence of dissociation probability on the rotational orientation of the hydrogen molecule [201–203]. In the theoretical study of Ref. [201], the rovibrational product state distributions of desorbed hydrogen from a Cu surface were calculated using a variational reactive scattering method. The calculated vibrational state distribution of desorbed hydrogen is very hot while the rotational distribution is Boltzmann-like [201]. These results are in good agreement with the desorption experiment [210]. In the following sections we present some results of theoretical calculation for dissociation of H_2 on Cu surfaces.

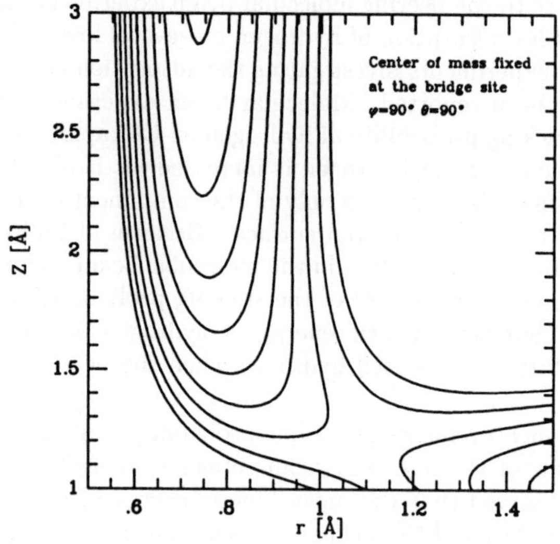

Figure 10.3: Contour plot of the potential energy surface for dissociative adsorption of H_2 on Cu(111).

10.3.2 Potential Energy Surface for H_2/Cu(111)

The accurate *ab initio* calculation of the PES for H_2 on Cu or other transition metals is a challenging task. Most dynamics studies have relied on empirical functional forms to model the H_2/Cu PES. Total energy *ab initio* calculations using density functional theory have recently been reported for the H_2/Cu system [211]. The barrier for dissociation is found to depend strongly on the impact site and the orientation of the diatom. Utilizing the available information from the *ab initio* calculation, a 6D empirical LEPS potential energy surface was constructed [203]

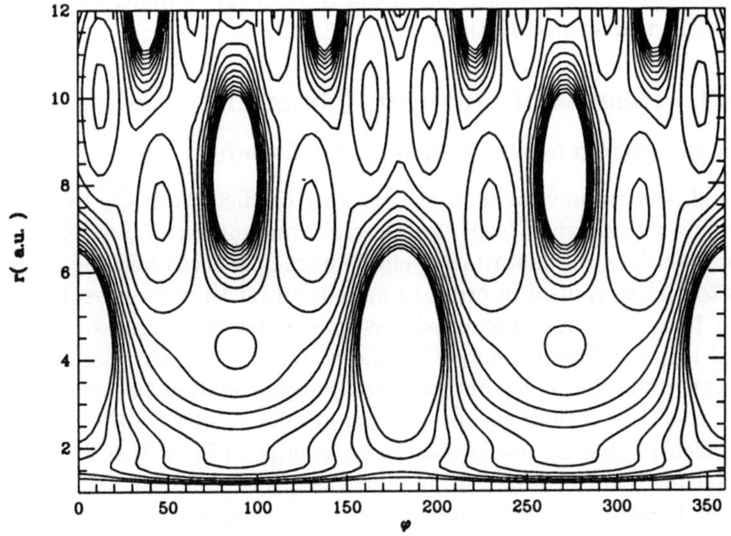

Figure 10.4: Contour plot of the $H_2/Cu(111)$ PES showing local surface corrugation and the dependence of the PES on r and ϕ coordinates. All other coordinates are fixed at the saddle point geometry.

$$V = U_1 + U_2 + U_3 - \left[Q_1^2 + (Q_2 + Q_3)^2 - Q_1(Q_2 + Q_3)\right]^{\frac{1}{2}} \tag{10.26}$$

where U_i and Q_i are defined by

$$U_i = \frac{1}{4(1+\Delta)} D_i \Big\{ (3 + \Delta_i) \exp[-2\alpha_i(q_i - q_{i0})] - (2 + 6\Delta_i) \exp[-\alpha_i(q_i - q_{i0})] \Big\} \tag{10.27}$$

$$Q_i = \frac{1}{4(1+\Delta)} D_i \Big\{ (1 + 3\Delta_i) \exp[-2\alpha_i(q_i - q_{i0})] - (6 + 2\Delta_i) \exp[-\alpha_i(q_i - q_{i0})] \Big\} \tag{10.28}$$

where U_1 and Q_1 describe the interaction between two hydrogen atoms and $U_{2,3}$ and $Q_{2,3}$ describe atom-surface interactions. $U_{2,3}$ and $Q_{2,3}$ are periodic functions of the lateral coordinates of the hydrogen atom. Figure

10.4 shows the local corrugation of the $H_2/Cu(111)$ potential surface with an explicit dependence on the azimuthal angle ϕ.

10.3.3 Results of Quantum Dynamics Studies

Effect of hydrogen vibration on dissociation

Although the enhancement of the sticking coefficient on Cu by hydrogen vibration is established from experimental measurements discussed in the previous section, its quantitative effect on the observed sticking coefficient needs to be determined in order to help interpret the experimental results. Figure 10.5 shows the calculated dissociation probabilities as functions of normal incident kinetic energy for the v=0 and v=1 vibrational states of hydrogen. The thermal vibration-averaged dissociation probability is also shown together with the experimental results of Ref. [207]. The figure shows clearly that at low nozzle beam energies (E_k < 0.4eV), the main contribution to adsorption is from vibrationally excited $H_2(v{=}1)$. At higher energies, however, the dissociation is dominated by the ground vibrational state of hydrogen. The thermal vibration-averaged theoretical results agree well with the experimental data of Ref. [207].

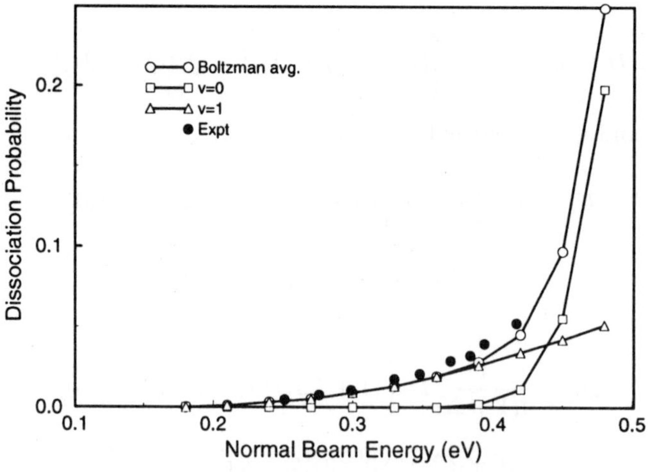

Figure 10.5: Vibrational state-specific dissociation probabilities P_v for H_2 on Cu(111). The vibrationally averaged dissociation probabilities are given by circles and the experimental results [207] are shown by solid diamonds.

Table 10.1: Vibrational population ratios of v=1 to v=0 in hydrogen desorbed from the Cu(111) surface.

T_s(K)	R_B^a	R_{theo} / R_B	R_{exp}^b/R_B
750	.000352	133.5	
850	.000897	75.8	93.6(33.4)
950	.00188	48.4	

[a]R_B is the Boltzmann ratio and [a]R_{exp} is the experimental ratio from Ref. [210]. The number in parenthesis is the experimental error estimate.

Desorption of hydrogen

The theoretical calculations of Ref. [201] also obtained the product rovibrational state distribution of hydrogen desorbed from the Cu(111) surface. Although the rotational state distribution of hydrogen is Boltzmann like, the vibrational state of H_2 is highly excited. Table 10.1 lists the calculated population ratios of v=1 to v=0 together with the Boltzmann ratio and the experimental measurements. As can be seen, the v=1 to v=0 population ratio of the desorbed hydrogen is much larger than the corresponding Boltzmann ratio at the surface temperature by a factor over 100! This vibrational enhancement factor decreases as the surface temperature is increased.

Effect of hydrogen rotation on dissociation

In contrast to the vibrational effect, the rotational effect on hydrogen dissociation on Cu is much less understood, until very recently. Most earlier 3D quantum calculations have used the plane rotor model, which is not appropriate for studying rotational effects. Using the correct 3D spherical rotor for hydrogen rotation, some important results on the effect of rotational orientation on hydrogen dissociation have been obtained [201,202,205]. Figure 10.6 shows the dependence of the dissociation probability of hydrogen on its magnetic quantum number m on Cu(111). As shown in the figure, the dissociation probability of H_2 in the $j = 5$ rotational state increases rapidly as the magnetic quantum number number m increases. If $P(jm)$ denotes the dissociation probability for a rotational state jm, then Fig. 10.6 shows that $P(jm)> P(jm')$ for $m > m'$, with $P(jj)$ being the largest. The same dependence of $P(jm)$ on the quantum number m for a given j is found for all other values of j as well. Thus hydrogen dissociation on Cu is dominated by those rotational states with large magnetic quantum numbers m. In particular, the $m = j$ state gives the largest dissociation probability for a given j-manifold.

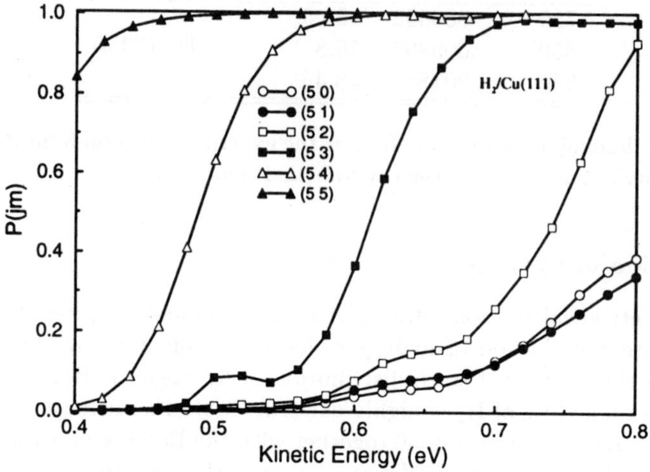

Figure 10.6: Dissociation probabilities of H_2(jm) on Cu(111) as a function of kinetic energy from initial rotational states j=5 and m=0–5.

Since rotational states with the same quantum number j but different m numbers do not differ in energy, this m-dependence of the dissociation for a given j is an unambiguous demonstration of the rotational orientation or steric effect on hydrogen dissociation. In analogy with classical mechanics, since the rotational wavefunction associated with the quantum number m=j has no nodes on the plane of the solid surface, the rotational motion of the m=j state resembles that of a helicopter and is thus called the helicopter mode. Similarly since the rotational motion of the m=j state resembles that of a cartwheel, it is thus called the cartwheel mode. This explicit m-dependence of the dissociation probability has also been found in theoretical calculations for H_2 on a Ni surface [205, 212]. A classical trajectory study [213] of associative desorption of H_2 from Cu also indicated that the helicopter mode (corresponding roughly to the m=j state) is preferred in hydrogen desorption from Cu. This is consistent with the common consensus that molecules with their rotational axes perpendicular to the metal surface (large quantum number m) can dissociate more easily than those with rotational axes parallel to the surface (small quantum number m).

Of course, the fundamental reason is that the molecule-surface interaction potential is least repulsive when the molecular axis of the diatom is parallel to the surface (θ=90 degrees) and becomes more repulsive as the axis tilts toward the surface.

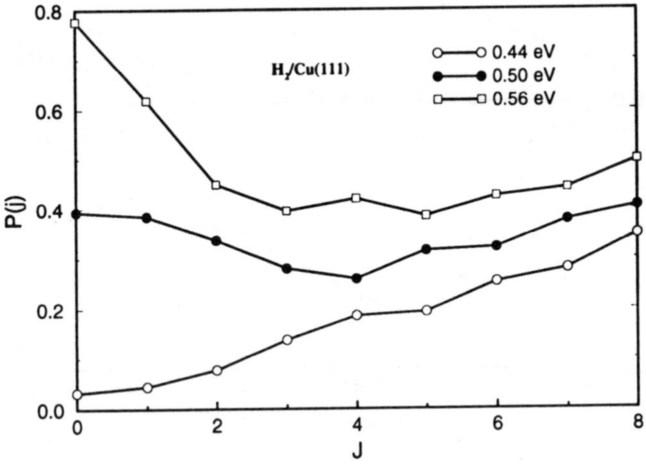

Figure 10.7: Degeneracy-averaged dissociation probability $\bar{P}(j)$ plotted as a function of rotational quantum number j at kinetic energies E_k=0.44, 0.50, and 0.56 eV for the dissociation of H_2 on Cu(111).

Since in nonpolarized experiments only degeneracy-averaged dissociation probabilities can be measured, one needs to average the dissociation probability $P(jm)$ over the $2j+1$ degeneracy. We can define a degeneracy-averaged dissociation probability for a given j as

$$\bar{P}(j) = \frac{1}{2j+1} \sum_{m=-j}^{m=j} P(jm) \tag{10.29}$$

Figure 10.7 plots the j-dependence of $\bar{P}(j)$ at several energies for H_2/Cu(111). At low kinetic energies, $\bar{P}(j)$ *increases* monotonically as j increases, as shown in the figure for E_k=0.44 eV. However, as the E_k increases, $\bar{P}(j)$ initially *decreases* as j increases and levels off at around j=5 before moving upward as j increases further. The latter behavior is consistent with experimental measurements of the mean kinetic energy of the rotational states of D_2 desorbed from Cu(111) [214].

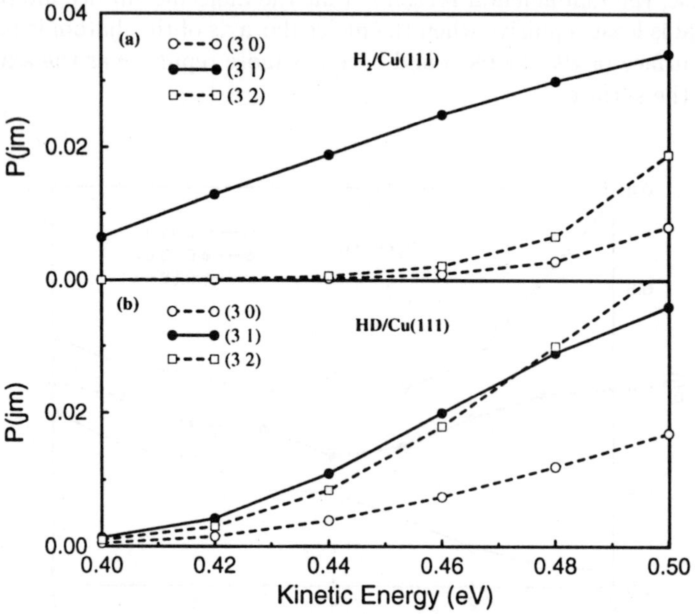

Figure 10.8: Dissociation probabilities P(jm) for j=3, m=0–2 as a function of kinetic energy for (a) H_2/Cu(111) and (b) HD/Cu(111).

Effect of homonuclear symmetry on hydrogen dissociation

Another important effect discovered in fixed site model studies of hydrogen dissociation on metals is a selection rule for dissociation of homonuclear diatoms on metals [201, 203, 205] as discussed in Sec. 10.2.4. This symmetry effect is displayed quite clearly in Fig. 10.8 where the rotation state-specific dissociation probability $P(jm)(j = 3)$ for hydrogen dissociation on Cu(111) as a function of collision energy is plotted for three values of $m (m = 0, 1, 2)$. The probability $P(33)$ is omitted because it is too large to fit in the figure. As is shown in the figure, the probability $P(31)$ ($j+m$=even) is significantly larger than the other two probabilities $P(30)$ and $P(32)$ ($j + m = odd$) for H_2/Cu. In fact at kinetic energies below 0.46 eV, the probabilities $P(30)$ and $P(32)$ are orders of magnitude smaller and essentially negligible compared with $P(31)$. This symmetry effect diminishes rather quickly as the energy increases and the reason for this is discussed in Sec. 10.2.4. However, for the dissociation of heteronuclear HD on Cu(111), the even and odd probabilities are of similar magnitude as shown in Fig. 10.8. As emphasized in Sec. 10.2.4, this symmetry effect only shows up at low kinetic energies at which the number of open product channels is severely limited due to the selection rule. Thus the effect is less pronounced if the total energy of the system is high or if the reaction is highly exothermic. This theoretical prediction of a symmetry effect remains to be verified experimentally.

Chapter 11

Semiclassical Descriptions of Quantum Mechanics

11.1 WKB Approximation

11.1.1 General Formalism

The semiclassical approximation in quantum mechanics was mainly developed for treating one-dimensional problems in which it has enjoyed great success. Despite many efforts, generalization of the semiclassical theory to practical applications in multidimensions has so far proven to be very difficult. Here we discuss the basic theory in the standard WKB approach to one-dimensional problems. We begin with the one-dimensional stationary Schrdinger equation

$$\frac{d^2\psi}{dx^2} + \frac{p^2(x)}{\hbar^2}\psi = 0 \tag{11.1}$$

where the momentum $p(x)$ is given by

$$p(x) = \sqrt{2m(E - V(x))} \tag{11.2}$$

If we start with the following *ansatz*

$$\psi(x) = e^{iS(x)/\hbar} \tag{11.3}$$

and substitute it into Eq. (11.1), we obtain an equation for the exponential function $S(x)$

$$-S'^2/\hbar^2 + iS''/\hbar + \frac{p^2(x)}{\hbar^2} = 0 \tag{11.4}$$

Equation (11.4) is nonlinear and no easier to solve than the original Schrödinger equation. However, in view of the semiclassical limit $\hbar \to 0$, we can expand the function $S(x)$ in powers of $\hbar$

$$S = S_0 + \hbar S_1 + \hbar^2 S_2 + \cdots \tag{11.5}$$

and group the terms in Eq. (11.4) according to powers of $\hbar$

$$[-S_0'^2 + p^2(x)] + \hbar[-2S_0'S_1' + iS_0''] + \hbar^2[-2S_0'S_2' - S_1'^2 + iS_1''] \cdots = 0 \tag{11.6}$$

By equating all the coefficients of $\hbar^n (n = 0, 1, 2, ...)$ to zero, we obtain the equations

$$S_0'^2 = p^2(x) \tag{11.7}$$

for the zeroth order and

$$S_1' = i(S_0''/2S_0') \tag{11.8}$$

for the first order of $\hbar$. These two equations are solved to yield the solutions

$$S_0 = \int p(x)dx \tag{11.9}$$

$$S_1 = \frac{i}{2}\ln|p(x)| \tag{11.10}$$

From the *ansatz* in Eq. (11.3), we obtain two linearly independent WKB solutions to the Schrödinger equation

$$\psi_{WKB} = \frac{1}{\sqrt{|p(x)|}} e^{\pm\frac{i}{\hbar}\int p(x)dx} \tag{11.11}$$

or their equivalent sine and cosine forms. In classically forbidden regions where $p^2(x) < 0$, the general solutions are exponentially increasing and decreasing functions

$$\psi_{WKB} = \frac{1}{\sqrt{|p(x)|}} e^{\pm\frac{1}{\hbar}\int |p(x)|dx} \tag{11.12}$$

The validity of the WKB approximation requires that the expansion in Eq. (11.5) is convergent and therefore satisfies the relation

$$|S_0| \gg |\hbar S_1| \tag{11.13}$$

By differentiating the above relation with respect to coordinate x, we obtain the validity condition for the WKB approximation

$$\left|\frac{d\lambda}{dx}\right| \ll 2\pi \tag{11.14}$$

where the de Broglie wavelength is defined as $\lambda(x) = 2\pi\hbar/p(x)$. This condition states that the variation of the de Broglie wavelength λ with coordinate has to be small in order for the WKB approximation to be applicable.

11.1.2 Turning Points and Connection Formulae

Direct application of the WKB solutions in Eqs. (11.11) and (11.12) is problematic because they break down at classical turning points where the classical momentum is zero or $p(x) = 0$. Thus one needs to remove singularities at the turning points by smoothly joining the WKB solutions on both sides of the turning points. In order to do that, let us investigate the correct behavior of the wavefunction near a turning point. The standard way to do this is to approximate the potential in the vicinity of a turning point by a linear function. Let's assume a single turning point at $x = x_0$ where the left side ($x < x_0$) is classically forbidden and the right side ($x > x_0$) is classically allowed as shown in Fig. 11.1. We can approximate the potential near x=x$_0$ by a linear form

$$V(x) \approx E - F(x - x_0) \tag{11.15}$$

where $F = -V'(x_0)$. The Schrödinger equation becomes

$$\left[\frac{d^2\psi(x)}{dx^2} + \frac{2m}{\hbar^2}F(x - x_0)\right]\psi(x) = 0 \tag{11.16}$$

The solutions to the Schrd̈inger equation with a linear potential are the well known Airy functions $A_i(z)$ and $B_i(z)$ [108] which are two linearly independent solutions of the equation

$$\left[\frac{d^2}{dz^2} - z\right]\left\{\begin{array}{c} A_i(z) \\ B_i(z) \end{array}\right\} = 0 \tag{11.17}$$

with $z = a(x_0 - x)$ and $a = -(2mF/\hbar^2)^{1/3}$. Thus near the region $x = x_0$ (region II), the correct wavefunction is given by a linear combination

$$\psi_{II} = \alpha A_i(z) + \beta B_i(z) \tag{11.18}$$

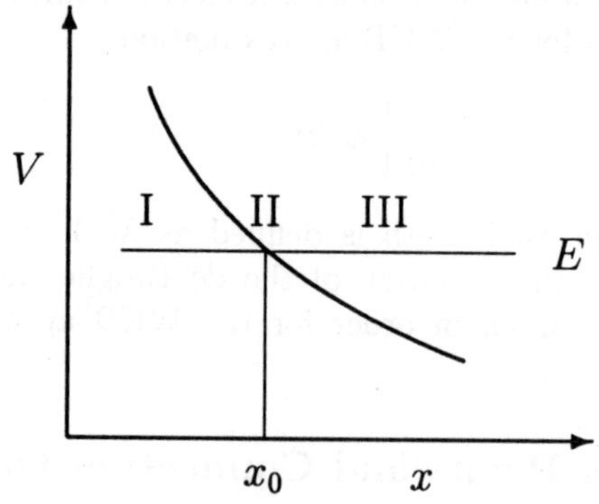

Figure 11.1: A single turning point at $x = x_0$.

The Airy functions have the asymptotic expansions in the limits of $z \to \infty (x \ll x_0)$ [108]

$$\begin{cases} A_i(z) \overset{z\to\infty}{\longrightarrow} \dfrac{1}{2\sqrt{\pi}} z^{-1/4} e^{-\frac{2}{3}z^{3/2}} \\ B_i(z) \overset{z\to\infty}{\longrightarrow} \dfrac{1}{\sqrt{\pi}} z^{-1/4} e^{\frac{2}{3}z^{3/2}} \end{cases} \tag{11.19}$$

and $z \to -\infty (x \gg x_0)$

$$\begin{cases} A_i(z) \overset{z\to-\infty}{\longrightarrow} \dfrac{1}{\sqrt{\pi}} (-z)^{-1/4} \sin\left(\dfrac{2}{3}(-z)^{3/2} + \dfrac{\pi}{4}\right) \\ B_i(z) \overset{z\to\infty}{\longrightarrow} \dfrac{1}{\sqrt{\pi}} (-z)^{-1/4} \cos\left(\dfrac{2}{3}(-z)^{3/2} + \dfrac{\pi}{4}\right) \end{cases} \tag{11.20}$$

In the classically forbidden region I ($x \ll x_0$), the WKB solution takes the exponential form

$$\psi_I = \frac{A}{\sqrt{|k(x)|}} e^{-\int_x^{x_0} |k(x)| dx} + \frac{B}{\sqrt{|k(x)|}} e^{\int_x^{x_0} |k(x)| dx} \tag{11.21}$$

with $k(x) = p(x)/\hbar$. There exists a common region in $x \ll x_0$ in which both solutions ψ_I of (11.21) and ψ_{II} of (11.18) are valid representations of the wavefunction. Therefore we can utilize the asymptotic expansions of the Airy functions in Eq. (11.19) and the relation

$$\int_x^{x_0} |k(x)| dx = \int_z^0 z^{1/2} dz = -\frac{2}{3} z^{3/2}, \tag{11.22}$$

to match the solutions of ψ_I and ψ_{II}. This gives rise to the following relations for the coefficients of the wavefunctions

$$\begin{cases} \alpha = 2\sqrt{\pi} a^{-1/6} A \\ \beta = \sqrt{\pi} a^{-1/6} B \end{cases} \tag{11.23}$$

In the classically allowed region III ($x \gg x_0$), the WKB solution can be written in the form

$$\psi_{III} = \frac{C}{\sqrt{k(x)}} \sin\left(\int_{x_0}^{x} k(x)dx + \frac{\pi}{4}\right) + \frac{D}{\sqrt{k(x)}} \cos\left(\int_{x_0}^{x} k(x)dx + \frac{\pi}{4}\right) \tag{11.24}$$

Again by using the asymptotic expansions of the Airy functions in Eq. (11.20), we can match the solutions of ψ_I and ψ_{II} to obtain

$$\begin{cases} \alpha = \sqrt{\pi} a^{-1/6} C \\ \beta = \sqrt{\pi} a^{-1/6} D \end{cases} \tag{11.25}$$

Combining equations (11.23) and (11.25), we obtain the connection formula between the WKB wavefunctions on both sides of a turning point x_0 in Fig. 11.1

$$\begin{aligned} C &\longleftrightarrow 2A \\ D &\longleftrightarrow B \end{aligned} \tag{11.26}$$

This connection formula can be equally applied to the case when the classically forbidden region is to the right of the turning point $x_0 (x > x_0)$ (right turning point) and the classically allowed region is to the left of $x_0 (x < x_0)$.

The special case of $B = D = 0$ is frequently encountered and gives a useful connection formula

$$\boxed{\frac{1}{\sqrt{|k(x)|}} e^{-\int_x^{x_0} |k(x)|dx} \longleftrightarrow \frac{2}{\sqrt{k(x)}} \sin\left(\int_{x_0}^{x} k(x)dx + \frac{\pi}{4}\right)} \tag{11.27}$$

For a turning point with the classically forbidden region to the right of x_0, one simply reverses the integration limit in Eq. (11.27) to obtain

$$\boxed{\frac{1}{\sqrt{|k(x)|}} e^{-\int_{x_0}^{x} |k(x)|dx} \longleftrightarrow \frac{2}{\sqrt{k(x)}} \sin\left(\int_{x}^{x_0} k(x)dx + \frac{\pi}{4}\right)} \tag{11.28}$$

The wavefunction on the right side of Eq. (11.27) can be rewritten as

$$\psi = \frac{e^{i\pi/4}}{\sqrt{k(x)}} \left[e^{-i\int_{x_0}^{x} k(x)dx} + e^{-i\pi/2} e^{i\int_{x_0}^{x} k(x)dx} \right] \tag{11.29}$$

which shows explicitly that the wavefunction has lost a phase of $\pi/2$ upon reflection from the turning point. This is a well known result in optics.

Semiclassical quantization

Using the connection formulae, the WKB approximation can be applied to a variety of one-dimensional problems. For example, for a bound state problem with two turning points a and b $(a < b)$, one can first apply the connection formula (11.27) to the left hand turning point a to obtain the WKB wavefunction in the classically allowed region $(a < x < b)$

$$\begin{aligned}\psi &= \frac{1}{\sqrt{k(x)}} \sin\left(\int_a^x k(x)dx + \frac{\pi}{4}\right) \\ &= \frac{1}{\sqrt{k(x)}} \sin\left[\int_a^b k(x)dx + \frac{\pi}{2} - \left(\int_x^b k(x)dx + \frac{\pi}{4}\right)\right] \end{aligned} \tag{11.30}$$

In order to match the wavefunction to the exponentially decaying solution in region $x > b$ using Eq. (11.28), we must have the relation

$$\int_a^b k(x)dx = \left(n + \frac{1}{2}\right)\pi \qquad (n = 0, 1, 2, \cdots) \tag{11.31}$$

which is the Sommerfield quantization relation. The more general formula is

$$\boxed{\oint k(x)dx = (2n + \delta)\pi} \qquad (n = 0, 1, 2, \cdots) \tag{11.32}$$

Equation (11.32) states that the total phase accumulated by the wavefunction after going through a round trip is equal to multiples of 2π. The δ is the total phase loss accumulated by the reflections of the wavefunction at both turning points a and b. For example, the reflection of the wavefunction from a soft wall causes a phase loss of $\pi/2$ as shown in Eq. (11.29). For a hard wall the phase loss is π.

11.2 Uniform Semiclassical Approximation

11.2.1 Comparison Equations

Since the structure of the turning points determines the shape of the wavefunction in the entire space, the existence of turning points makes the WKB solution complicated. The wavefunction takes different forms on different sides of a turning point and they need to be matched by a connection formula. In view of this problem, a general method was suggested originally by Langer [215] to obtain a single WKB solution that is valid in the entire space across the turning points. This is called the uniform semiclassical approximation. The basic idea is to map the solution of the original Schrödinger equation to the solution of a reference equation which can be solved analytically. The basic approach is discussed in the following.

Let $\psi(x)$ be a solution of the equation

$$\left[\frac{d^2}{dx^2} + k^2(x)\right]\psi(x) = 0 \tag{11.33}$$

with a single turning point at x_0. We seek a solution $\phi(x)$ in the form of

$$\phi(x) = A(x)\Psi[z(x)] \tag{11.34}$$

where the new variable z is a function of x and $\Psi(z)$ is the solution of a comparison equation

$$\left[\frac{d^2}{dz^2} + K^2(z)\right]\Psi(z) = 0 \tag{11.35}$$

with a corresponding turning point at $z_0 = z(x_0)$.

From the primitive WKB expression

$$\psi(x) = k^{-1/2}(x)\exp\left(i\int_{x_0}^{x} k(x)dx\right) \tag{11.36}$$

for ψ and

$$\Psi(z) = K^{-1/2}(z)\exp\left(i\int_{z_0}^{z} K(z)dz\right) \tag{11.37}$$

for Ψ and Eq. (11.34), we should define

$$A(x) = \left(\frac{K(z)}{k(x)}\right)^{1/2} \tag{11.38}$$

and impose the relation between the variables z and x by

$$\boxed{\int_{z_0}^{z} K(z)dz = \int_{x_0}^{x} k(x)dx} \tag{11.39}$$

which leads to the equation

$$\frac{dz}{dx} = \frac{k(x)}{K(z)} \tag{11.40}$$

Thus the uniform wavefunction in Eq. (11.34) is written as

$$\boxed{\phi(x) = \left(\frac{K(z)}{k(x)}\right)^{1/2} \Psi[z(x)]} \tag{11.41}$$

By substituting the expression of (11.41) in Eq. (11.33) and utilizing Eqs. (11.35) and (11.40), we can derive an equation for $\phi(x)$

$$\left[\frac{d^2}{dx^2} + k^2(x) + \gamma(x)\right] \phi = 0 \tag{11.42}$$

where $\gamma(x)$ is given by

$$\begin{aligned} \gamma(x) &= -\left(\frac{dz}{dx}\right)^{1/2} \frac{d^2}{dx^2} \left(\frac{dz}{dx}\right)^{-1/2} \\ &= -\left(\frac{k(x)}{K(z)}\right)^{1/2} \frac{d^2}{dx^2} \left(\frac{k(x)}{K(z)}\right)^{-1/2} \end{aligned} \tag{11.43}$$

Thus the newly defined function $\phi(x)$ will satisfy the same equation as the original function $\psi(x)$, if the extra term $\gamma(x)$ vanishes from Eq. (11.42) or if $\gamma(x)$ is very small relative to $k^2(x)$, i.e.,

$$\left|\frac{\gamma(x)}{k^2(x)}\right| \ll 1 \tag{11.44}$$

Under such conditions, the function $\phi(x)$ should be a good approximation to $\psi(x)$. The validity condition (11.44) is usually satisfied if the quantity $k(x)/K(z)$ is a slowly varying function of x. Even at turning points the condition (11.44) can be satisfied, provided that both $k(x)$ and $K(z)$ approach zero simultaneously. Since $k(x)$ vanishes at the turning points, $K(z)$ must be chosen to vanish also at the turning points. In order words, the turning point structure of $K(z)$ in z space has to be the same as that of $k(x)$ in x space in order to satisfy the uniform condition (11.44) everywhere (including all turning points). If $K(z)$ is chosen to be a constant, the uniform approximation is equivalent to the WKB approximation. Thus, the standard WKB approximation can be considered as a special case of the uniform approximation with a constant reference potential.

11.2.2 Applications of the Uniform Approximation

An isolated turning point

For an isolated turning point at x_0 with classically forbidden region to its left ($x < x_0$) such as shown in Fig. 11.1, we can choose $K^2(z) = z$ to define a comparison equation

$$\left[\frac{d^2}{dz^2} + z\right] \Psi(z) = 0 \tag{11.45}$$

where the turning point is chosen at z=0 in the z space. The solution of this equation is the Airy function defined by Eq. (11.17), i.e., $\Psi(z) = A_i(-z)$. From Eq. (11.39), we have

$$\int_{x_0}^{x} |k(x)|d = \int_0^z |K(z)|^{1/2} dz = \pm\frac{2}{3}|z|^{3/2} \tag{11.46}$$

which is valid in both the classically forbidden and classically allowed regions. Equation (11.46) can be solved to yield the equation

$$z = \pm\left(\frac{3}{2}\int_{x_0}^{x} |k(x)|dx\right)^{2/3} \tag{11.47}$$

Thus the uniform semiclassical wavefunction for a single turning point defined in Eq. (11.41) is now given by

$$\begin{aligned}\phi(x) &= \left[\frac{K(z)}{k(x)}\right]^{1/2} \Psi(z) \\ &= \boxed{\frac{z^{1/4}}{\sqrt{k(x)}} A_i[-z(x)]}\end{aligned} \tag{11.48}$$

which is a global approximate solution valid everywhere.

Using the uniform semiclassical wavefunction of Eq. (11.48) and the asymptotic expansion of $A_i(z)$, we can easily derive the connection formula for an isolated turning point. For example, in the classically allowed region, we use the asymptotic expansion for $A_i(z)$ in Eq. (11.20) to obtain the WKB wavefunction

$$\begin{aligned}\phi_R(x \gg x_0) \longrightarrow &\quad \frac{z^{1/4}}{k^{1/2}(x)}\frac{1}{\sqrt{\pi}}(-z)^{-1/4}\sin\left(\frac{2}{3}(-z)^{3/2} + \frac{\pi}{4}\right) \\ &= \frac{1}{\sqrt{\pi k(x)}}\sin\left(\int_{x_0}^{x} k(x)dx + \frac{\pi}{4}\right)\end{aligned} \tag{11.49}$$

Similarly using the asymptotic expansion of $A_i(z)$ in Eq. (11.19), we obtain the WKB wavefunction in the classically forbidden region

$$\phi_L(x \ll x_0) \longrightarrow \quad \frac{(-z)^{1/4}}{|k(x)|^{1/2}}\frac{1}{2\sqrt{\pi}}(-z)^{-1/4}e^{-\frac{2}{3}(-z)^{3/2}}$$

$$= \frac{1}{2\sqrt{\pi|k(x)|}} e^{-\int_{x_0}^{x} |k(x)|dx} \tag{11.50}$$

Equations (11.49) and (11.50) give exactly the connection formula of (11.27).

Potential well

In the case of a potential well with two turning points a and b ($a < b$) as shown in Fig. 11.2, the comparison equation can be chosen to be that of

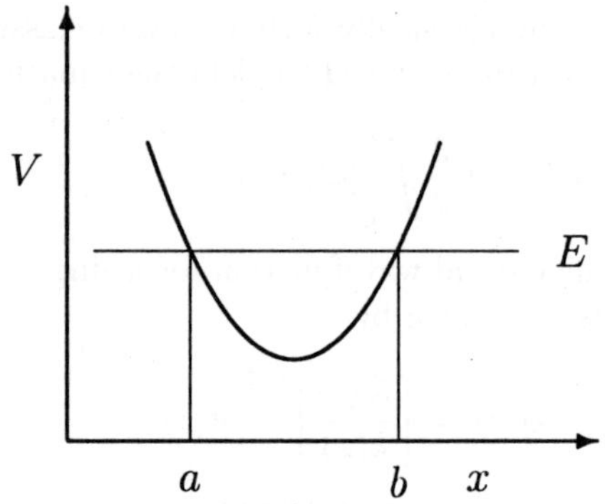

Figure 11.2: Potential well with two turning points a and b.

a harmonic oscillator

$$\left[\frac{d^2}{dz^2} + (2n+1-z^2)\right]\Psi(z) = 0 \tag{11.51}$$

with $K^2(z) = (2n+1-z^2)$ and the two turning points at $z_0 = \pm\sqrt{(2n+1)}$. The solution $\Psi(z)$ is the well known harmonic oscillator wavefunction

$$\Psi_{HO}(z) = \left(\frac{1}{\sqrt{\pi}2^n n!}\right)^{1/2} H_n(z) e^{-\frac{1}{2}z^2} \tag{11.52}$$

where $H_n(z)$ are the Hermite polynomials defined in the Appendix. The constant n is determined from Eq. (11.38)

$$\int_a^b k(x)dx = \int_{-z_0}^{z_0} (z_0^2 - z^2)^{1/2} dz$$
$$= \frac{z_0^2\pi}{2} = (n + \frac{1}{2})\pi \tag{11.53}$$

which is just the quantization condition of (11.31). Therefore n should take the values $0, 1, 2, \cdots$. The uniform semiclassical wavefunction for a potential well is thus given by

$$\boxed{\phi(x) = \frac{(2n+1-z^2)^{1/4}}{\sqrt{k(x)}}\Psi_{HO}(z)} \tag{11.54}$$

where the implicit relation between x and z is given by the equation

$$\begin{aligned}\int_a^x k(x)dx &= \int_{-z_0}^{z} (z_0^2 - z^2)^{1/2} dz \\ &= \frac{1}{2}z(z_0^2 - z^2)^{1/2} + \frac{z_0^2}{2}\left[\frac{\pi}{2} + \sin^{-1}\left(\frac{z}{z_0}\right)\right]\end{aligned} \tag{11.55}$$

Potential barrier

Since the semiclassical result for the case of a potential barrier is very important and useful, such as in barrier penetration or tunneling, we use the uniform semiclassical method to derive the result here. However, the mathematics involved is more complicated and tedious than in previous examples. For energies below the top of the barrier, there are two turning points a and b ($a < b$) as shown in Fig. 11.3.

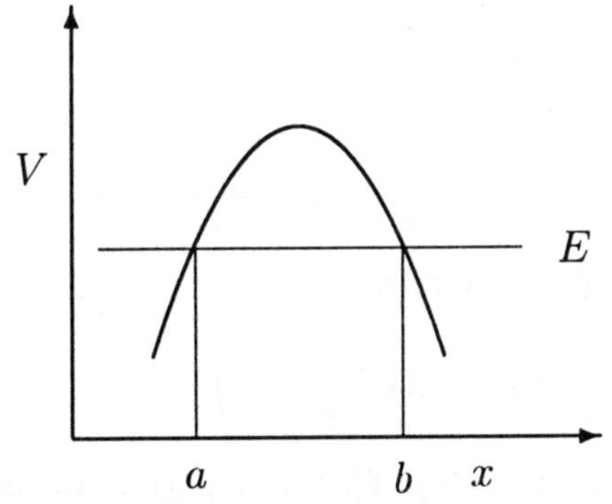

Figure 11.3: Potential barrier with two turning points a and b.

We can define a comparison equation

$$\left[\frac{d^2}{dz^2} + z^2 - t\right]\Psi(z) = 0 \quad (t \geq 0) \tag{11.56}$$

which describes the motion of a unit-mass particle in an inverted harmonic potential of $V = -\frac{1}{2}z^2$ with the energy $\epsilon = t/2$. The two turning points in the z space are $z_0 = \pm\sqrt{t}$.

The general solution to the comparison Eq. (11.56) is a linear combination of two independent parabolic cylinder functions $D(t, \pm z)$ [108], i.e.,

$$\Psi(z) = \alpha D(t, z) + \beta D(t, -z) \tag{11.57}$$

Thus the uniform semiclassical wavefunction is given by

$$\phi(x) = \left|\frac{z^2 - t}{k^2(x)}\right|^{1/4} [\alpha D(t, z) + \beta D(t, -z)] \tag{11.58}$$

The parabolic cylinder function has the following asymptotic expansions [108]

$$\begin{cases} D(t, z) \overset{z\to\infty}{\longrightarrow} \left(\dfrac{2}{z}\right)^{1/2} e^{i\varphi(t,z)} \\ D(t, -z) \overset{z\to\infty}{\longrightarrow} \dfrac{1}{2i\kappa}\left(\dfrac{2}{z}\right)^{1/2} \left[(1-\kappa^2)e^{i\varphi(t,z)} - (1+\kappa^2)e^{-i\varphi(t,z)}\right] \end{cases}$$

In the above function, κ is defined as [108]

$$\kappa = \left(1 + e^{\pi t}\right)^{1/2} - e^{\pi t/2} \tag{11.59}$$

the phase function $\varphi(t, z)$ is defined by

$$\varphi(t, z) = \frac{z^2}{2} - \frac{t}{2}\ln(\sqrt{2}z) + \frac{1}{2}\phi_2 + \frac{\pi}{4} \tag{11.60}$$

and the phase factor ϕ_2 is given by

$$\phi_2 = \arg\Gamma\left(\frac{1}{2} + i\frac{t}{2}\right) \tag{11.61}$$

where Γ is the Gamma function [216]. It is important to note in Eq. (11.59) that the asymptotic expansion of $D(t, -z)$ cannot be obtained by simply replacing z by $-z$ in the asymptotic expansion of $D(t, z)$ and vice versa. This phenomenon is called the Stokes phenomenon in mathematics [217]. The fundamental reason for the Stokes phenomenon is the fact that the region of convergence for the asymptotic expansion of $D(t, z)$ in the limit of $z \to \infty$ is different from that in the limit of $z \to -\infty$.

Using the asymptotic expansion of $D(t,z)$ in Eq. (11.59), we can derive an important and useful connection formula between the WKB wavefunctions on the two sides of a quadratic barrier. From Eqs. (11.57), (11.58) and (11.59), we obtain asymptotic expansions for the uniform wavefunction in the outer regions of the two turning points of the barrier

$$\begin{cases} \phi(x) \overset{x\gg b}{\longrightarrow} \dfrac{2}{\sqrt{k(x)}}\left(\alpha+\beta\dfrac{1-\kappa^2}{2i\kappa}\right)e^{i\varphi(t,z)} - \dfrac{2\beta}{\sqrt{k(x)}}\dfrac{1+\kappa^2}{2i\kappa}e^{-i\varphi(t,z)} \\ \phi(x) \overset{x\ll a}{\longrightarrow} \dfrac{2}{\sqrt{k(x)}}\left(\beta+\alpha\dfrac{1-\kappa^2}{2i\kappa}\right)e^{i\varphi(t,-z)} - \dfrac{2\alpha}{\sqrt{k(x)}}\dfrac{1+\kappa^2}{2i\kappa}e^{-i\varphi(t,-z)} \end{cases} \tag{11.62}$$

On the other hand, the standard WKB wavefunctions in the outer regions of the turning points are given by

$$\begin{cases} \psi(x) \overset{x\gg b}{\longrightarrow} \dfrac{A}{\sqrt{k(x}}e^{i\int_b^x k(x)dx} + \dfrac{B}{\sqrt{k(x)}}e^{-i\int_b^x k(x)dx} \\ \psi(x) \overset{x\ll a}{\longrightarrow} \dfrac{C}{\sqrt{k(x}}e^{i\int_a^x k(x)dx} + \dfrac{D}{\sqrt{k(x)}}e^{-i\int_a^x k(x)dx} \end{cases} \tag{11.63}$$

Since the indefinite integral $\int K(z)dz$ is known analytically [216]

$$\int K(z)dz = \int \sqrt{z^2-t}\,dz = \frac{z\sqrt{z^2-t}}{2} - \frac{t}{2}ln(\sqrt{z^2-t}+z) \tag{11.64}$$

we use Eq. (11.64) to obtain (for $x \gg b$)

$$\begin{aligned} \int_b^x k(x)dx &= \int_{\sqrt{t}}^z K(z)dz \\ &\overset{z\gg\sqrt{t}}{\longrightarrow} \frac{z^2}{2} - \frac{t}{2}\ln z + \frac{t}{4}\ln\left(\frac{t}{e}\right) - \frac{t}{2}\ln 2 \\ &= \varphi(t,z) + \frac{t}{4}\ln\left(\frac{t}{2e}\right) - \frac{1}{2}\phi_2 - \frac{\pi}{4} \\ &= \varphi(t,z) + w \end{aligned} \tag{11.65}$$

where the phase w is defined as

$$w = \frac{t}{4}\ln\left(\frac{t}{2e}\right) - \frac{1}{2}\phi_2 - \frac{\pi}{4} \tag{11.66}$$

Similarly for $x \ll a$, we obtain

$$\int_x^a k(x)dx = \int_z^{-\sqrt{t}} K(z)dz \overset{z\ll-\sqrt{t}}{\longrightarrow} \varphi(t,-z) + w \tag{11.67}$$

Using Eqs. (11.65) and (11.67) and equating (11.62) with (11.63), we obtain the relations between the various coefficients

$$\begin{cases} A = \left(\alpha + \beta\dfrac{1-\kappa^2}{2i\kappa}\right)e^{-iw} \\ B = -\beta\dfrac{1+\kappa^2}{2i\kappa}e^{iw} \end{cases} \tag{11.68}$$

and

$$\begin{cases} C = -\alpha\dfrac{1+\kappa^2}{2i\kappa}e^{iw} \\ D = \left(\beta + \alpha\dfrac{1-\kappa^2}{2i\kappa}\right)e^{-iw} \end{cases} \tag{11.69}$$

By eliminating α and β, and utilizing Eq. (11.59) for κ, we obtain the connection formulae

$$\begin{cases} C = \left(1+e^{\pi t}\right)^{1/2} e^{i\frac{\pi}{2}} e^{2iw} A + ie^{\frac{\pi}{2}t} B \\ D = -ie^{\frac{\pi}{2}t} A + e^{-i\frac{\pi}{2}} e^{-2iw}\left(1+e^{\pi t}\right)^{1/2} B \end{cases} \tag{11.70}$$

We further define a new phase factor

$$\bar{w} = 2w + \frac{\pi}{2} = 2\left[\frac{t}{4}\ln\left(\frac{t}{2e}\right) - \frac{1}{2}\phi_2 - \frac{\pi}{4}\right] + \frac{\pi}{2} \tag{11.71}$$

and an action integral

$$\theta = \int_a^b |k(x)|dx = \int_{-\sqrt{t}}^{\sqrt{t}} \sqrt{|z^2 - t|}dz = \frac{\pi t}{2} \tag{11.72}$$

where the result of Eq. (11.53) has been used in the last equation.

Using the above definitions for $\bar{w}$ and θ, the connection formulae of Eq. (11.70) can be written in matrix form

$$\begin{bmatrix} C \\ D \end{bmatrix} = \begin{bmatrix} \sqrt{(1+p^2)}e^{i\bar{w}} & ip \\ -ip & \sqrt{(1+p^2)}e^{-i\bar{w}} \end{bmatrix}\begin{bmatrix} A \\ B \end{bmatrix} \tag{11.73}$$

with $p = e^{\theta}$. The inverse of Eq. (11.73) is

$$\begin{bmatrix} A \\ B \end{bmatrix} = \begin{bmatrix} \sqrt{(1+p^2)}e^{-i\bar{w}} & -ip \\ ip & \sqrt{(1+p^2)}e^{i\bar{w}} \end{bmatrix}\begin{bmatrix} C \\ D \end{bmatrix} \tag{11.74}$$

Equation (11.73) is the general formula that connects the WKB wavefunctions on both sides of a barrier.

The connection formula (11.74) can be applied to obtain the semiclassical tunneling probability under a potential barrier. For a quantum particle incident from the left of the barrier, there is only transmitted wave on the right side of the barrier. Therefore the coefficient B in the wavefunction expression of Eq. (11.63) is zero. From Eq. (11.74), we thus obtain the transmission coefficient

$$T = \frac{A}{C} = \frac{\exp(-i\bar{w})}{[1+e^{2\theta}]^{1/2}} \tag{11.75}$$

which gives the transmission probability

$$\boxed{|T|^2 = \frac{1}{1+e^{2\theta}}} \tag{11.76}$$

Note that the transmission probability given by Eq. (11.76) is the correct (uniform) semiclassical result. The standard WKB result $|T|^2 = e^{-2\theta}$ is only valid when the condition $e^{2\theta} \gg 1$ is satisfied as can be seen clearly from Eq. (11.76). Equation (11.76) also gives the very neat result $|T|^2 = 1/2$ when $\theta = 0$ or when the energy of the particle is equal to the top of the barrier.

11.2.3 Langer Modification

In many applications, the potential is singular at the origin $r = 0$, such that

$$k^2(r) \overset{r\to 0}{\longrightarrow} \frac{\alpha}{r^n} (n > 0) \tag{11.77}$$

In the widely encountered radial Schrödinger equation, the radial coordinate r is in the range of $[0, \infty]$ and the potential is singular near the origin due to either the presence of the Coulomb potential or the centrifugal potential. In addition, the interaction potential itself may be singular. In these situations, the standard WKB approximation leads to incorrect results because the WKB wavefunction has the wrong behavior near the origin. To understand this problem, let us examine the validity condition of (11.44). From Eq. (11.43), it is clear that $\gamma(x)$ will be singular at the origin unless $K(z)$ is chosen to be singular also at the origin to cancel out the singularity of $k(x)$, which is in general difficult to implement in practice. In the case of a constant $K(z)$ (corresponding to the WKB approximation), $\gamma(r)$ becomes

$$\gamma(r) \overset{r\to 0}{\longrightarrow} \frac{n(n-4)}{16r^2} \tag{11.78}$$

Thus near the origin, the validity condition (11.44) becomes

$$\left|\frac{n(n-4)r^{n-2}}{16\alpha}\right| \ll 1 \tag{11.79}$$

The above inequality always holds near the origin if $n > 2$. However for $0 < n \leq 2$, the condition (11.79) is in general not satisfied near the origin.

To remove the singularity at r=0, Langer made the substitution [215]

$$\begin{cases} r = e^x \\ \psi(r) = e^{\frac{x}{2}}\Psi(x) \end{cases} \qquad (-\infty \leq x \leq \infty) \tag{11.80}$$

The equation for the new wavefunction $\Psi(x)$ can be shown to satisfy the equation

$$\left[\frac{d^2}{dx^2} + \Gamma^2(x)\right]\Psi(x) = 0 \tag{11.81}$$

where

$$\Gamma^2(x) = e^{2x}k^2(e^x) - \frac{1}{4} = e^{2x}k_m^2(r) \tag{11.82}$$

where the modified momentum is

$$k_m^2(r) = k^2(r) - \frac{1}{4r^2} \tag{11.83}$$

Since

$$k(r) = \frac{2\mu}{\hbar^2}(E - V) \tag{11.84}$$

the modification of Eq. (11.83) is identical to the addition of an extra potential

$$\boxed{V_m(r) = V(r) + \frac{\hbar^2}{8\mu r^2}} \tag{11.85}$$

It can be shown that the validity condition (11.44) for $\Gamma(x)$ with constant $K(z)$ is satisfied near origin. Thus we can safely apply the WKB approximation to obtain the wavefunction from Eq. (11.81)

$$\begin{aligned} \Psi(x) &= \frac{1}{\sqrt{\Gamma(x)}} e^{\pm i \int \Gamma(x) dx} \\ &= e^{-\frac{x}{2}} \frac{1}{\sqrt{k_m(r)}} e^{\pm i \int k_m(r) dr} \end{aligned} \tag{11.86}$$

From Eq. (11.80), we obtain the WKB solution to the original problem

$$\psi(r) = \frac{1}{\sqrt{k_m(r)}} e^{\pm i \int k_m(r) dr} \tag{11.87}$$

which is simply the WKB solution to the modified Schrödinger equation

$$\left[\frac{d^2}{dr^2} + k_m^2(r)\right] \psi(r) = 0 \tag{11.88}$$

Thus, using the Langer modification of (11.83), the WKB approximation can be directly applied to solve Eq. (11.88). A common application of the Langer modification arises in solving the radial equation with a centrifugal potential

$$k^2(r) = \frac{2m}{\hbar^2}\Big(E - V(r)\Big) - \frac{l(l+1)}{r^2} \tag{11.89}$$

The modified potential becomes

$$\begin{aligned} k_m^2(r) &= \frac{2m}{\hbar^2}(E - V(r)) - \frac{l(l+1)}{r^2} - \frac{1}{4r^2} \\ &= \frac{2m}{\hbar^2}(E - V(r)) - \frac{(l+\frac{1}{2})^2}{r^2} \end{aligned} \tag{11.90}$$

which is the well known Langer modification of

$$\boxed{l(l+1) \longrightarrow (l+\frac{1}{2})^2} \tag{11.91}$$

for centrifugal potentials.

The validity condition of (11.79) shows that if the potential is more singular than quadratic ($n > 2$), there is generally no need to apply the Langer modification because the singularity of the potential will automatically dominate the behavior of the wavefunction near the origin.

11.3 Semiclassical Scattering

11.3.1 Phase Shift in Elastic Scattering

For elastic scattering in a central potential $V(r)$, the Langer modification is needed to calculate the scattering phase shift. One thus solves the modified radial Schrödinger equation

$$\left[\frac{d^2}{dr^2} + \frac{2m}{\hbar^2}\left(E - V - \frac{\hbar^2}{2m}\frac{(l+\frac{1}{2})^2}{r^2}\right)\right] \psi(r) = 0 \tag{11.92}$$

The WKB solution can be written as

$$\psi(r) = A\sin\left(\int_{r_0}^{r} k_m(r)dr + \frac{\pi}{4}\right) \tag{11.93}$$

where the lower integration limit r_0 is the turning point determined by the equation

$$E = V(r_0) + \frac{\hbar^2}{2m}\frac{(l+\frac{1}{2})^2}{r_0^2} \tag{11.94}$$

Comparing the radial solution Eq. (11.93) with the standard asymptotic form in Eq. (4.139), we obtain the phase shift

$$\begin{aligned}\delta_l &= \int_{r_0}^{\infty} [k_m(r) - k_0(r)]\, dr - k_0 r_0 + (l + \frac{1}{2})\frac{\pi}{2} \\ &= \int_{r_0}^{\infty} \left[\left(\frac{2m}{\hbar^2}(E-V) - \frac{(l+\frac{1}{2})^2}{r^2}\right)^{1/2} - \sqrt{\frac{2mE}{\hbar^2}}\right] dr \\ &\quad -\sqrt{\frac{2mE}{\hbar^2}}r_0 + (l+\frac{1}{2})\frac{\pi}{2}\end{aligned} \tag{11.95}$$

It can be shown that the phase shift δ_l in Eq. (11.95) vanishes if the potential $V(r)$ is zero. However, without the Langer modification, δ_l does not vanish when $V(r)$ does.

From Eq. (11.95), we can write

$$\begin{aligned}\Theta(l) &= 2\frac{d\delta_l}{dl} \\ &= \boxed{\pi - (l+\frac{1}{2})\int_{r_0}^{\infty} \frac{\hbar dr}{r^2\sqrt{2m[E-V(r)] - \hbar^2(l+\frac{1}{2})^2/r^2}}}\end{aligned} \tag{11.96}$$

which is the exact deflection function $\Theta(l)$ in classical scattering.

11.3.2 Elastic Cross Sections

In elastic scattering, the differential cross section is given by

$$d\sigma = 2\pi \sin\theta |f(\theta)|^2 d\theta \tag{11.97}$$

where the scattering amplitude is expressed as a sum over partial waves (cf. 4.162)

$$\begin{aligned} f(\theta) &= \frac{1}{2ik}\sum_{l=0}^{\infty}(2l+1)\left[e^{i2\delta_l}-1\right]P_l(\cos\theta) \\ &= \frac{1}{2ik}\sum_{l=0}^{\infty}(2l+1)e^{i2\delta_l}P_l(\cos\theta) \end{aligned} \tag{11.98}$$

where the second term in Eq. (4.162) vanishes because

$$\sum_{l=0}^{\infty}(2l+1)P_l(\cos\theta) = 0 \tag{11.99}$$

for $\theta \neq 0$ (cf. Eqs. (A.29) and (A.32)).

In the semiclassical limit, many partial waves contribute to the scattering cross section and the dominant contribution is from certain large l values. Thus we can approximate the discrete sum over l in Eq. (4.162) by an integration over a continuous variable l by the prescription

$$\sum_{l=0}^{\infty}(2l+1) \longrightarrow \int_0^{\infty} 2(l+1)\ dl \simeq \int_0^{\infty} 2\lambda d\lambda \tag{11.100}$$

where $\lambda = l + 1/2$. Equation (11.98) then becomes

$$f_{sc}(\theta) \simeq \frac{1}{2ik}\int_0^{\infty} 2\lambda e^{i2\delta(\lambda)}P_l(\cos\theta)d\lambda \tag{11.101}$$

For not too small scattering angles, the condition $\sin\theta \geq 1/l$ is satisfied and we can use the asymptotic expansion for Legendre polynomials given by Eq. (A.18) in the appendix to simplify the integral

$$f_{sc}(\theta) = \int_0^{\infty} g(\theta,\lambda)\left[e^{i\phi_+(\theta,\lambda)} - e^{i\phi_-(\theta,\lambda)}\right]d\lambda \tag{11.102}$$

where

$$g(\theta,\lambda) = -\sqrt{\frac{\lambda}{2\pi k^2\sin\theta}} \tag{11.103}$$

and

$$\phi_{\pm}(\theta,\lambda) = 2\delta(\lambda) \pm (\lambda\theta + \pi/4) \tag{11.104}$$

Applying the stationary phase approximation (see Sec. 11.4.1), the dominant contribution to the integral in Eq. (11.102) is given by the stationary point λ_s

$$\frac{d\phi_\pm}{d\lambda_s} = 0 \tag{11.105}$$

which yields the equation

$$\Theta(\lambda_s) = 2\frac{d\delta(\lambda_s)}{d\lambda_s} = \mp\theta \tag{11.106}$$

where $\Theta(\lambda)$ is just the classical deflection function given in Eq. (11.96).

The scattering amplitude in the stationary phase approximation is then given by

$$\begin{aligned} f_{sc}(\theta) &\simeq g(\theta, \lambda_s)\sqrt{\frac{2\pi i}{\phi''_\pm}}\exp\left[i\phi_\pm(\theta, \lambda_s)\right] \\ &= \sqrt{\frac{\lambda_s}{k^2 \sin\theta |\Theta'(\lambda_s)|}}\exp\left[i\phi_\pm(\theta, \lambda_s) + (\text{sign})i\pi/4\right] \end{aligned} \tag{11.107}$$

where

$$\phi_\pm(\theta, \lambda_s) = 2\delta(\lambda_s) \pm (\lambda_s\theta + \pi/4) \tag{11.108}$$

and the sign is either plus or minus depending on the sign of the second derivative $d^2\phi_\pm/d\lambda_s^2$. If more than one stationary point contributes to the integral, Eq. (11.107) involves the sum over all of them

$$f_{sc}(\theta) \simeq \sum_s \sqrt{\frac{\lambda_s}{k^2 \sin\theta |\Theta'(\lambda_s)|}}\exp\left[i\phi_\pm(\theta, \lambda_s) + (\text{sign})i\pi/4\right] \tag{11.109}$$

It is important to note that Eq. (11.109) looks identical to the classical result by using the correspondence relation $\lambda = l + 1/2 \longrightarrow kb$ where b is the classical impact parameter. However, there is a phase factor in the semiclassical result of (11.109) which is responsible for interference patterns in quantum scattering.

11.4 Path Integral and Stationary Phase Approximation

11.4.1 Stationary Phase Approximation

One dimensional case

At the heart of the semiclassical approximation is the stationary phase approximation (SPA). To see how the SPA works, let us examine the integral

$$I = \int_{-\infty}^{\infty} dx \; g(x) e^{\frac{i}{\hbar} f(x)} \tag{11.110}$$

in the limit $\hbar \to 0$. Because of the rapidly oscillating phase in the integrand as $\hbar \to 0$, the most important contribution to the integral comes from the integrand in a small region near the stationary point x_0 for which the stationary phase condition $f'(x_0) = 0$ is satisfied. The contribution from the integrand in other regions of space is negligibly small as $\hbar \to 0$. Thus we can expand the function $f(x)$ around the stationary phase point x_0 to second order

$$f(x) \simeq f(x_0) + \frac{1}{2}(x - x_0)^2 f''(x_0) \tag{11.111}$$

Utilizing the well known Gaussian integral result

$$\int_{-\infty}^{\infty} dx e^{iax^2} dx = \left(\frac{\pi i}{a}\right)^{1/2} = \left(\frac{\pi}{|a|}\right)^{1/2} e^{\mathrm{sign}[a] i\pi/4} \tag{11.112}$$

where $\mathrm{sign}[a] = \pm 1$ depending on whether a is positive or negative, the integral in Eq. (11.110) can be evaluated by the stationary phase approximation (in the limit of $\hbar \to 0$)

$$\begin{aligned} I &= e^{\frac{i}{\hbar} f(x_0)} \int_{-\infty}^{\infty} dx g(x) e^{\frac{i}{\hbar}\frac{1}{2}(x-x_0)^2 f''(x_0)} \\ &= g(x_0) \left(\frac{2\pi i\hbar}{f''(x_0)}\right)^{1/2} e^{\frac{i}{\hbar} f(x_0)} \\ &= g(x_0) \left(\frac{2\pi\hbar}{|f''(x_0)|}\right)^{1/2} e^{\mathrm{sign}[f''(x_0)]\frac{i\pi}{4}} e^{\frac{i}{\hbar} f(x_0)} \end{aligned} \tag{11.113}$$

where $\mathrm{sign}[f''(x_0)] = \pm 1$ depending on whether $f''(x_0)$ is positive or negative.

In general, there can be more than one stationary phase point $x_k(k = 1, 2, \cdots)$. Assuming these points are sufficiently separated, the SPA integral in Eq. (11.113) can be generalized to

$$\boxed{I = \sum_k g(x_k) \left(\frac{2\pi i\hbar}{f''(x_k)}\right)^{1/2} e^{\frac{i}{\hbar}f(x_k)}} \tag{11.114}$$

Multidimensional case

Denoting $\mathbf{x}$ as an N-dimensional vector with components $x_i(i = 1, 2, \cdots, N)$, the Gaussian integral is a generalization of Eq. (11.112)

$$I = \int_{-\infty}^{\infty} e^{i\mathbf{x}^T\mathbf{A}\mathbf{x}}\, d\mathbf{x} = \left[\frac{(\pi i)^N}{|\mathbf{A}|}\right]^{1/2} e^{-iM\pi/2} \tag{11.115}$$

where $|\mathbf{A}|$ is the absolute value of the determinant of the matrix $\mathbf{A}$ and M is the number of negative eigenvalues of the matrix $\mathbf{A}$ which is also called the Maslov index. Utilizing Eq. (11.115), the multidimensional SPA integral becomes

$$\int_{-\infty}^{\infty} d\mathbf{x}\; g(\mathbf{x}) e^{\frac{i}{\hbar}f(\mathbf{x})} = \sum_\alpha g(\mathbf{x}_\alpha) \left[\frac{(2\pi i\hbar)^N}{|\mathbf{f}''(\mathbf{x}_\alpha)|}\right]^{1/2} e^{-iM\pi/2} e^{\frac{i}{\hbar}f(\mathbf{x}_\alpha)} \tag{11.116}$$

where the second derivative matrix is defined as

$$\mathbf{f}''_{ij}(\mathbf{x}_\alpha) = \left.\frac{\partial^2 f(\mathbf{x})}{\partial x_i \partial x_j}\right|_{\mathbf{x}=\mathbf{x}_\alpha} \tag{11.117}$$

We will see how the stationary phase approximation is intimately connected with the semiclassical approximation of quantum mechanics in the following subsection.

11.4.2 Path Integral Representation of Quantum Mechanics

One degree of freedom

Let us examine the matrix element of the evolution operator in one dimension

$$K(x, y, t) = <x|e^{-\frac{i}{\hbar}Ht}|y> \tag{11.118}$$

where the Hamiltonian is given by

$$H = T + V = -\frac{\hbar^2}{2m}\frac{d^2}{dx^2} + V(x) \tag{11.119}$$

If we split the propagator into many small time slices $\epsilon = t/N$, we can rewrite Eq. (11.118) as

$$K(x, y, t) = \lim_{\epsilon\to 0}\int dx_1 dx_2 \cdots dx_{N-1} <x|e^{-\frac{i}{\hbar}H\epsilon}|x_1>$$
$$\times <x_1|e^{-\frac{i}{\hbar}H\epsilon}|x_2><x_2|\cdots|x_{N-1}><x_{N-1}|e^{-\frac{i}{\hbar}H\epsilon}|y> \tag{11.120}$$

By splitting the short time propagator $e^{-\frac{i}{\hbar}H\epsilon}$ as (Trotter product) [218]

$$e^{-\frac{i}{\hbar}H\epsilon} = e^{-\frac{i}{\hbar}T\epsilon}e^{-\frac{i}{\hbar}V\epsilon} + O(\epsilon^2) \tag{11.121}$$

the matrix element of the short time propagator can easily be calculated

$$\begin{aligned}
&<x_1|e^{-\frac{i}{\hbar}H\epsilon}|x_2>= \\
&= \int <x_1|e^{-\frac{i}{\hbar}T\epsilon}|p><p|e^{-\frac{i}{\hbar}V\epsilon}|x_2>\, dp \\
&= \frac{1}{2\pi\hbar}\int \exp\left[-\frac{i}{\hbar}\left(\frac{p^2}{2m}\epsilon + p(x_2 - x_1)\right)\right] e^{-\frac{i}{\hbar}V(x_2)\epsilon} dp \\
&= \sqrt{\frac{m}{2\pi i\hbar\epsilon}}\exp\left[\frac{i}{\hbar}\left(\frac{m}{2}\left(\frac{x_2 - x_1}{\epsilon}\right)^2 - V(x_2)\right)\epsilon\right]
\end{aligned} \tag{11.122}$$

where the momentum wavefunction

$$<x|p>= \frac{1}{\sqrt{2\pi\hbar}}e^{\frac{i}{\hbar}px} \tag{11.123}$$

and the Gaussian integral result

$$\int_{-\infty}^{\infty} e^{-ax^2+bx}dx = \sqrt{\frac{\pi}{a}}\exp\left(\frac{b^2}{4a}\right) \tag{11.124}$$

have been used to obtain the final result in Eq. (11.122).

It is recognized that the exponent in Eq. (11.122) is just the classical action $S(x_1, x_2, \epsilon)$ [219], viz.

$$\begin{aligned}
S(x_1, x_2, \epsilon) &= \left[\frac{m}{2}\left(\frac{x_2 - x_1}{\epsilon}\right)^2 - V(x_2)\right]\epsilon \\
&= \int_0^{\epsilon}\left(\frac{1}{2}mv^2 - V\right)dt \\
&= \int_0^{\epsilon} L(x_1, x_2, t)dt
\end{aligned} \tag{11.125}$$

where L is the Lagrangian

$$L(x_1, x_2, t) = \frac{m}{2}v^2 - V = T - V \tag{11.126}$$

Equation (11.122) can thus be rewritten as

$$\boxed{<x_1|e^{-\frac{i}{\hbar}H\epsilon}|x_2> \stackrel{\epsilon\to 0}{\longrightarrow} \left(\frac{1}{2\pi i\hbar}\left|\frac{\partial^2 S}{\partial x_i \partial x_2}\right|\right)^{1/2} e^{\frac{i}{\hbar}S(x_1,x_2,\epsilon)}} \tag{11.127}$$

Using Eq. (11.127), the finite time propagator in Eq. (11.118) can be written in the form

$$K(x, y, t) = \left(\frac{m}{2\pi i\hbar\epsilon}\right)^{\frac{N}{2}} \int dx_1 dx_2 \cdots dx_{N-1} e^{\frac{i}{\hbar}S(x,x_1,x_2,\cdots,x_{N-1},y,t)} \tag{11.128}$$

where

$$\begin{aligned} S(x, y, t) &= \sum_{i=1}^{N-1} S(x_{i-1}, x_i, \epsilon) \\ &= \int_0^t L(x, x_1, x_2, \cdots, x_{N-1}, y, t')dt' \end{aligned} \tag{11.129}$$

is the classical action evaluated through a particular broken line path connecting $x = x_0, x_1, x_2, \cdots, x_{N-1}, x_N = y$. Thus the integration over all the intermediate coordinates in Eq. (11.128) can be interpreted as summing over all possible broken line paths connecting x and y in time t. In the limit of $N \to \infty$, this summation becomes an integration over all continuous paths

$$K(x, y, t) = C \int D[x] e^{\frac{i}{\hbar}S(x,y,t)} \tag{11.130}$$

where the symbol $\int D[x]$ denotes integration over all possible paths and C is a normalization constant given by

$$C = \lim_{N\to\infty} \left(\frac{m}{2\pi i\hbar\epsilon}\right)^{\frac{N}{2}} \tag{11.131}$$

From the derivation of the short time propagator in Eq. (11.122), the free particle propagator with the Hamiltonian $H_0 = p^2/2m$ is easily obtained by replacing ϵ by t and setting $V(x)$=0

$$<x_1|e^{-\frac{i}{\hbar}H_0 t}|x_2> = \left(\frac{m}{2\pi i\hbar t}\right)^{1/2} \exp\left[\frac{i}{\hbar}\frac{m}{2}\frac{(x_2 - x_1)^2}{t}\right] \tag{11.132}$$

Multidegrees of freedom

In the case of n degrees of freedom, the Hamiltonian is written as

$$H = \sum_{\alpha=1}^{n} T_\alpha + V$$
$$= \sum_{\alpha=1}^{n} \left(-\frac{\hbar^2}{2m_\alpha}\frac{d^2}{dx_\alpha^2}\right) + V(\mathbf{x}) \tag{11.133}$$

The one-dimensional short time propagator of Eq. (11.122) is easily generalized to the multidimensional case

$$K(\mathbf{x}_1, \mathbf{x}_2, \epsilon) = <\mathbf{x}_1|e^{-\frac{i}{\hbar}H\epsilon}|\mathbf{x}_2> \prod_{\alpha=1}^{n}\left(\frac{m_\alpha}{2\pi i\hbar\epsilon}\right)^{1/2}$$
$$\times \exp\left[\frac{i}{\hbar}\left(\sum_{\alpha=1}^{n}\frac{\alpha}{2}\left(\frac{x_{\alpha 2}-x_{\alpha 1}}{\epsilon}\right)^2 - V(\mathbf{x}_2)\right)\epsilon\right] \tag{11.134}$$

Similarly, the multidimensional generalization of the one-dimensional finite time propagator of Eq. (11.130) is

$$K(\mathbf{x}, \mathbf{y}, t) = C\int D[\mathbf{x}]e^{\frac{i}{\hbar}S(\mathbf{x},\mathbf{y},t)} \tag{11.135}$$

11.4.3 Stationary Phase Approximation of the Path Integral

In the semiclassical limit of $\hbar \to 0$, the phase in the path integral such as Eq. (11.128) is a rapidly oscillating function of the path. The most important contributions to the integral are those of classical paths for which the classical actions S_c are stationary,

$$\delta S_c(x, y, t) = \delta\int_x^y L(x, y, t)dt = 0 \tag{11.136}$$

Thus, we can apply the $N-1$ dimensional stationary phase approximation of Eq. (11.116) to evaluate the integral

$$<x|e^{-\frac{i}{\hbar}Ht}|y> = \left(\frac{m}{2\pi i\hbar\epsilon}\right)^{\frac{N}{2}}\int dx_1 dx_2\cdots dx_{N-1}e^{\frac{i}{\hbar}S(x,x_1,x_2,\cdots,x_{N-1},y,t)}$$
$$\overset{\hbar\to 0}{\longrightarrow} \sum_c A_c(x, y, t)e^{-i\frac{\pi}{4}}e^{\frac{i}{\hbar}S_c(x,y,t)}e^{-iM_c\frac{\pi}{2}} \tag{11.137}$$

where the summation is over all classical paths. The positive definite amplitude $A_c(x, y, t)$ is given by

$$
\begin{aligned}
A_c(x,y,t) &= \left(\frac{m}{2\pi\hbar\epsilon}\right)^{\frac{N}{2}} \left[\frac{(2\pi\hbar)^{N-1}}{|\mathbf{S}''_{N-1}|}\right]^{1/2} \\
&= \left[\frac{1}{2\pi\hbar}\left(\frac{m}{\epsilon}\right)^{N} \frac{1}{|S''_{N-1}|}\right]^{1/2}
\end{aligned}
\tag{11.138}
$$

which can be simplified to

$$
A_c(x,y,t) = \left[\frac{1}{2\pi\hbar}\left|\frac{\partial^2 S_c(x,y,t)}{\partial x \partial y}\right|\right]^{1/2} \tag{11.139}
$$

Thus the explicit expression of the propagator in the stationary phase approximation can be written as

$$
\boxed{<x|e^{-\frac{i}{\hbar}Ht}|y> \xrightarrow{SP} \sum_c \left[\frac{1}{2\pi i\hbar}\left|\frac{\partial^2 S_c(x,y,t)}{\partial x \partial y}\right|\right]^{1/2} e^{\frac{i}{\hbar}S_c(x,y,t)} e^{-iM_c\frac{\pi}{2}}} \tag{11.140}
$$

where the phase factor $e^{-i\frac{\pi}{4}}$ in Eq. (11.137) has been included in the amplitude for simplicity.

Equation (11.139) gives a relation between the phase and amplitude of the unitary propagator $K(x, y, t)$. This is in fact a general property of any unitary operator in the stationary phase approximation [158]. For any unitary propagator $U(x, y)$ written as

$$
U(x,y) = A(x,y)e^{\frac{i}{\hbar}\phi(x,y)} \tag{11.141}
$$

where $\phi(x, y)$ is real, the unitary condition is

$$
\begin{aligned}
<y|U^\dagger U|y'> &= \int U^*(x,y)U(x,y)dx \\
&= \int A^*(x,y)A(x,y')e^{\frac{i}{\hbar}[\phi(x,y')-\phi(x,y)]}dx \\
&= \delta(y-y')
\end{aligned}
\tag{11.142}
$$

Since the integral is dominated by the integrand near $y'=y$ in the limit $\hbar \to 0$, we can expand

$$
\phi(x,y') - \phi(x,y) \approx (y'-y)\frac{\partial\phi(x,y)}{\partial y} \tag{11.143}
$$

and approximate $A(x, y')$ by $A(x, y)$. Assuming $z = \frac{\partial \phi(x,y)}{\partial y}$ is a monotonic function of x, the integral in Eq. (11.142) can be written as

$$\int |A(x,y)|^2 \left| \frac{\partial^2 \phi(x,y)}{\partial x \partial y} \right|^{-1} e^{\frac{i}{\hbar}(y-y')z} dz = \delta(y - y') \qquad (11.144)$$

which determines the prefactor $A(x, y)$ apart from a phase factor by the equation

$$|A(x,y)| = \left(\frac{1}{2\pi\hbar} \left| \frac{\partial^2 \phi(x,y)}{\partial x \partial y} \right| \right)^{1/2} \qquad (11.145)$$

The quantum short time propagator of Eq. (11.127) is exactly equivalent to the result of the stationary phase approximation with a single classical path, as well as the free propagator of Eq. (11.132). In fact, it can be shown that for any quadratic Lagrangian [218], the quantum propagator is exactly given by the stationary phase result involving only classical paths [220]. Equation (11.140) can be straightforwardly generalized to the N-dimensional case

$$\begin{aligned} K(\mathbf{x}, \mathbf{y}, t) &= <\mathbf{x}|e^{-\frac{i}{\hbar}Ht}|\mathbf{y}> \\ &= \sum_c \left[\frac{1}{(2\pi i\hbar)^n} \left| \frac{\partial^2 S_c(\mathbf{x}, \mathbf{y})}{\partial \mathbf{x} \partial \mathbf{y}} \right| \right]^{1/2} e^{\frac{i}{\hbar}S_c(\mathbf{x},\mathbf{y},t)} e^{-iM_c\frac{\pi}{2}} \end{aligned} \qquad (11.146)$$

From the multidimensional Gaussian integral in (11.115) and the stationary phase approximation, we know that the Maslov index M is the number of negative eigenvalues of the second derivative matrix of the action S. This, however, does not give a practical way to determine the value of M. Fortunately, it is well known that the Maslov index M can actually be determined by the number of caustics (or turning points) that the system encounters along the classical path. This can be easily understood from the fact that the wavefunction loses a phase of $\pi/2$ when reflected from a turning point as shown by the semiclassical connection formula in section (11.1.2) in one dimension. This notation can be generalized to multidimensions as well [221].

11.4.4 Initial Value Representation

The semiclassical propagator in Eq. (11.140) can be used to calculate the matrix element between two wavepackets $\Phi_f(x)$ and $\Phi_i(x)$

$$K_{fi}(t) = \iint dx_1 dx_2 \Phi_f^*(x_2) <x_2|e^{-\frac{i}{\hbar}Ht}|x_1> \Phi_i(x_1)$$

$$= \sum_c \iint dx_1 dx_2 \Phi_f^*(x_2) \left(2\pi i\hbar \left| \frac{\partial x_2(x_1, p_1, t)}{\partial p_1} \right| \right)^{-1/2}$$

$$\times e^{\frac{i}{\hbar} S(x_2, x_0, t)} e^{-iM_c \frac{\pi}{2}} \Phi_i(x_1) \tag{11.147}$$

where the momentum p_1 is defined by

$$p_1 = -\frac{\partial S(x_2, x_1, t)}{\partial x_1} \tag{11.148}$$

The determination of the classical action $S(x_2, x_1, t)$ involves a root searching problem because one only knows the two end points x_1 and x_2. In order to find $S(x_2, x_1, t)$, one needs to run trajectories with the initial position x_1 and trial momentum p_1 until the choice of p_1 enables the trajectory to reach x_2 in the given time t. In addition, the propagator is singular at caustics or turning points where

$$\frac{\partial x_2(x_1, p_1, t)}{\partial p_1} = 0 \tag{11.149}$$

In order to avoid the complicated and often intractable root searching problem, it is suggested [222–224] to transform the integration variable from x_2 to p_1

$$dx_2 = \left| \frac{\partial x_2(x_1, p_1, t)}{\partial p_1} \right| dp_1 \tag{11.150}$$

Thus, Eq. (11.147) becomes

$$K_{fi}(t) = \sum_c \iint dx_1 dp_1 \Phi_f^*(x_2) \left(\frac{1}{2\pi i\hbar} \left| \frac{\partial x_2(x_1, p_1, t)}{\partial p_1} \right| \right)^{1/2}$$

$$\times e^{\frac{i}{\hbar} S(x_1, p_1, t)} e^{-iM_c \frac{\pi}{2}} \Phi_i(x_1) \tag{11.151}$$

Since $x_2(x_1, p_1, t)$ is given by the trajectory running forward with the initial position x_1 and momentum p_1, one can simply integrate over all the initial coordinates and momenta without the need of root searching. In addition, the singularities at caustics have been removed because the integrand vanishes at caustics of (11.149). However, singularities now appear at new places satisfied by the condition

$$\frac{\partial p_1(x_2, x_1, t)}{\partial x_2} = 0 \tag{11.152}$$

Equation (11.151) can be applied to propagate the time-dependent wavefunction as well. If we set $|\Phi_f\rangle = \Phi$, $|\Phi_i\rangle = |x\rangle$, $p_1 = p$, $x_2 = x'$, and $t = -t$, in Eq. (11.151), we have

$$\begin{aligned}\langle x|e^{-\frac{i}{\hbar}Ht}|\Phi\rangle &= \langle\Phi|e^{\frac{i}{\hbar}Ht}|x\rangle^* \\ &= \sum_c \int dp\Phi(x')\left(\frac{i}{2\pi\hbar}\left|\frac{\partial x'(x,p,-t)}{\partial p}\right|\right)^{1/2} \\ &\quad\times e^{-\frac{i}{\hbar}S(x,p,-t)}e^{iM_c\frac{\pi}{2}}\end{aligned} \tag{11.153}$$

where the trajectory needs to run backwards from 0 to $-t$ (or equivalently t to 0) with the initial conditions (x,p). Equations (11.151) and (11.153) can be generalized to multidimensions in a straightforward fashion as

$$\begin{aligned}\langle\mathbf{x}|e^{-\frac{i}{\hbar}Ht}|\mathbf{y}\rangle &= \sum_c \left[\frac{1}{(2\pi i\hbar)^n}\left|\frac{\partial\mathbf{x}_2(\mathbf{x}_1,\mathbf{p}_1,t)}{\partial\mathbf{p}_1}\right|\right]^{1/2} \\ &\quad\times e^{\frac{i}{\hbar}S_c(\mathbf{x}_1,\mathbf{y}_1,t)}e^{-iM_c\frac{\pi}{2}}\end{aligned} \tag{11.154}$$

and

$$\begin{aligned}\langle\mathbf{x}|e^{-\frac{i}{\hbar}Ht}|\Phi\rangle &= \sum_c \int d\mathbf{p}\Phi(\mathbf{x}')\left[\left(\frac{i}{2\pi\hbar}\right)^n\left|\frac{\partial x'(\mathbf{x},\mathbf{p},-t)}{\partial\mathbf{p}}\right|\right]^{1/2} \\ &\quad\times e^{-\frac{i}{\hbar}S(x,p,-t)}e^{iM_c\frac{\pi}{2}}\end{aligned} \tag{11.155}$$

Further reading

Refs. [158,225–232] are good sources for further reading on the topic of the semiclassical approximation.

Appendix A

Orthogonal Polynomials and Special Functions

A.1 Hermite Polynomials

Definition

The Hermite polynomials H_n can be defined by the differential form

$$H_n(z) = (-)^n e^{z^2} \left(\frac{d^n}{dz^n} e^{-z^2} \right) \qquad (n = 0, 1, 2, \cdots, \infty) \tag{A.1}$$

It is easy to see from Eq. (A.1) that H_n is a polynomial of degree n with the property $H_n(-z) = (-)^n H_n(z)$. It can also be shown that H_n satisfies the differential equation

$$\left[\frac{d^2}{dz^2} - 2z\frac{d}{dz} + 2n \right] H_n(z) = 0 \tag{A.2}$$

The generating function of H_n is

$$e^{-s^2+2sz} = \sum_{n=0}^{\infty} \frac{s^n}{n!} H_n(z) \tag{A.3}$$

Orthogonality

The Hermite polynomials satisfy the orthogonality relation

$$\int_{-\infty}^{\infty} e^{-z^2} H_m(z) H_n(z) dz = \left(\sqrt{\pi} 2^n n! \right) \delta_{mn} \tag{A.4}$$

Recursion relation

Instead of directly calculating H_n from the definition in Eq. (A.1)), the Hermite polynomials can be generated by the recursion relation

$$H_{n+1} = 2zH_n - 2nH_{n-1} \tag{A.5}$$

starting from the two lowest order polynomials, $H_0 = 1$ and $H_1 = 2z$.

Eigenfunctions of harmonic oscillator

The eigenfunctions of the harmonic oscillator ϕ_n satisfy the Schrödinger equation

$$\left[-\frac{\hbar^2}{2m}\frac{\partial^2}{\partial x^2} + \frac{1}{2}m\omega^2 x^2\right]\phi_n(x) = (n + \frac{1}{2})\hbar\omega\phi_n(x) \tag{A.6}$$

where the integer n takes the values $0, 1, 2, \cdots$. $\phi_n(x)$ is expressed in terms of the Hermite polynomials

$$\phi_n(x) = \sqrt{\frac{\alpha}{\sqrt{\pi}2^n n!}}e^{-\frac{1}{2}(\alpha x)^2}H_n(\alpha x) \tag{A.7}$$

where the constant α is defined by

$$\alpha = \sqrt{\frac{m\omega}{\hbar}} \tag{A.8}$$

From the normalization relation of Eq. (A.4) for the Hermite polynomials, the wavefunction ϕ_n is normalized

$$\int_{-\infty}^{\infty} \phi_m(x)\phi_n(x)dx = \delta_{mn} \tag{A.9}$$

and has the parity $\phi_n(-x) = (-)^n\phi_n(x)$. From the recursion relation of (A.5) for H_n, it is straightforward to verify that ϕ_n satisfies the recursion relation

$$\sqrt{\frac{n+1}{2}}\phi_{n+1} = \alpha x\phi_n - \sqrt{\frac{n}{2}}\phi_{n-1} \tag{A.10}$$

It is useful to note that the ground state wavefunction of the harmonic oscillator is a Gaussian function

$$\phi_0(x) = \sqrt{\frac{\alpha}{\sqrt{\pi}}}e^{-\frac{1}{2}(\alpha x)^2} \tag{A.11}$$

A.2 Legendre Polynomials

Definition

The Legendre polynomials $P_l(x)$ $(l = 0, 1, 2, \cdots, \infty)$ can be defined by the differential form

$$P_l(x) = \frac{1}{2^l l!} \frac{d^l}{dx^l}(x^2 - 1)^l \qquad x \in [-1, 1] \tag{A.12}$$

P_l is a polynomial of degree l and has the parity $P_l(-x) = (-)^l P_l(x)$. The Legendre polynomial P_l is a special case (m=0) of the associated Legendre polynomial P_l^m which is defined as

$$P_l^m(x) = (1 - x^2)^{m/2} \frac{d^m}{dx^m} P_l(x) = \frac{(1 - x^2)^{m/2}}{2^l l!} \frac{d^{l+m}}{dx^{l+m}} \left(x^2 - 1\right)^l \tag{A.13}$$

where the integer m is confined to the range $0 \le m \le l$. It can be shown that P_l^m satisfies the relation [139]

$$P_l^{-m}(x) = (-1)^m P_l^m(x) \tag{A.14}$$

and has the parity

$$P_l^m(-x) = (-)^{l-m} P_l^m(x) \tag{A.15}$$

Obviously when $m = 0$, $P_l^0 = P_l$, and thus we can focus on discussing the properties of the associated Legendre polynomials P_l^m.

Orthogonality

P_l^m is a solution of the differential equation

$$\left[(1 - x^2)\frac{d^2}{dx^2} - 2x\frac{d}{du} + l(l + 1) - \frac{m^2}{1 - x^2}\right] P_l^m(x) = 0 \tag{A.16}$$

and satisfies the orthogonality relation

$$\int_{-1}^{1} P_k^m(x) P_l^m(x) dx = \frac{2}{2l + 1} \frac{(l + m)!}{(l - m)!} \delta_{kl} \tag{A.17}$$

Asymptotic expansion

The Legendre polynomial has the following asymptotic expansions valid for large $l (l \gg 1)$:

1) $\sin\theta \geq 1/l$

$$P_l(\cos\theta) \longrightarrow \left(\frac{1}{2}(l+\frac{1}{2})\pi\sin\theta\right)^{-1/2} \sin\left[(l+\frac{1}{2})\theta + \frac{\pi}{4}\right] \tag{A.18}$$

2) $\sin\theta \leq 1/l$

$$P_l(\cos\theta) \longrightarrow (\cos\theta)^l J_0\left((l+\frac{1}{2})\theta\right) \tag{A.19}$$

where J_0 is the zeroth order Bessel function.

Recursion relations

The recursion relations for P_l^m are

$$(2l+1)xP_l^m = (l+1-m)P_{l+1}^m + (l+m)P_{l-1}^m \tag{A.20}$$

$$(1-x^2)\frac{d}{dx}P_l^m = (l+1)xP_l^m - (l+1-m)P_{l+1}^m \tag{A.21}$$

The following values of Legendre polynomials are useful:

$$P_0 = 1, \quad P_1 = x, \quad P_l(1) = 1$$

A.3 Spherical Harmonics

Definition

The spherical harmonics Y_l^m are defined in terms of the associated Legendre polynomials P_l^m and an exponential function

$$Y_l^m(\theta,\phi) = (-)^m\sqrt{\frac{(2l+1)}{4\pi}\frac{(l-m)!}{(l+m)!}}P_l^m(\cos\theta)e^{im\phi} \tag{A.22}$$

The Y_l^m are simultaneous eigenfunctions of the angular momentum operator $\mathbf{L}^2$ and the z component L_z of $\mathbf{L}$

$$\begin{cases} \mathbf{L}^2 Y_l^m &= l(l+1)\hbar^2 Y_l^m \\ L_z Y_l^m &= m\hbar Y_l^m \end{cases} \tag{A.23}$$

where the coordinate representations of the angular momentum operators are given by

$$\mathbf{L}^2 = -\hbar^2 \left[-\frac{1}{\sin\theta} \frac{\partial}{\partial\theta} \left(\sin\theta \frac{\partial}{\partial\theta} \right) + \frac{1}{\sin^2\theta} \frac{\partial^2}{\partial\phi^2} \right] \tag{A.24}$$

$$L_z = -i\hbar \frac{\partial}{\partial\phi} \tag{A.25}$$

Under the coordinate inversion $(\theta, \phi) \rightarrow (\pi - \theta, \phi + \pi)$, it is easy to verify that

$$Y_l^m(\pi - \theta, \phi + \pi) = (-)^l Y_l^m(\theta, \phi) \tag{A.26}$$

Thus Y_l^m has parity $(-)^l$.

Orthogonality

Using the orthogonality property of Eq. (A.17) for the associated Legendre polynomials P_l^m, one can see that the spherical harmonics are orthonormal

$$\begin{aligned} <Y_l^m | Y_{l'}^{m'}> &= \int_0^{2\pi} d\phi \int_0^{\pi} \sin\theta \, d\theta \; Y_l^{m*}(\theta, \phi) Y_{l'}^{m'}(\theta, \phi) \\ &= \delta_{ll'} \delta_{mm'} \end{aligned} \tag{A.27}$$

Since the Y_l^m form a complete basis set in the two-dimensional space (θ, ϕ), we have the closure relation

$$\sum_{l=0}^{\infty} \sum_{m=-l}^{m=l} |Y_l^m><Y_l^m| = \mathbf{I} \tag{A.28}$$

or in coordinate representation

$$\sum_{l=0}^{\infty} \sum_{m=-l}^{l} Y_l^{m*}(\theta, \phi) Y_l^m(\theta', \phi') = \delta(\cos\theta - \cos\theta') \delta(\phi - \phi') \tag{A.29}$$

A formula that involves the integral over three spherical harmonics is quite useful

$$\begin{aligned} \int d\Omega Y_{j_3}^{*m_3} Y_{j_2}^{m_2} Y_{j_1}^{m_1} &= \left[\frac{(2j_1 + 1)(2j_2 + 1)}{4\pi(2j_3 + 1)} \right]^{1/2} \\ &\quad \times <j_1 m_1 j_2 m_2 | j_3 m_3><j_1 0 j_2 0 | j_3 0> \end{aligned} \tag{A.30}$$

Recursion relation

The spherical harmonics Y_l^m satisfy the recursion relation

$$\cos\theta Y_l^m = \sqrt{\frac{(l+1+m)(l+1-m)}{(2l+1)(2l+3)}} Y_{l+1}^m + \sqrt{\frac{(l+m)(l-m)}{(2l+1)(2l-1)}} Y_{l-1}^m \quad (A.31)$$

In general, one can use the recursion relation to generate the associated Legendre polynomial first and then use Eq. (A.22) to construct the spherical harmonics.

Addition theorem

The addition theorem establishes the relation between the Legendre polynomials and the spherical harmonics

$$\frac{2l+1}{4\pi} P_l(\cos\gamma) = \sum_{m=-l}^{l} Y_l^{m*}(\theta_1, \phi_1) Y_l^m(\theta_2, \phi_2) \quad (A.32)$$

where γ is the angle between the directions of the unit vectors $\hat{\mathbf{r}}_1 = (\theta_1, \phi_1)$ and $\hat{\mathbf{r}}_2 = (\theta_2, \phi_2)$ or explicitly

$$\cos\gamma = \hat{\mathbf{r}}_1 \cdot \hat{\mathbf{r}}_2 = \sin\theta_1 \sin\theta_2 \cos(\phi_1 - \phi_2) + \cos\theta_1 \cos\theta_2 \quad (A.33)$$

A.4 Chebychev Polynomials

Definition

The Chebychev polynomials are simply defined as cosine functions

$$T_n(x) = \cos(n\theta) \qquad x \in [-1, 1] \quad (A.34)$$

where $x = \cos\theta$. $T_n(x)$ is the solution of the differential equation

$$\left[(1-x^2)\frac{d^2}{dx^2} - x\frac{d}{dx} + n^2\right] T_n(x) = 0 \quad (A.35)$$

From the definition in Eq. (A.34), $T_n(x)$ has n zeros located at

$$x = \cos\left(\frac{\pi(k-1/2)}{n}\right) \qquad (k = 1, 2, \cdots, n) \quad (A.36)$$

Also from Eq. (A.34), the magnitude of $T_n(x)$ is bounded between ± 1 which makes it very attractive in the stable polynomial approximation of functions.

Orthogonality

The Chebychev polynomials are orthogonal

$$\int_{-1}^{1} \frac{T_i(x)T_j(x)}{\sqrt{1-x^2}}dx = \begin{cases} 0 & i \neq j \\ \pi/2 & i = j \neq 0 \\ \pi & i = j = 0 \end{cases} \tag{A.37}$$

Recursion relation

The recursion relation for the Chebychev polynomials is

$$T_{n+1}(X) = 2XT_n(X) - T_{n-1}(X) \tag{A.38}$$

with values $T_0(x) = 1,\ T_1(x) = x,\ T_2(x) = 2x^2 - 1,\ \cdots$.

A.5 Spherical Bessel Functions

Definition

The spherical Bessel functions $\tilde{j}_l(kr)$ are solutions of the equation

$$\left[\frac{1}{r}\frac{d^2}{dr^2}r - \frac{l(l+1)}{r^2} + k^2\right]\tilde{j}_l(kr) = 0 \tag{A.39}$$

In molecular dynamics, it is often more convenient to directly deal with the Ricatti-Bessel function defined as $j_l(kr) = kr\tilde{j}_l(kr)$. It is easy to verify that $j_l(kr)$ satisfies the equation

$$\left[\frac{d^2}{dr^2} - \frac{l(l+1)}{r^2} + k^2\right] j_l(kr) \tag{A.40}$$

which is the familiar asymptotic radial equation in quantum dynamics.

The Ricatti-Bessel function can be represented by an infinite series expansion

$$\begin{aligned} j_l(z) &= \left(\frac{\pi z}{2}\right)^{1/2} J_{l+\frac{1}{2}} \\ &= z^{l+1}\sum_{n=0}^{\infty}\frac{(-z^2/2)^n}{n!(2l+2n+1)!!} \end{aligned} \tag{A.41}$$

where $J_{l+1/2}$ is the regular Bessel function of order $l + \frac{1}{2}$ The Ricatti-Bessel function j_l is a regular solution to Eq. (A.40) and vanishes near the origin

$$j_l(z) \xrightarrow{z\to 0} \frac{z^{l+1}}{(2l+1)!!} \tag{A.42}$$

The second order differential Eq. (A.40) has two linearly independent solutions. Since j_l is a regular solution ($j_l(0) \to 0$), it is a unique solution of Eq. (A.40). The other independent solution is irregular and non-unique. In fact, any nontrivial linear combination of two independent solutions gives rise to another irregular solution. The ones that are commonly used in scattering are the so-called Ricatti-Neumann functions n_l and the Ricatti-Henkel functions $h_l^{\pm}$. The Ricatti-Neumann functions n_ls are given by a series relation

$$\begin{aligned} n_l(z) &= (-)^l \left(\frac{\pi z}{2}\right)^{1/2} J_{-l-1/2} \\ &= z^{-l}(2l-1)!! \sum_{n=0}^{\infty} \frac{(-z^2/2)^n}{n!(2n-2l-1)!!} \end{aligned} \tag{A.43}$$

At the origin n_l behaves as

$$n_l(z) \overset{z\to 0}{\longrightarrow} z^{-l}(2l-1)!! \tag{A.44}$$

For the special case of l=0, j_l and n_l are simply the sine and cosine functions.

The widely used Ricatti-Hankel functions $h_l^{\pm}(z)$ are linear combinations of j_j and n_l in the same fashion as are $\exp(\pm iz)$ of the sine and cosine

$$h_l^{\pm}(z) = n_l(z) \pm i j_l(z) \tag{A.45}$$

Orthogonality

The Ricatti-Bessel function $j_l(z)$ satisfies the orthogonality relation

$$<j_l(k)|j_l(k')>= \int_0^{\infty} j_l(kr) j_l(k'r) dr = \frac{\pi}{2}\delta(k-k') \tag{A.46}$$

Asymptotic Expansion

In the asymptotic limit of $z \to \infty$, j_l becomes a sine function

$$j_l(kr) \overset{r\to\infty}{\longrightarrow} \sin(kr - \frac{l\pi}{2}) \tag{A.47}$$

and n_l becomes a cosine function

$$n_l(kr) \overset{r\to\infty}{\longrightarrow} \cos(kr - \frac{l\pi}{2}) \tag{A.48}$$

The Ricatti-Henkel functions ($h_l^{\pm}$) are outgoing and incoming waves asymptotically

$$h_l^{\pm}(kr) \overset{r\to\infty}{\longrightarrow} e^{\pm i(kr - \frac{l\pi}{2})} \tag{A.49}$$

Thus we can view j_l, n_l, and $h_l^{\pm}$ as the generalizations of sine, cosine and $\exp(\pm ikr)$ functions.

From their asymptotic limits, it is straightforward to verify that the Wronskian between j_l and n_l is given by

$$W[j_l, n_l] = j_l'(z)n_l(z) - j_l(z)n_l'(z) = k \tag{A.50}$$

Recursion relations

If $\tilde{f}_l(z)$ is any arbitrary spherical Bessel function, it satisfies the following recursion relation

$$(2l+1)\tilde{f}_l(z) = z\tilde{f}_{l+1} + \tilde{f}_{l-1} \tag{A.51}$$

and

$$\tilde{f}_l(z) = \left[z^l\left(-\frac{1}{z}\frac{d}{dz}\right)^l\right]\tilde{f}_0 \tag{A.52}$$

The Ricatti-Bessel function, for example, can be simply obtained by the relation $j_l(z) = z\tilde{j}_l(z)$.

A.6 Useful Mathematical Formulae

The following mathematical formulae are frequently encountered in quantum dynamics theory:

$$\frac{1}{|\mathbf{r}_1 - \mathbf{r}_2|} = \sum_{l=0}^{\infty} \frac{r_<^l}{r_>^{l+1}} P_l(\cos\gamma) \tag{A.53}$$

$$\frac{e^{ik|\mathbf{r}_1 - \mathbf{r}_2|}}{|\mathbf{r}_1 - \mathbf{r}_2|} = k\sum_{l=0}^{\infty}(2l+1)\tilde{j}_l(kr_<)\tilde{h}_l^{(+)}(kr_>)P_l(\cos\gamma) \tag{A.54}$$

Here γ is the angle between the vectors $\mathbf{r}_1$ and $\mathbf{r}_2$ and $r_<(r_>)$ is the smaller (larger) of the lengths r_1 and r_2. Another useful formula is the plane wave expansion

$$\begin{aligned}\exp(i\mathbf{k}\cdot\mathbf{r}) &= \exp(ikr\cos\gamma)\\ &= \sum_{l=0}^{\infty}(2l+1)i^l\frac{j_l(kr)}{kr}P_l(\cos\gamma)\\ &= \sum_{l=0}^{\infty}\sum_{m=-l}^{l}4\pi i^l\frac{j_l(kr)}{kr}Y_l^{m*}(\hat{\mathbf{k}})Y_l^m(\hat{\mathbf{r}}) \qquad \text{(A.55)}\end{aligned}$$

Here $\tilde{j}_l(kr)$ and $\tilde{h}_l^{(+)}(kr)$ are the spherical-Bessel and spherical-Hankel functions defined in Sec. A.5.

Appendix B

Gaussian Quadrature

General definition

In the numerical integration of a function $f(x)$ weighted by $W(x)$, one usually approximates the integral by an N-term finite quadrature

$$\int_a^b W(x)f(x)dx \approx \sum_{i=1}^{N} w_i f(x_i) \tag{B.1}$$

with N nodes x_i and N weights $w_i (i = 1, 2, \cdots, N)$. In standard quadrature methods such as the Simpson or Newton-Cotes formulae, the nodes x_i are predetermined while the weights are often chosen to render the discrete sum in (B.1) exact for some low order polynomial integrand $f(x)$ such as a quadratic or cubic function. For example, the 3-point Simpson formula gives the exact result for a quadratic integrand. In Gaussian quadratures, however, one chooses both the nodes and weights ($2N$ in total) such that the N-term discrete sum in Eq. (B.1) is *exact* for any polynomial integrand $f(x)$ whose order is no higher than $2N - 1$.

To illustrate more clearly how Gaussian quadrature works, let us examine a N-term quadrature for the power integral

$$I_k = \int_a^b W(x)x^k dx = \sum_{i=1}^{N} w_i x_i^k \qquad (k = 0, 1, 2, \cdots, 2N - 1) \tag{B.2}$$

If we require the above quadrature to be *exact* for $k = 0, 1, 2, \cdots, 2N - 1$, then the nodes $x_i (i = 0, 1, 2, \cdots, N)$ and weights $w_i (i = 0, 1, 2, \cdots, N)$ will be ***uniquely*** determined by $2N$ equations. Since any polynomial function

$F(x)$ up to the power of $2N-1$ can always be written as a superposition

$$F(x) = \sum_{k=0}^{2N-1} C_k x^k \tag{B.3}$$

Gaussian quadrature gives the *exact* integral for such polynomial integrands. It should be emphasized here that for a given weighting function $W(x)$, the Gaussian quadrature nodes and weights are *unique*.

The methods for finding Gaussian nodes and weights are simplified if orthogonal polynomials are employed. For a given weighting function $W(x)$, we can always find orthogonal polynomials $P_i(x)(i = 1, 2, \cdots,)$ such that

$$<P_i|P_j>= \int_a^b W(x)P_i(x)P_j(x)dx = \delta_{ij} \tag{B.4}$$

The Gaussian quadrature theorem states that the N Gaussian quadrature nodes $x_i(i = 1, 2, \cdots, N)$ are given by the N roots of the Nth order polynomial $P_N(x)$. Thus the problem of finding Gaussian nodes becomes that of finding the zeros of orthogonal polynomials. For any given set of polynomials $Q_k(x)(k = 0, 1, 2, \cdots, N)$, we can always form a new set of orthogonal polynomials $P_k(x)(k = 0, 1, 2, \cdots, N)$ by linear combination

$$P_i = \sum_m Q_m C_{mi} \tag{B.5}$$

By requiring the P_i to be orthogonal, we obtain an eigenvalue equation

$$\mathbf{C}^T\mathbf{O}\mathbf{C} = \mathbf{I} \tag{B.6}$$

which is solved by diagonalizing the overlap matrix $\mathbf{O}$ defined by

$$\mathbf{O}_{mn} =<Q_i|Q_j>= \int_a^b W(x)Q_i(x)Q_j(x)dx \tag{B.7}$$

Once the Gaussian nodes x_i are known, the Gaussian weights can be straightforwardly determined. For example, we can solve linear algebraic equations for the weights w_i as follows

$$\begin{aligned} I_0 &= \sum_{i=1}^{N} P_0(x_i) w_i \\ I_1 &= \sum_{i=1}^{N} P_1(x_i) w_i \\ \cdots &= \cdots \\ I_{N-1} &= \sum_{i=1}^{N} P_{N-1}(x_i) w_i \end{aligned} \tag{B.8}$$

where the integral I_k is defined as

$$I_k = \int_a^b W(x) P_k(x) dx = I_0 \delta_{k0} \qquad (k = 0, 1, 2, \cdots, N-1) \tag{B.9}$$

with I_0 given by

$$I_0 = \int_a^b W(x) P_0(x) dx \tag{B.10}$$

Eq. (B.8) can be inverted to solve for the Gaussian weights w_i.

Gauss-Legendre quadrature

Since the Legendre polynomials are orthogonal with a constant weighting function $W(x) = 1$,

$$<P_l|P_{l'}>= \int_{-1}^{1} P_l(x) P_{l'}(x) dx = \frac{2}{2l+1} \delta_{ll'} \tag{B.11}$$

one can determine Gauss-Legendre nodes by finding the zeros of Legendre polynomials and subsequently determine the weights. Thus for integrals with a constant weighting function, one can use Gauss-Legendre quadrature to carry out the numerical integrals

$$\int_{-1}^{1} f(x) dx \approx \sum_{i=1}^{N} w_i f(x_i) \tag{B.12}$$

where the quadrature nodes x_i are given by the zeros of the Legendre polynomial $P_N(x)$. Equation (B.12) is exact if $f(x)$ is *any* polynomial whose order is no higher than $2N-1$.

Gauss-Hermite quadrature

Similarly, since the Hermite polynomials are orthogonal with a weighting function e^{-x^2}

$$\int_{-\infty}^{\infty} e^{-x^2} H_m(x) H_n(x) dx = \left(\sqrt{\pi} 2^n n!\right) \delta_{mn} \tag{B.13}$$

one can use Gauss-Hermite quadrature to evaluate numerical integral

$$\int_{-\infty}^{\infty} e^{-x^2} f(x) dx \approx \sum_{i=1}^{N} w_i f(x_i) \tag{B.14}$$

where the quadrature nodes x_i are zeros of the Hermite polynomial $H_N(x)$.

Gauss-Chebychev quadrature

Since the Chebychev polynomials are orthogonal with a weighting function $1/\sqrt{1-x^2}$

$$\int_{-1}^{1} \frac{T_i(x) T_j(x)}{\sqrt{1-x^2}} dx = \frac{\pi}{2}(1+\delta_{i0})\delta_{ij} \tag{B.15}$$

one can use Gauss-Chebychev quadratures to evaluate numerical integrals

$$\int_{-1}^{1} \frac{f(x)}{\sqrt{1-x^2}} dx \approx \sum_{i=1}^{N} w_i f(x_i) \tag{B.16}$$

where the quadrature nodes x_i are zeros of the Chebychev polynomial $T_N(x)$.

Appendix C

Clebsch-Gordon Coefficients

Definition

Let $\mathbf{j}_1$, $\mathbf{j}_2$ be the angular momenta of systems 1 and 2, respectively, and $\mathbf{J}$ the angular momentum of the total system

$$\mathbf{J} = \mathbf{j}_1 + \mathbf{j}_2 \tag{C.1}$$

The angular momentum eigenstates of $\mathbf{j}_1$ and $\mathbf{j}_2$ are, respectively,

$$\begin{cases} |j_1 m_1\rangle & (m_1 = -j_1, \cdots, j_1) \\ |j_2 m_2\rangle & (m_2 = -j_2, \cdots, j_2) \end{cases} \tag{C.2}$$

and their direct product state can be denoted $|j_1 m_1 j_2 m_2\rangle = |j_1 m_1\rangle |j_2 m_2\rangle$. The simultaneous eigenvectors of $\mathbf{j}_1^2, \mathbf{j}_2^2, \mathbf{J}^2, J_z$ can be obtained through a unitary transformation from the direct product state $|j_1 m_1 j_2 m_2\rangle$ as

$$|j_1 j_2 JM\rangle = \sum_{m_1 m_2} |j_1 m_1 j_2 m_2\rangle\langle j_1 m_1 j_2 m_2 | j_1 j_2 JM\rangle \tag{C.3}$$

where the coefficients of the transformation $\langle j_1 m_1 j_2 m_2 | j_1 j_2 JM\rangle$ are called Clebsch-Gordon (CG) coefficients. The eigenvectors of $(\mathbf{j}_1^2, \mathbf{j}_2^2, \mathbf{J}^2, J_z)$ are basis vectors of the *coupled* angular momentum representation while those of $|j_1 m_1 j_2 m_2\rangle$ are basis vectors of the *uncoupled* angular momentum representation. The CG coefficients are defined to be real, viz.,

$$\langle j_1 m_1 j_2 m_2 | j_1 j_2 JM\rangle = \langle j_1 m_1 j_2 m_2 | j_1 j_2 JM\rangle^* \tag{C.4}$$

Orthogonality relation

Since the CG coefficients form a unitary (orthogonal) transformation between the coupled and uncoupled angular momentum representations, they obviously satisfy the orthogonality conditions

$$\sum_{m_1 m_2} \langle j_1 m_1 j_2 m_2 | j_1 j_2 J M \rangle \langle j_1 m_1 j_2 m_2 | j_1 j_2 J' M' \rangle = \delta_{JJ'} \delta_{MM'} \quad \text{(C.5)}$$

and

$$\sum_{JM} \langle j_1 m_1 j_2 m_2 | j_1 j_2 J M \rangle \langle j_1 m_1' j_2 m_2' | j_1 j_2 J' M' \rangle = \delta_{m_1 m_1'} \delta_{m_2 m_2'} \quad \text{(C.6)}$$

Methods of calculation

The explicit expressions for the CG coefficients were derived by Racah [233, 234]

$$\begin{aligned} &\langle j_1 m_1 j_2 m_2 | j_1 j_2 j_3 m_3 \rangle = \\ &\quad \delta_{m_1+m_2, m_3} \left[(2j_3+1) \frac{(s-2j_3)!(s-2j_2)!(s-2j_2)!!(j_1+m_1)!}{(s+1)!} \right. \\ &\quad \left. \times (j_1-m_1)!(j_2+m_2)!(j_2-m_2)!(j_3+m_3)!(j_3-m_3)! \right]^{1/2} \\ &\quad \times \sum_\nu (-)^\nu / [\nu!(j_1+j_2-j_3-\nu)!(j_1-m_1-\nu)!(j_2+m_2-\nu)! \\ &\quad \times (j_3-j_2+m_1+\nu)!(j_3-j_1-m_2+\nu)!] \end{aligned} \quad \text{(C.7)}$$

However, in practice, the CG coefficients are often more conveniently calculated by using recursion relations. By applying the raising and lowering operators $J_\pm$ to Eq. (C.3), we can derive the following recursion relation for fixed values of j_1, j_2, J, and M

$$\begin{aligned} &\sqrt{J(J+1)-M(M\pm 1)} \langle j_1 m_1 j_2 m_2 | j_1 j_2 J M \rangle = \\ &\quad \sqrt{j_1(j_1+1)-m_1(m_1\pm 1)} \langle j_1 m_1 \pm 1 j_2 m_2 | j_1 j_2 J M \pm 1 \rangle \\ &\quad + \sqrt{j_2(j_2+1)-m_2(m_2\pm 1)} \langle j_1 m_1 j_2 m_2 \pm 1 | j_1 j_2 J M \pm 1 \rangle \end{aligned} \quad \text{(C.8)}$$

The 3j Symbol

The introduction of the Wigner "3j" symbol displays the symmetry properties of the CG coefficients more explicitly

$$\begin{bmatrix} j_1 & j_2 & j_3 \\ m_1 & m_2 & m_3 \end{bmatrix} \equiv \frac{(-)^{j_1-j_2-m_3}}{\sqrt{2j_3+1}} <j_1 m_1 j_2 m_2 | j_3 - m_3> \tag{C.9}$$

From the properties of the C-G coefficients, the 3j symbol is zero unless the following conditions are satisfied:

1) $m_1 + m_2 + m_3 = 0$

2) $|j_1 - j_2| \leq j_3 \leq j_1 + j_2$

The 3j symbol has the following symmetry properties.

1) It is invariant to a circular permutation of the three columns:

$$\begin{bmatrix} j_1 & j_2 & j_3 \\ m_1 & m_2 & m_3 \end{bmatrix} \equiv \begin{bmatrix} j_2 & j_3 & j_1 \\ m_2 & m_3 & m_1 \end{bmatrix} \equiv \begin{bmatrix} j_3 & j_1 & j_2 \\ m_3 & m_1 & m_2 \end{bmatrix} \tag{C.10}$$

2) Permutation of any two columns is equivalent to multiplying by a factor $(-)^{j_1+j_2+j_3}$:

$$\begin{bmatrix} j_2 & j_1 & j_3 \\ m_2 & m_1 & m_3 \end{bmatrix} \equiv \begin{bmatrix} j_1 & j_2 & j_3 \\ m_1 & m_2 & m_3 \end{bmatrix} (-)^{j_1+j_2+j_3} \tag{C.11}$$

3) Simultaneously changing the sign of m_1, m_2, and m_3 is equivalent to multiplying by a factor $(-)^{j_1+j_2+j_3}$.

$$\begin{bmatrix} j_1 & j_2 & j_3 \\ -m_1 & -m_2 & -m_3 \end{bmatrix} \equiv \begin{bmatrix} j_1 & j_2 & j_3 \\ m_1 & m_2 & m_3 \end{bmatrix} (-)^{j_1+j_2+j_3} \tag{C.12}$$

Appendix D

Space-Fixed and Body-Fixed Frames

D.1 Coordinate Rotations and Angular Momentum

To conform to standard notation for coordinate transformations, we employ the active view in which the physical system is rotated in our discussion. If the physical system in space (X, Y, Z) is rotated by an infinitesimal angle ϵ along the direction of a unit vector $\hat{\mathbf{u}}$ to a new point in space (X', Y', Z'), then a vector $\mathbf{A}$ is transformed to the new one via the relation

$$\begin{aligned} \mathbf{A}' &= R_u(\epsilon)\mathbf{A} \\ &= \mathbf{A} - \epsilon(\hat{\mathbf{u}} \times \mathbf{A}) \end{aligned} \tag{D.1}$$

In particular, the coordinate vector $\mathbf{r}$ also transforms by this relation.

When the physical system is rotated from the point $\mathbf{r}$ to the new point $\mathbf{r}'$ in space, the value of the old function at $\mathbf{r}$ must be the same as the new function at $\mathbf{r}'$ or

$$\psi'(\mathbf{r}') = \psi(\mathbf{r}) \tag{D.2}$$

If we define a rotation operator R for the transformation of the function by

$$\psi'(\mathbf{r}) = R\psi(\mathbf{r}) \tag{D.3}$$

then we have the relation

$$R\psi(\mathbf{r}') = \psi(\mathbf{r}) = \psi[R_u^{-1}(\epsilon)\mathbf{r}'] \tag{D.4}$$

By using Eq. (D.1) for the coordinate vector $\mathbf{r}$, we obtain

$$
\begin{aligned}
R\psi(\mathbf{r}') &= \psi[\mathbf{r}' - \epsilon(\hat{\mathbf{u}} \times \mathbf{r}')] \\
&= \sum_{n=0}^{\infty} \frac{[-\epsilon(\hat{\mathbf{u}} \times \mathbf{r}') \cdot \nabla]^n}{n!} \psi(\mathbf{r}') \\
&= \sum_{n=0}^{\infty} \frac{\left(-\frac{i}{\hbar}\epsilon \hat{\mathbf{u}} \cdot \mathbf{J}\right)^n}{n!} \psi(\mathbf{r}')
\end{aligned} \tag{D.5}
$$

where $\mathbf{J} = -i\hbar \mathbf{r} \times \nabla$ is the angular momentum operator. By replacing $\mathbf{r}'$ by $\mathbf{r}$, we arrive at the transformation relation

$$
\begin{aligned}
\psi'(\mathbf{r}) &= R\psi(\mathbf{r}) \\
&= \boxed{\exp\left(-\frac{i}{\hbar}\epsilon J_u\right) \psi(\mathbf{r})}
\end{aligned} \tag{D.6}
$$

where $J_u = \hat{\mathbf{u}} \cdot \mathbf{J}$ is the angular momentum component along the direction $\hat{\mathbf{u}}$. Although we have derived the transformation relation (D.6) for the orbital angular momentum, Eq. (D.6) in fact also serves as the definition for the *total* angular momentum of the system.

D.2 Rotation Matrix

Definition

Let (X, Y, Z) denote the space-fixed (SF) coordinate system and (x, y, z) denote the rotated (body-fixed or molecular) coordinate system which is obtained through rotation of the SF coordinate system by three Euler angles as shown in Fig. D.1. The rotation of the physical system from the SF $\rightarrow$ BF frame can be represented by the product of three successive rotations (cf. Fig. D.1)

$$
\mathbf{R}(\phi, \theta, \chi) = e^{-i\chi J_z} e^{-i\theta J_{Y'}} e^{-i\phi J_Z} \tag{D.7}
$$

where $J_{Y'}$ is the angular momentum along the intermediate Y' axis. Using the transformation relation for operators under unitary transformation, we can write

$$
e^{-i\chi J_z} = e^{-i\theta J_{Y'}} e^{-i\chi J_Z} e^{i\theta J_{Y'}} \tag{D.8}
$$

and

$$
e^{-i\theta J_{Y'}} = e^{-i\phi J_Z} e^{-i\theta J_{Y'}} e^{i\phi J_Z} \tag{D.9}
$$

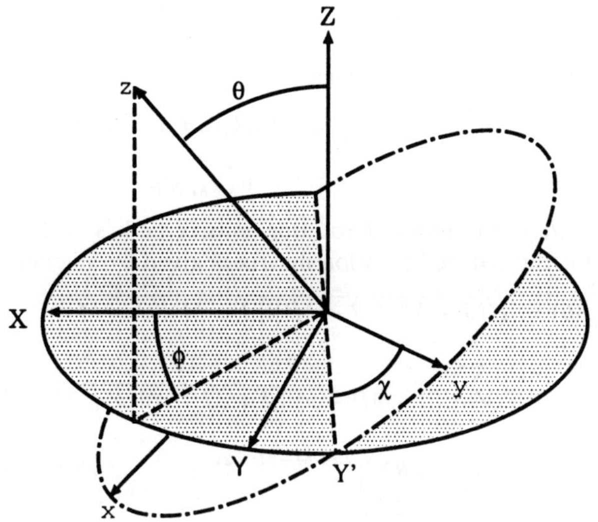

Figure D.1: Space-fixed (X,Y,Z) and body-fixed (x,y,z) coordinate systems specified by three Euler angles (ϕ, θ, χ). Y' is the Y-axis of the intermediate coordinate system.

By applying the above equations, we can rewrite the rotation matrix in Eq. (D.7) as

$$\boxed{\mathbf{R}(\phi, \theta, \chi) = e^{-i\phi J_Z} e^{-i\theta J_Y} e^{-i\chi J_Z}} \tag{D.10}$$

where J_Y and J_Z are angular momentum projections on the Y and Z axes in the "old" (SF) coordinate system. Equation (D.10) states that the rotation operation depicted in Fig. D.1 can be equally achieved by three successive rotations defined in the same (SF) coordinate system but with a reversal of the order of rotation [139].

Upon rotation, the new state vector $|\psi\rangle'$ is related to old one $|\psi\rangle$ through a unitary transformation

$$|\psi\rangle' = \mathbf{R}(\phi, \theta, \chi)|\psi\rangle \tag{D.11}$$

If $|\psi\rangle$ is an angular momentum eigenstate $|JM\rangle$, we then have

$$|JM\rangle' = \sum_{M'} |JM'\rangle\langle JM'|\mathbf{R}(\phi, \theta, \chi)|JM\rangle$$

$$= \sum_{M'} |JM'> D^J_{M'M}(\phi, \theta, \chi) \tag{D.12}$$

where the rotation matrix is defined as

$$D^J_{M'M}(\phi, \theta, \chi) = <JM'|\mathbf{R}(\phi, \theta, \chi)|JM>$$
$$= e^{-i\phi M'} d^J_{M'M}(\theta) e^{-i\chi M} \tag{D.13}$$

where $d^J_{M'M}(\theta)$ is the reduced rotation matrix and is real. The explicit expression for the reduced rotation matrix is given by Wigner [235]

$$d^J_{M'M}(\theta) = \sqrt{(J+M')!(J-M')!(J+M)!(J-M)!}$$
$$\times \sum_\nu \frac{(-1)^\nu}{(J-M'-\nu)!(J+M-\nu)!(\nu+M'-M)!\nu!}$$
$$\times \left(\cos\left(\frac{\theta}{2}\right)\right)^{2J+M-M'-2\nu} \left(-\sin\left(\frac{\theta}{2}\right)\right)^{M'-M+2\nu} \tag{D.14}$$

where the sum over ν is for all integers for which the factorial arguments are nonnegative.

Special properties

The rotation matrix $d^J_{M'M}(\theta)$ has the following symmetry properties:

$$d^J_{M'M}(\theta) = d^J_{-M-M'}(\theta) \tag{D.15}$$

$$d^J_{M'M}(-\theta) = d^J_{MM'}(\theta) = (-1)^{M'-M} d^J_{M'M}(\theta) \tag{D.16}$$

In the case of $M' = 0$, $d^J_{MM'}(\theta)$ is related to the associated Legendre polynomial

$$d^J_{M0}(\theta) = (-1)^M \sqrt{\frac{(L-M)!}{(L+M)!}} P_{JM}(\cos\theta) \tag{D.17}$$

Orthogonality

The rotation matrix $D^J_{M'M}$ is orthogonal

$$<D^{J'}_{M'N'}|D^J_{MN}> = \int_0^{2\pi} d\phi \int_0^\pi \sin\theta\, d\theta \int_0^{2\pi} d\chi\, D^{J'*}_{M'N'}(\phi, \theta, \chi) D^J_{MN}(\phi, \theta, \chi)$$
$$= \frac{8\pi^2}{2J+1} \delta_{J'J} \delta_{M'M} \delta_{N'N} \tag{D.18}$$

Since the rotation operator is unitary, the inverse of the rotation matrix is given by its hermitian adjoint

$$[D^J]^{-1}_{MM'} = D^{J*}_{M'M} \tag{D.19}$$

D.3 Transformation Between SF and BF Basis

If the total angular momentum of a molecular system **J** is the sum of two independent angular momenta **j** and **l** (**l** is usually the orbital angular momentum), the SF angular momentum eigenstate in the coupled representation $|JMjl\rangle$ is obtained through the orthogonal transformation from the *uncoupled* representation by Eq. (C.3)

$$|JMjl\rangle = \sum_m \langle jmlM-m|JM\rangle\, |jm\rangle\, |lM-m\rangle \tag{D.20}$$

which can also be reversed to yield

$$|jm\rangle\, |lM-m\rangle = \sum_J \langle jmlM-m|JM\rangle\, |JMjl\rangle \tag{D.21}$$

where the summation includes all allowed values of m or J. It is easy to verify that the SF eigenfunction Y^{JM}_{jl} is also an eigenfunction of the parity

$$\hat{p}Y^{JM}_{jl}(\hat{\mathbf{R}}, \hat{\mathbf{r}}) = Y^{JM}_{jl}(-\hat{\mathbf{R}}, -\hat{\mathbf{r}}) = pY^{JM}_{jl}(\hat{\mathbf{R}}, \hat{\mathbf{r}}) \tag{D.22}$$

with the parity p given by

$$p = (-1)^{j+l} \tag{D.23}$$

In the space-fixed (SF) coordinate system, the total angular momentum eigenfunction can be written explicitly

$$Y^{JM}_{jl}(\hat{\mathbf{R}}, \hat{\mathbf{r}}) = \sum_m \langle jmlM-m|JM\rangle\, Y^m_j(\hat{r})Y^{M-m}_l(\hat{R}) \tag{D.24}$$

where the unit vectors denote spatial orientation $\hat{\mathbf{R}} = (\theta_R, \phi_R)$ and $\hat{\mathbf{r}} = (\theta_r, \phi_r)$

If the molecular system is rotated from the SF frame to the BF frame by the three Euler angles (ϕ, θ, χ) in space, then the rotated angular momentum wavefunction Y^{JM}_{jl} (in the SF frame) is obtained from the unrotated one $\widetilde{Y}_{jl}^{JK}$ by the transformation relation

$$Y_{jl}^{JM}(\hat{\mathbf{R}},\hat{\mathbf{r}}) = \sum_K D_{KM}^J(\phi,\theta,\chi)\widetilde{Y}_{jl}^{JK}(\hat{\mathbf{R}},\hat{\mathbf{r}}) \tag{D.25}$$

where $\widetilde{Y}_{jl}^{JK}(\hat{\mathbf{R}},\hat{\mathbf{r}})$ is the unrotated wavefunction of $Y_{jl}^{JK}(\hat{\mathbf{R}},\hat{\mathbf{r}})$ in the BF frame. If we choose the z axis in the BF frame to coincide with the direction of the unit vector $\hat{R}$, then we have $\hat{R} = (0,0)$ and $\hat{r} = (\gamma, 0)$. Thus the unrotated wavefunction simplifies to

$$\begin{aligned}\widetilde{Y}_{jl}^{JK}(00,\gamma 0) &= \sum_m \langle jmlK-m|JK\rangle\, Y_j^m(\gamma,0)Y_l^{K-m}(00)\\ &= \sqrt{\frac{2l+1}{4\pi}}\,\langle jKl0|JK\rangle\, Y_j^K(\gamma,0)\end{aligned} \tag{D.26}$$

Equation (D.25) can thus be rewritten as

$$\begin{aligned}Y_{jl}^{JM}(\hat{\mathbf{R}},\hat{\mathbf{r}}) &= \langle \hat{R}\hat{r}|JMjl\rangle\\ &= \sqrt{\frac{2l+1}{4\pi}}\sum_K \langle jKl0|JK\rangle\, D_{KM}^J(\phi,\theta,\chi)Y_j^K(\gamma,0)\end{aligned} \tag{D.27}$$

We can now define a body-fixed angular momentum function by

$$\mathcal{Y}_{jK}^{JM} = \widetilde{D}_{KM}^J P_{jK} \tag{D.28}$$

where $P_{jK} = \sqrt{2\pi}Y_j^K(\gamma,0)$ are normalized associated Legendre polynomials and $\widetilde{D}_{KM}^J$ are normalized rotation matrices defined as

$$\widetilde{D}_{KM}^J = \sqrt{\frac{2J+1}{8\pi^2}}D_{KM}^J \tag{D.29}$$

Using Eq. (D.28), we can rewrite Eq. (D.27) as an orthogonal transformation between SF and BF angular momentum functions

$$Y_{jl}^{JM} = \sum_K C_{lK}\mathcal{Y}_{jK}^{JM} \tag{D.30}$$

where the orthogonal transformation matrix is

$$C_{lK} = \sqrt{\frac{2l+1}{2J+1}}\,\langle jKl0|JK\rangle \tag{D.31}$$

Equation (D.30) can be inverted to yield

$$\mathcal{Y}_{jK}^{JM} = \sum_{l} C_{lK} Y_{jl}^{JM} \tag{D.32}$$

By performing the parity operation on $\mathcal{Y}_{jK}^{JM}$ we obtain

$$\begin{aligned}
\hat{p}\mathcal{Y}_{jK}^{JM} &= \sum_{l} C_{lK} \hat{p} Y_{jl}^{JM} \\
&= \sum_{l} C_{lK} (-1)^{j+l} Y_{jl}^{JM} \\
&= (-1)^{J} \sum_{l} C_{l-K} Y_{jl}^{JM} \\
&= (-1)^{J} \mathcal{Y}_{j-K}^{JM}
\end{aligned} \tag{D.33}$$

where the symmetry relation of C.12 for the CG coefficient has been used. Thus, the BF function $\mathcal{Y}_{jK}^{JM}$ is not an eigenfunction of parity. It is thus often convenient to group the K and $-K$ terms together to define a parity-adapted BF angular momentum function as

$$\begin{aligned}
\mathcal{Y}_{jK}^{JMp} &= \frac{1}{\sqrt{2(1+\delta_{K0})}} \left[\mathcal{Y}_{jK}^{JM} + \hat{p}\mathcal{Y}_{jK}^{JM}\right] \\
&= \frac{1}{\sqrt{2(1+\delta_{K0})}} \left[\mathcal{Y}_{jK}^{JM} + (-1)^{P} \mathcal{Y}_{j-K}^{JM}\right]
\end{aligned} \tag{D.34}$$

where the total parity P is defined as $P = (-1)^{J+p}$ with parity p given by the definition of (D.23). We note that for K=0, only *even* total parity exists. With the definition of Eq. (D.34), we only need to deal with quantum numbers $K \geq 0$,

Using $\mathcal{Y}_{jK}^{JMp}$ as the body-fixed angular momentum eigenfunction, the transformation relation of Eq. (D.30) can be rewritten as

$$Y_{jl}^{JM} = \sum_{K \geq 0} C_{lK}^{p} \mathcal{Y}_{jK}^{JMp} \tag{D.35}$$

where the C_{lK}^{p} are given by

$$\begin{aligned}
C_{lK}^{p} &= \sqrt{2-\delta_{K0}} C_{lK} \\
&= \sqrt{2-\delta_{K0}} \sqrt{\frac{2l+1}{2J+1}} <jKl0|JK>
\end{aligned} \tag{D.36}$$

Orthogonality relation

From the orthogonality properties of Y_{jl}^{JM} and $\mathcal{Y}_{jK}^{JMp}$, we obtain the orthogonality relation for the transformation matrix

$$\begin{aligned}\sum_{l(p)} C_{lK}^{p} C_{lK'}^{p} &= \sum_{l(p)} \frac{2l+1}{2J+1}\sqrt{2-\delta_{K0}}\sqrt{2-\delta_{K'0}} \\ &\quad \times <jKl0|JK><jK'l0|JK'> \\ &= \delta_{KK'} \end{aligned} \tag{D.37}$$

where the summation over l is for a fixed parity block. The inverse relation is given by

$$\begin{aligned}\sum_{K\geq 0} C_{lK}^{p} C_{l'K}^{p} &= \sum_{K\geq 0} \frac{\sqrt{(2l+1)(2l'+1)}}{2J+1}(2-\delta_{K0}) \\ &\quad \times <jKl0|JK><jKl'0|JK> \\ &= \delta_{ll'} \end{aligned} \tag{D.38}$$

Similar orthogonality relations can be written out for the coefficients C_{lK} by simply removing the restrictions on l and K in the summations and setting the factor $(2-\delta_{K0})$ to unity in the above equations.

D.4 Total Angular Momentum in the BF Frame

It was discovered [236–238] that the projection of the total angular momentum in a rotating frame obeys an anomalous commutation relation. If $\hat{J}_i (i = x, y, z)$ denote the projections of the total angular momentum $\mathbf{J}$ along the body-fixed axes, they satisfy the *anomalous* commutation relation

$$[\hat{J}_i, \hat{J}_j] = \hat{J}_i\hat{J}_j - \hat{J}_j\hat{J}_i = -i\sum_k \epsilon_{ijk}\hat{J}_k \tag{D.39}$$

where ϵ_{ijk} is the standard antisymmetric tensor. As a result of this, the raising and lowering operators of the total angular momentum $\hat{J}_{\pm} = \hat{J}_x \pm i\hat{J}_y$ in the body-fixed frame behave like the lowering and raising operators $\hat{J}_{\mp} = \hat{J}_X \mp i\hat{J}_Y$ in the space-fixed frame! The following is a formal proof of Eq. (D.39) by Van Vleck [238] using direction cosine vectors.

Let $\mathbf{u}^{\alpha} (\alpha = x, y, z)$ and $\mathbf{e}_i$ $(i = X, Y, Z)$ be unit vectors along the body-fixed axes and space-fixed axes, respectively. Their projections on the space-fixed axis $u_i^\alpha = \mathbf{u}^\alpha \cdot \mathbf{e}_i$ are called direction cosines. The vectors $\mathbf{u}^\alpha$ obey the vector multiplication rule

$$\sum_{\alpha\beta} \epsilon_{\alpha\beta\gamma} \mathbf{u}^\alpha \mathbf{u}^\beta = \mathbf{u}^\gamma \tag{D.40}$$

By the definition of a vector, the commutation relation of $\mathbf{u}^\alpha$ with the total angular momentum $\mathbf{J}$ (space-fixed) is given by

$$[\hat{J}_i, u_j^\alpha] = i \sum_k \epsilon_{ijk} u_k^\alpha \tag{D.41}$$

Using Eqs. (D.40) and (D.41) along with the standard commutation relation of $\mathbf{J}$, it is not difficult to show that the body-fixed angular momentum $\hat{J}_\alpha = \mathbf{u}^\alpha \cdot \mathbf{J} = \sum_i u_i^\alpha \hat{J}_i$ satisfies the anomalous commutation relation (D.39).

It should be emphasized here that it is the *total* angular momentum whose projection onto the body-fixed representation satisfies the anomalous commutation relation. However, the component angular momenta still satisfy the normal commutation relation [238]. The effect of this anomalous commutation relation is reflected in the action of the angular momentum operators $J_\pm$ in the BF frame on BF angular momentum eigenfunctions [239]. The simplest approach is to represent the total angular momentum operator in the BF coordinate system in terms of the three Euler angles and to explicitly work out the following relation [240]

$$\begin{aligned} \hat{J}_\pm D^J_{KM} &= \sqrt{J(J+1) - K(K \mp 1)} D^J_{K\mp 1 M} \\ &= \lambda^\mp_{JK} D^J_{K\mp 1 M} \end{aligned} \tag{D.42}$$

or use the normalized rotation matrix

$$\hat{J}_\pm \widetilde{D}^J_{KM} = \lambda^\mp_{JK} \widetilde{D}^J_{K\mp 1 M} \tag{D.43}$$

This is an important relation that has been used to derive the coupling terms of the centrifugal potential for molecular Hamiltonian matrices in Sec. 3.2.2 and a number of other places. For example, we can evaluate the following expression in the BF coordinate system

$$\begin{aligned} \hat{J}_\pm \hat{\jmath}_\mp \mathcal{Y}^{JM}_{jK} &= (\hat{J}_\pm \widetilde{D}^J_{KM})(\hat{\jmath}_\mp P_{jK}) \\ &= (\lambda^\mp_{JK} \widetilde{D}^J_{K\mp 1 M})(\lambda^\mp_{jK} P_{jK\mp 1}) \\ &= \lambda^\mp_{JK} \lambda^\mp_{jK} \mathcal{Y}^{JM}_{jK\mp 1} \end{aligned} \tag{D.44}$$

The action of the angular momentum operator in the BF frame on parity-adapted BF angular momentum functions yields

$$\begin{aligned}\hat{J}_z\hat{\jmath}_z\mathcal{Y}_{jK}^{JMp} &= \hat{J}_z\frac{\hbar K}{\sqrt{2(1+\delta_{K0})}}\left[\mathcal{Y}_{jK}^{JM} - (-1)^P\mathcal{Y}_{j-K}^{JM}\right] \\ &= \frac{\hbar^2 K^2}{\sqrt{2(1+\delta_{K0})}}\left[\mathcal{Y}_{jK}^{JM} + (-1)^P\mathcal{Y}_{j-K}^{JM}\right] \\ &= \hbar^2 K^2 \mathcal{Y}_{jK}^{JMp} \end{aligned} \qquad \text{(D.45)}$$

and

$$\begin{aligned}\hat{J}_\pm\hat{\jmath}_\mp\mathcal{Y}_{jK}^{JMp} &= \frac{1}{\sqrt{2(1+\delta_{K0})}}\Big[\lambda_{JK}^{\mp}\lambda_{jK}^{\mp}\mathcal{Y}_{jK\mp1}^{JM} \\ &\quad +(-1)^P\lambda_{JK}^{\pm}\lambda_{jK}^{\pm}\mathcal{Y}_{j-K\mp1}^{JM}\Big]\end{aligned} \qquad \text{(D.46)}$$

Matrix elements of the $\mathbf{L}^2$ operator in the BF Basis

Using results derived in the preceding section for angular momentum operators in the BF frame, we can derive the matrix elements of the orbital angular momentum $\mathbf{L}^2$ in BF basis that are needed to construct the matrix elements of the centrifugal potential. Using the parity-adapted BF angular momentum basis, the matrix elements in question are defined by

$$W_{K'K}^{JMpj} = <\mathcal{Y}_{jK'}^{JMp}|\mathbf{L}^2|\mathcal{Y}_{jK}^{JMp}> \qquad \text{(D.47)}$$

where the parity-adapted BF angular momentum basis $\mathcal{Y}_{jK}^{JMp}$ is defined in Eq. (D.34).

To begin the derivation, we first write the orbital angular momentum operator as

$$\mathbf{L}^2 = \mathbf{J}^2 - \mathbf{j}^2 = (J^2 + j^2 - 2\hat{J}_z\hat{\jmath}_z) - (\hat{J}_+\hat{\jmath}_- + \hat{J}_-\hat{\jmath}_+) \qquad \text{(D.48)}$$

where $\mathbf{J}$ and $\mathbf{j}$ are, respectively, the total and internal angular momentum operators. Since $\mathbf{L}^2$ is a scalar operator, it is invariant with respect to the reference frame in which the angular momentum operators are projected. Thus we can project the angular momentum operators $\mathbf{J}$ and $\mathbf{j}$ in the BF frame and keep in mind that the total angular momentum operator $\mathbf{J}$ satisfies the anomalous commutation relation. Using the result of Eq. (D.45), the matrix element in Eq. (D.47) is then written as

$$\begin{aligned}W_{K'K}^{JMpj} &= \left[J(J+1) + j(j+1) - 2K^2\right]\delta_{K'K} \\ &\quad - <\mathcal{Y}_{jK'}^{JMp}|\hat{J}_+\hat{\jmath}_- + \hat{J}_-\hat{\jmath}_+|\mathcal{Y}_{jK}^{JMp}>\end{aligned} \qquad \text{(D.49)}$$

The matrix element in the second bracket in Eq. (D.49) can be calculated as follows. First, we use the result of Eq. (D.46) to write

$$(\hat{J}_+\hat{\jmath}_- + \hat{J}_-\hat{\jmath}_+)\mathcal{Y}_{jK}^{JMp} = \frac{1}{\sqrt{2(1+\delta_{K0})}}\Big\{\lambda_{JK}^+\lambda_{jK}^+\left[Y_{jK+1}^{JM} + (-1)^P Y_{j-K-1}^{JM}\right]$$
$$+\lambda_{JK}^-\lambda_{jK}^-\left[Y_{jK-1}^{JM} + (-1)^P Y_{j-K+1}^{JM}\right]\Big\} \qquad \text{(D.50)}$$

Since $K \geq 0$, we can separate the $K = 0$ case from the second bracket in the preceding equation and use definition of (D.34) to rewrite the equation as

$$(\hat{J}_+\hat{\jmath}_- + \hat{J}_-\hat{\jmath}_+)\mathcal{Y}_{jK}^{JMp} = \lambda_{JK}^+\lambda_{jK}^+\frac{1}{\sqrt{1+\delta_{K0}}}\mathcal{Y}_{jK+1}^{JMp} + \lambda_{JK}^-\lambda_{jK}^-\frac{\delta_{K0}}{\sqrt{1+\delta_{K0}}}$$
$$\times\mathcal{Y}_{jK+1}^{JMp} + \lambda_{JK}^-\lambda_{jK}^-(1-\delta_{K0})\sqrt{\frac{1+\delta_{K-1,0}}{1+\delta_{K0}}}\mathcal{Y}_{jK-1}^{JMp}$$
$$= \lambda_{JK}^+\lambda_{jK}^+\sqrt{1+\delta_{K0}}\mathcal{Y}_{jK+1}^{JMp} + \lambda_{JK}^-\lambda_{jK}^-(1-\delta_{K0})$$
$$\times\sqrt{1+\delta_{K1}}\mathcal{Y}_{jK-1}^{JMp} \qquad \text{(D.51)}$$

Thus we have determined the matrix element in Eq. (D.47) to be

$$W_{K'K}^{JMpj} = \left[J(J+1) + j(j+1) - 2K^2\right]\delta_{K'K} - \lambda_{JK}^+\lambda_{jK}^+\sqrt{1+\delta_{K0}}\,\delta_{K',K+1}$$
$$-\lambda_{JK}^-\lambda_{jK}^-\sqrt{1+\delta_{K1}}\,\delta_{K',K-1} \qquad \text{(D.52)}$$

One could also calculate the matrix element of Eq. (D.47) by transforming the BF basis to the SF basis. Using Eqs. (D.32) and (D.36) we can calculate the matrix element by

$$W_{K'K}^{JMpj} = \sum_{ll'(p)} C_{l'K'}^p C_{lK}^p <Y_{jl'}^{JM}|\mathbf{L}^2|Y_{jl}^{JM}>$$
$$= \sum_{l(p)} l(l+1) C_{lK'}^p C_{lK}^p$$
$$= \sum_{l(p)} \sqrt{2-\delta_{K'0}}\sqrt{2-\delta_{K0}}\frac{l(l+1)(2l+1)}{2J+1}$$
$$\times <jK'l0|JK'><jKl0|JK> \qquad \text{(D.53)}$$

where the summation over l, l' is for fixed parity only. By equating (D.52) with (D.53), we obtain a relation between the CG coefficients

$$\sum_{l(p)} \sqrt{2-\delta_{K'0}}\sqrt{2-\delta_{K0}}\frac{l(l+1)(2l+1)}{2J+1} <jK'l0|JK'><jKl0|JK>$$
$$= \left[J(J+1)+j(j+1)-2K^2\right]\delta_{K'K} - \lambda^+_{JK}\lambda^+_{jK}\sqrt{1+\delta_{K0}}\delta_{K',K+1}$$
$$- \lambda^-_{JK}\lambda^-_{jK}\sqrt{1+\delta_{K1}}\delta_{K',K-1} \qquad \text{(D.54)}$$

If we do not separate the parity and allow both positive and negative values of K, we can obtain a similar relation

$$\sum_{l} \frac{l(l+1)(2l+1)}{2J+1} <jKl0|JK><jK'l0|JK'>=$$
$$\left[J(J+1)+j(j+1)-2K^2\right]\delta_{K'K} - \lambda^+_{JK}\lambda^+_{jK}\delta_{K',K+1}$$
$$- \lambda^-_{JK}\lambda^-_{jK}\delta_{K',K-1} \qquad \text{(D.55)}$$

Equation (D.54) and (D.55) seem to be unexplored so far.

Bibliography

[1] J. C. Tully, in *Dynamics of Molecular Collisions, Part B*, edited by W. H. Miller (Plenum Press, New York, 1976).

[2] M. Baer, in *Theory of Chemical Reaction Dynamics Vol. II*, edited by M. Baer (CRC, Boca Raton, FL, 1985).

[3] M. S. Child, in *Atom Molecule Collision Theory*, edited by R. B. Bernstein (Plenum Press, New York, 1979).

[4] H. Nakamura, in *Dynamics of Molecules and Chemical Reactions*, edited by R. E. Wyatt and J. Z. H. Zhang (Marcel Dekker, New York, 1996).

[5] H. F. Schaefer III ed., *Methods of Electronic Structure Theory* (Plenum, New York, 1977).

[6] A. Szabo and N. S. Ostlund, *Modern Quantum Chemistry: Introduction to Advanced Electronic Structure Theory* (MacMillan, New York, 1982).

[7] S. Wilson, *Electron Correlation in Molecules* (Clarendon Press, Oxford, 1984).

[8] I. N. Levine, *Quantum Chemistry*, 4th Ed. (Prentice Hall, Englewood Cliffs, N. J., 1991)).

[9] T. H. Dunning and L. B. Harding, in *Theory of Chemical Reaction Dynamics Vol. I*, edited by M. Baer (CRC, Boca Raton, FL, 1985).

[10] M. Morse, *Calculus of Variations in the Large,* American Mathematical Society Colloquium Publications, Vol 18 (1934).

[11] J. V. Lill, G. A. Parker, and J. C. Light, *Chem. Phys. Lett.* **89**, 483 (1982); Z. Bacic and J. C. Light, *Ann. Rev. Phys. Chem.* **40**, 469 (1989).

[12] P. McGuire and D. J. Kouri, *J. Chem. Phys.* **60**, 2488 (1974).

[13] R. T. Pack, *J. Chem. Phys.* **60**, 633 (1974).

[14] W. D. Davison, Chem. Soc. Faraday Disc. **33**, 71 (1962).

[15] K. Takayanagi, Adv. Mol. Phys. **1**, 149, (1965).

[16] G. Zazur and H. Rabitz, *J. Chem. Phys.* **60**, 2057 (1974).

[17] S. Green, *J. Chem. Phys.* **62**, 2271 (1975).

[18] M. H. Alexander and A. P. DePristo, *J. Chem. Phys.* **66**, 2166 (1977).

[19] D. H. Zhang and J. Z. H. Zhang, *J. Chem. Phys.* **98**, 6276 (1993); *ibid.* **99** 6624 (1993).

[20] M. D. Marshall, P. Jensen, and P. R. Bunker, *Chem. Phys. Lett.* **176**, 255 (1991).

[21] S. C. Althorpe, D. C. Clary, and P. R. Bunker, *Chem. Phys. Lett.* **187**, 345 (1991).

[22] D. H. Zhang and J. Z. H. Zhang, *J. Chem. Phys.* **98**, 5978 (1993); *ibid.* **99** 6624 (1993).

[23] H. Sun and R. O. Watts, *J. Chem. Phys.* **92**, 603 (1990).

[24] M. Quack and M. A. Suhm, Mol. Phys. **69**, 791 (1990); (b) *ibid. J. Chem. Phys.* **95**, 28 (1991).

[25] W. C. Necoechea and D. G. Truhlar, *Chem. Phys. Lett.* **224**, 297 (1994).

[26] D. H. Zhang, Q. Wu, J. Z. H. Zhang, M.v. Dirke, and Z. Bacic, *J. Chem. Phys.* **102**, 2315 (1995).

[27] Q. Wu, D. H. Zhang, and J. Z. H. Zhang, *J. Chem. Phys.* **103**, 2548 (1995).

[28] M. Quack and M. A. Suhm, *J. Chem. Phys.* **95**, 28 (1991).

[29] R. E. Wyatt and C. Iung, in *Dynamics of Molecules and Chemical Reactions*, edited by R. E. Wyatt and J. Z. H. Zhang (Marcel Dekker, New York, 1996).

[30] J. Dai and J. Z. H. Zhang, *J. Chem. Phys.* **103**, 1491 (1995).

[31] D. Neuhauser, *J. Chem. Phys.* **100**, 5076 (1994).

[32] S. K. Gray and C. E. Wozny, *J. Chem. Phys.* **91**, 7671 (1989); **94**, 2817 (1991).

[33] A. Isele, C. Meier, V. Engel, N. Fahrer, and Ch. Schlier, *J. Chem. Phys.* **101**, 5919 (1994).

[34] J. Dai and J. Z. H. Zhang, *J. Chem. Phys.* **104**, 3664 (1995).

[35] V. A. Mandelshtam, T. P. Grozdanov, and H. S. Taylor, *J. Chem. Phys.* **103**, 10074 (1995).

[36] G. Herzberg, *Molecular Spectra and Molecular Structure: I. Spectra of Diatomic Molecules, 2nd ed.* (Krieger, FL, 1950).

[37] D. H. Zhang and J. Z. H. Zhang, *J. Chem. Phys.* **101**, 3671 (1994).

[38] A. J. Dobbyn, M. Stumpf, H-M. Keller, and R. Schinke, *J. Chem. Phys.* **103**, (1995).

[39] B. Kendrick and R. T. Pack, *Chem. Phys. Lett.* **235**, 291 (1995); *J. Chem. Phys.* **104**, 7475,7502 (1996).

[40] F. Calogero, *The Variable Phase Approach to Potential Scattering* (Academic Press, New York, 1967).

[41] A. M. Arthurs and A. Dalgarno, *Proc. R. Soc. London, Ser. A*, **256** 540 (1960).

[42] A. M. Lane and R. G. Thomas, *Rev. Mod. Phys.* **30**, 257 (1958).

[43] C. Bloch, *Nucl. Phys. A* **4**, 503 (1957).

[44] P. J. A. Buttle, *Phys. Rev.* **160**, 719 (1967).

[45] J. C. Light and R. B. Walker, *J. Chem. Phys.* **65**, 4272 (1976); E. B. Stechel, R. B. Walker, and J. C. Light, *J. Chem. Phys.* **69**, 3518 (1978).

[46] B. R. Johnson, *J. Comput. Phys.* **13** 4459 (1973).

[47] D. E. Manolopoulos, *J. Chem. Phys.* **85**, 6425 (1986).

[48] J. R. Taylor, *The Quantum Theory of Nonrelativistic Collisions* (Wiley, New York, 1972).

[49] P. Roman, *Advanced Quantum Theory* (Addison-Wesley, Reading, MA, 1965)

[50] R. G. Newton, *Scattering Theory of Waves and Particles* (Springer-Verlag, New York, 1982)

[51] C. J. Joachain, *Quantum Collision Theory* (North-Holland, Amsterdam, 1976)

[52] G. C. Schatz and A. Kuppermann, *J. Chem. Phys.* **65**, 4642 (1976).

[53] L. M. Hubbard, S. H. Shi, and W. H. Miller, *J. Chem. Phys.* **78**, 2381 (1983); G. C. Schatz, L. M. Hubbard, P. S. Dardi, and W. H. Miller, *J. Chem. Phys.* **81**, 231 (1984).

[54] A. K. Kuppermann and P. G. Hipes, *J. Chem. Phys.* **84**, 5962 (1986).

[55] R. T. Pack and G. A. Parker, *J. Chem. Phys.* **87**, 3888 (1987).

[56] W. H. Miller, *J. Chem. Phys.* **50**, 407 (1969).

[57] D. J. Kouri, in *Theory of Chemical Reaction Dynamics Vol. I*, edited by M. Baer (CRC, Boca Raton, FL, 1985).

[58] J. Z. H. Zhang, D. J. Kouri, K. Haug, D. W. Schwenke, Y. Shima, and D. G. Truhlar, *J. Chem. Phys.* **88**, 2492 (1988).

[59] W. Kohn, *Phys. Rev.*, **74**, 1763 (1948).

[60] T. -Y. Wu and T. Ohmura, *Quantum Theory of Scattering* (Prentice-Hall, Englewood Cliffs, NJ, 1962).

[61] J. Z. H. Zhang, S. I. Chu, and W. H. Miller, *J. Chem. Phys.* **88**, 6233 (1988).

[62] W. H. Miller, *Ann. Rev. Phys. Chem.* **41**, 245 (1990), and references therein.

[63] (a) K. Haug, D. W. Schwenke, Y. Shima, D. G. Truhlar, J. Z. H. Zhang, and D. J. Kouri, *J. Phys. Chem.* **90**, 6757 (1986); (b) J. Z. H. Zhang, D. J. Kouri, K. Haug, D. W. Schwenke, Y. Shima, and D. G. Truhlar, *J. Chem. Phys.* **88**, 2492 (1988); (c) D. W. Schwenke, K. Haug, M. Zhao, D. G. Truhlar, Y. Sun, J. Z. H Zhang, and D. J. Kouri, *J. Phys. Chem.* **92**, 3202 (1988); (d) M. Mladenovic, M. Zhao, D. G. Truhlar, D. W. Schwenke, Y. Sun, and D. J. Kouri, *J. Phys. Chem.* **92**, 7035 (1988).

[64] (a) J. Z. H Zhang and W. H. Miller, *Chem. Phys. Lett.* **153**, 465 (1988); (b) J. Z. H. Zhang, S-I Chu, and W. H. Miller, *J. Chem. Phys.* **88**, 6233 (1988); (c) J. Z. H. Zhang and W. H. Miller, *J. Chem. Phys.* **91**, 1528 (1989); (d) *ibid.* **92**, 1811 (1990); (e) *ibid.* **94**, 7785 (1990).

[65] (a) D. E. Manolopoulos and R. E. Wyatt, *Chem. Phys. Lett.* **152**, 23 (1988); (b) *ibid. J. Chem. Phys.* **91**, 6096 (1989); (c) D. E. Manolopoulos, M. D'Mello, and R. E. Wyatt, *J. Chem. Phys.* **93**, 403 (1990).

[66] (a) G. A. Parker, R. T. Pack, B. J. Archer, and R. B. Walker, *Chem. Phys. Lett.* **137**, 564 (1987); (b) R. T. Pack and G. A. Parker, *J. Chem. Phys.* **87**, 3888 (1987); (c) J. D. Kress, Z. Bacic, G. A. Parker, and R. T. Pack, *Chem. Phys. Lett.* **157**, 484 (1989).

[67] (a) A. Kuppermann and P. G. Hipes, *J. Chem. Phys.* **84**, 5962 (1986); (b) A. Kuppermann, in *Advances in Molecular Vibrations and Collision Dynamics, Vol. 2B*, edited by J. M. Bowman (JAI Press, Greenwich, CT, 1994).

[68] Schatz, G. C. *Chem. Phys. Lett.* **150**, 92 (1988).

[69] J. Linderberg, *Int. J. Quantum Chem. Symp.* **19**, 467 **1986**.

[70] (a) J. M. Launay and M. L. Dourneuf, *Chem. Phys. Lett.* **163**, 178 (1989); (b) *ibid. 169*, 473 **1990**.

[71] (a) J. Z. H. Zhang, *J. Chem. Phys.* **94**, 6047 (1991); (b) *ibid Chem. Phys. Lett.* **181**, 63 (1991).

[72] (a) Y. S. M. Wu, A. Kuppermann, and B. Lepetit, *Chem. Phys. Lett.* **186**, 319 (1991); (b) *ibid.* **201**, 178 (1993).

[73] J. F. Castillo, D. E. Manolopoulos, K. Stark and H.-J. Werner, *J. Chem. Phys.* **104**, 6531 (1996).

[74] B. Liu, *J. Chem. Phys.* **58**, 1924 (1973); P. Siegbahn and B. Liu, *J. Chem. Phys.* **68**, 2457 (1978); D. G. Truhlar and C. J. Horowitz, *J. Chem. Phys.* **68**, 2466 (1978).

[75] J. Z. H. Zhang and W. H. Miller, *J. Chem. Phys.* **91**, 1528 (1989).

[76] D. A. V. Kliner, K. -D. Rinnen, and R. N. Zare, *Chem. Phys. Lett.* **166**, 107 (1990).

[77] R. E. Continetti, B. A. Balko, and Y. T. Lee, *J. Chem. Phys.* **93**, 5719 (1990); R. E. Continetti, Ph.D Thesis, Univ. Calif., Berkeley, 1989.

[78] A. Kuppermann, in *Dynamics of Molecules and Chemical Reactions*, edited by R. E. Wyatt and J. Z. H. Zhang (Marcel Dekker, New York, 1996).

[79] D. M. Newmark, A. M. Wodtke, G. N. Robinson, C. C. Hayden and Y. T. Lee, ACS Symp. Ser. **263**, 479 (1984); *J. Chem. Phys.* **82**, 3045 (1985).

[80] J. Z. H. Zhang and W. H. Miller, *J. Chem. Phys.* **88**, 4549 (1988); *ibid.* **92**, 1811 (1990).

[81] J. Z. H. Zhang, *Chem. Phys. Lett.* **181**, 63 (1991).

[82] J. Z. H. Zhang, W. H. Miller, Alexandra Weaver, and Daniel M. Neumark, *Chem. Phys. Lett.* **182**, 283 (1991).

[83] C. Yu, Y. Sun, D. J. Kouri, P. Halvick, D. G. Truhlar and D. W. Schwenke, *J. Chem. Phys.* **90**, 7608 (1989); C. Yu, D. J. Kouri, M. Zhao, D. G. Truhlar and D. W. Schwenke, *Chem. Phys. Lett.* **157**, 491 (1989).

[84] J. D. Kress, Z. Bacic, G. A. Parker and R. T Pack, *Chem. Phys. Lett.* **157**, 484 (1989); Z. Bacic, J. D. Kress, G. A. Parker, and R. T Pack, *J. Chem. Phys.* **92**, 2344 (1990). *Chem. Phys. Lett.* **169**, 372 (1990).

[85] M. D'Mello, D. E. Manolopoulous, and R. E. Wyatt, *Chem. Phys. Lett.* **168**, 113 (1990); D. E. Manolopoulous, M. D'Mello, and R. E. Wyatt, *Chem. Phys. Lett.* **169**, 482 (1990).

[86] J. M. Launay and M. L. Dourneuf, *Chem. Phys. Lett.* **169**, 473 (1990).

[87] D. Neuhauser, M. Baer, R. S. Judson, and D. J. Kouri, *Chem. Phys. Lett.* **169**, 372 (1990).

[88] R. Steckler, D. G. Truhlar, and B. C. Garrett, *J. Chem. Phys.* **82**, 5499 (1985).

[89] K. Stark and H.-J. Werner, *J. Chem. Phys.* **104**, 6515 (1996).

[90] M. S. Child, *Molecular Collision Dynamics* (Academic Press, London, 1974).

[91] W. H. Miller, ed., *Dynamics of Molecular Collisions, Vol. I-II* (Plenum Press, New York, 1976)

[92] J. M. Bowman, ed., *Molecular Collision Dynamics* (Springer-Verlag, Berlin 1983).

[93] M. Baer, ed., *Theory of Chemical Reaction Dynamics, Vol. I–IV,* (CRC, Boca Raton, FL, 1985).

[94] R. E. Wyatt and J. Z. H. Zhang, ed., *Dynamics of Molecules and Chemical Reactions* (Marcel Dekker, New York, 1996).

[95] J. Z. H. Zhang, J. Dai, and W. Zhu, *J. Phys. Chem.* **101**, A, 2746 (1997), and references therein.

[96] D. Neuhauser and M. Baer, *J. Chem. Phys.* **90**, 4351 (1989).

[97] J. Z. H. Zhang, *J. Chem. Phys.* **92**, 324 (1990); *Chem. Phys. Lett.* **160**, 417 (1989).

[98] J. Z. H. Zhang, Comput. Phys. Commun. **63**, 28 (1991).

[99] S. Das and D. J. Tannor, *J. Chem. Phys.* **92**, 3403 (1990)

[100] C. J. Williams, J. Qian and D. J. Tannor, *J. Chem. Phys.* **95**, 1721 (1991)

[101] M. Founargiotakis and J. C. Light, *J. Chem. Phys.* **93**, 633 (1990).

[102] (a) J. A. Fleck, Jr. J. R. Morris, and M. D. Feit, Appl. Phys. **10**, 129 (1976); (b) M. D. Feit, J. A. Fleck, Jr., and A. Steiger, J. Comput. Phys. **47**, 412 (1982).

[103] H. Tal-Ezer and D. Kosloff, *J. Chem. Phys.* **81**, 3967 (1984).

[104] C. Lanczos, J. Res. Natl. Bur. Stand. **45**, 255 (1950).

[105] T. J. Park and J. C. Light, *J. Chem. Phys.* **85**, 5870 (1986)

[106] C. Leforestier, R. Bisseling, C. Cerjan, M. D. Feit, R. Friesner, A. Guldberg, A. Hammerich, G. Jolicard, W. Karrlein, H. D. Meyer, N. Lipkin, O. Roncero, and R. Kosloff, *J. Comput. Phys.* **94**, 59 (1991)

[107] S. K. Gray, *J. Chem. Phys.* **96**, 6543 (1992).

[108] H. Abramowitz and I. A. Stegun, *Handbook of Mathematical Functions, with Formulas, Graphs, and Mathematical Tables,* (Dover, New York, 1972).

[109] V. Mandelshtam and H. S. Taylor, *J. Chem. Phys.* **102**, 7390(1995); *J. Chem. Phys.* **103**, 2903 (1995).

[110] R. Kosloff and D. Kosloff, *J. Chem. Phys.* **79**, 1823 (1983).

[111] E. J. Heller, *J. Chem. Phys.* **62**, 1544 (1975).

[112] J. A. Miller, R. J. Kee, and C. K. Westbrook, *Ann. Rev. Phys. Chem.* **41**, 345 (1990).

[113] C. Leforestier and W. H. Miller, *J. Chem. Phys.* **100**, 733 (1994).

[114] J. Dai and J. Z. H. Zhang, *J. Phys. Chem.* **100**, 6898 (1996).

[115] A. J. H. M. Meijer, and E. M. Goldfield, *J. Chem. Phys.* **108**, 5404 (1998).

[116] M. R. Pastrana, L. A. M. Quintales, J. Brandas, and A. J. C. Varandas, *J. Phys. Chem.* **94**, 8037 (1990).

[117] (a) Q. Sun and J. M. Bowman, *J. Chem. Phys.* **92**, 5201 (1990); (b) Q. Sun, D. L. Yan, N. S. Wang, J. M. Bowman, and M. C. Lin, *J. Chem. Phys.* **93**, 4730 (1990); (d) J. M. Bowman and D. S. Wang, *J. Chem. Phys.* **96**, 7852 (1992); (e) D. Wang and J. M. Bowman, *J. Chem. Phys.* **96**, 8906 (1992); (f) *ibid.* **98**, 6235 (1993); (g) *ibid.Chem. Phys. Lett.* **207**, 227 (1993).

[118] (a) A. N. Brook and D. C. Clary, *J. Chem. Phys.* **92**, 4178 (1990); (b) D. C. Clary, *J. Chem. Phys.* **95**, 7298 (1991); (c) *ibid.* **96**, 3656 (1992); (d) D. C. Clary, *Chem. Phys. Lett.* **192**, 34 (1992); (e) G. Nyman and D. C. Clary, *J. Chem. Phys.* **99**, 7774 (1993).

[119] (a) H. Szichman, I. Last, A. Baram, and M. Baer, *J. Phys. Chem.* **97**, 6436 (1993); (b) H. Szichman and M. Baer, *J. Chem. Phys.* **101**, 2081 (1994); *Chem. Phys. Lett.* **285**, 242 (1995).

[120] N. Balakrishnan and G. D. Billing, *J. Chem. Phys.* **101**, 2785 (1994). *Chem. Phys.* **499**, 189 (1994).

[121] J. Echave and D. C. Clary, *J. Chem. Phys.* **100**, 402 (1994).

[122] W. H. Thompson and W. H. Miller, *J. Chem. Phys.* **101**, 8620 (1994).

[123] E. M. Goldfield, S. K. Gray, G. C. Schatz, *J. Chem. Phys.* **102**, 8807 (1995).

[124] D. H. Zhang and J. Z. H. Zhang, *J. Chem. Phys.* **99**, 5615 (1993). *J. Chem. Phys.* **100**, 2697 (1994).

[125] J. Z. H. Zhang and D. H. Zhang, *J. Chem. Phys.* **101**, 1146 (1994).

[126] U. Manthe, T. Seideman, and W. H. Miller, *J. Chem. Phys.* **99**, 10078 (1993); *J. Chem. Phys.* **101**, 4759 (1994).

[127] D. Neuhauser, *J. Chem. Phys.* **100**, 9272 (1994).

[128] (a) J. Z. H. Zhang and D. H. Zhang, *Chem. Phys. Lett.* **232**, 370 (1995); (b) D. H. Zhang, J. Z. H. Zhang, Y. C. Zhang, D. Wang, Q. Zhang, *J. Chem. Phys.* **102**, 7400 (1995).

[129] Y. Zhang, D. Zhang, W. Li, Q. Zhang, D. Wang, D. H. Zhang, and J. Z. H. Zhang, *J. Phys. Chem.* **99**, 16824 (1995).

[130] D. H. Zhang and J. Z. H. Zhang, *J. Chem. Phys.* **103**, 6512 (1995).

[131] W. Zhu, J. Dai, and J. Z. H. Zhang, *J. Chem. Phys.* **105**, 4881 (1996).

[132] J. Dai, W. Zhu, and J. Z. H. Zhang, *J. Phys. Chem.* **100**, 13901 (1996).

[133] D. H. Zhang and J. C. Light, *J. Chem. Phys.* **104**, 4544 (1996).

[134] D. H. Zhang and J. C. Light, *J. Chem. Phys.* **105**, 1291 (1996).

[135] D. Neuhauser, M. Baer, R. S. Judson, and D. J. Kouri, *Computer Phys. Commun. 63*, 460 (1991),

[136] R. C. Mowrey, *J. Chem. Phys.* **94**, 7098 (1991).

[137] D. H. Zhang and Zhang, J. Z. H. *J. Chem. Phys.* **98**, 6276 (1993).

[138] D. H. Zhang and J. Z. H. Zhang, *J. Chem. Phys.* **98**, 5978 (1993); *J. Chem. Phys.* **99**, 6624 (1993).

[139] M. E. Rose, *Elementary Theory of Angular Momentum,* (Wiley, New York, 1957).

[140] (a) S. P. Walch and T. H. Dunning, *J. Chem. Phys.* **72**, 1303 (1980); (b) G. C. Schatz and H. Elgersma, *Chem. Phys. Lett.* **73**, 21 (1980).

[141] M. Alagia, N. Balucani, P. Casavecchia, D. Strangers, and G. G. Volpi, *J. Chem. Soc. Faraday. Trans.* **91**, 575 (1995).

[142] I. W. M. Smith and R. Zellner, *J. Chem. Soc. Faraday. Trans.* **69**, 1617 (1973); C. D. Jonah, Chem. Res. *10*, 326 (1977); R. R. Rabishankara and R. L. Thopmson, *Chem. Phys. Lett.* **99**, 377 (1983); J. Brunning, D. W. Derbyshire, I. W. M. Smith, and M. D. Williams, J. Chem. Soc. **84**, 105 (1988); J. E. Spencer, H. Endo, and G. P. Glass, *Proceedings of the Sixteenth International Symposium on Combustion* (The Combustion Institute, Pittsburgh, 1977), P. 829; D. L. Bauch, R. A. Cox, R. F. Hampson, Jr., J. A. Kerr, J. Troe, and R. T. Watson, *J. Phys. Chem. Ref. Data* **13**, 1259 (1984), and references therein.

[143] J. Wolfrum, *Faraday Diss. Chem. Soc.* **84**, 191 (1987); Spectrochim. Acta Part A **46** 567 (1990); M. J. Frost, P. Sharkey, and I. W. M. Smith, *Faraday Diss. Chem. Soc.* **91**, 305 (1991).

[144] J. K. Rice and A. P. Baronavsky, *J. Chem. Phys.* **94**, 1006 (1991); S. L. Nickolaisen, H. E. Cartland, and C. Wittig, *ibid.* **96**, 4378 (1992); C. R. Quick and J. J. Tiee, *Chem. Phys. Lett.* **100**, 223 (1983).

[145] N. F. Scherer, L. R. Khundkar, R. B. Bernstein, and A. H. Zewail, *J. Chem. Phys.* **87**, 1451 (1987); N. F. Scherer, C. Sipes, R. B. Bernstein, and A. H. Zewail, *ibid.* **92**, 5239 (1990); C. Wittig, S. Sharpe, and R. A. Beaudet, Acc. Chem. Res. **21** 341 (1988); S. I. Ionov, G. A. Brucker, C. Jaques, L. Valachovic, and C. Wittig, *J. Chem. Phys.* **97**, 9486 (1992).

[146] Schatz, G. C.; Fitzcharles; Harding, L. B. *Faraday Discus. Chem. Soc.* **84**, 359 (1987).

[147] M. Alagia, N. Balucani, P. Casavecchia, D. Stranges, and G. G. Volpi, *J. Chem. Phys.* **98**, 8341 (1993).

[148] K. Kudla, G. C. Schatz, and A. F. Wagner, *J. Chem. Phys.* **95**, 1635 (1991).

[149] G. C. Schatz and J. Dyke, *Chem. Phys. Lett.* **188**, 11 (1992).

[150] A. N. Brook and D. C. Clary, *J. Chem. Phys.* **92**, 4178 (1990).

[151] D. C. Clary and G. C. Schatz, *J. Chem. Phys.* **99**, 4578 (1992).

[152] E. M. Goldfield, S. K. Gray, and G. C. Schatz, *J. Chem. Phys.* **102**, 8807(1995).

[153] N. Balakrishnan and G. D. Billing, *J. Chem. Phys.* **104**, 4004 (1996).

[154] (a) D. J. Kouri, M. Arnold, and D. K. Hoffman, *Chem. Phys. Lett.* **203**, 166 (1993); (b) D. J. Kouri, Y. Huang, W. Zhu, and D. K. Hoffman, *J. Chem. Phys.* **100**, 3662 (1994).

[155] D. J. Tannor and D. E. Weeks, *J. Chem. Phys.* **98**, 3884 (1993).

[156] G. G. Balint-Kurti, R. N. Dixon, and C. C. Martson, *J. Chem. Soc. Faraday. Trans.* **86**, 1741 (1990).

[157] D. H. Zhang, Q. Wu, and J. Z. H. Zhang, *J. Chem. Phys.* **102**, 124 (1995).

[158] W. H. Miller, *Adv. Chem. Phys.,* **25**, 69 (1974)

[159] T. Peng and J. H. Z. Zhang, *J. Chem. Phys.* **105**, 6072 (1996).

[160] W. Zhu, T. Peng and J. H. Z. Zhang, *J. Chem. Phys.* **106**, 1742 (1997).

[161] J. Dai and J. H. Z. Zhang, *J. Chem. Soc. Faraday. Trans.* **93**, 699 (1997).

[162] F. T. Smith, *Phys. Rev.* **118**, 349 (1960).

[163] R. B. Metz, T. Kitsopoulos, A. Weaver, and D. M. Neumark, *J. Chem. Phys.* **88**, 1463 (1988); A. Weaver, R. B. Metz, S. E. Bradforth, and D. M. Neumark, *J. Phys. Chem.* **92**, 5558 (1988); *J. Chem. Phys.* **93**, 5352 (1990).

[164] C. L. Shoemaker, and R. E. Wyatt, *Advances in Quantum Chemistry*, **14**, 169 (1981).

[165] V. A. Mandelshtam and H. S. Taylor, in *Dynamics of Molecules and Chemical Reactions*, edited by R. E. Wyatt and J. Z. H. Zhang (Marcel Dekker, New York, 1996).

[166] See *Resonances*, ACS Symp. Ser. 263, edited by D. G. Truhlar (1984).

[167] J. Qi and J. M. Bowman, *J. Chem. Phys.* **100**, 15165 (1996); D. Wang and J. M. Bowman, *Chem. Phys. Lett.* **235**, 277 (1995).

[168] W. H. Miller, in *Dynamics of Molecules and Chemical Reactions*, edited by R. E. Wyatt and J. Z. H. Zhang (Marcel Dekker, New York, 1996).

[169] D. C. Chatfield, *et al*, in *Dynamics of Molecules and Chemical Reactions*, edited by R. E. Wyatt and J. Z. H. Zhang (Marcel Dekker, New York, 1996).

[170] R. G. Gilbert and S. C. Smith, in *Theory of Unimolecular and Recombination Reactions* (Blackwell Scientific, Oxford, 1990).

[171] G. C. Schatz and M. A. Ratner, *Quantum Mechanics in Chemistry* (Prentice-Hall, Englewood Cliffs, NJ, 1993).

[172] R. D. Levine, *Quantum Mechanics of Molecular Rate Processes* (Clarendon Press, Oxford, 1969).

[173] R. D. Levine and R. B. Berstein, *Molecular Reaction Dynamics and Chemical Reactivity* (Oxford, New York, 1987).

[174] W. Heitler, *The Quantum Theorey of Radiation, 3rd ed.* (Clarendon Press, Oxford, 1954).

[175] J. J. Sakurai, *Advanced Quantum Mechanics* (Addison-Wesley, Reading, MA, 1967)

[176] Y. Aharonov and D. Bohm, *Phys. Rev.* **115**, 485 (1959).

[177] H. A. Kramers and W. Heisenberg, *Zs. f. Phys.* **31**, 681 (1925).

[178] E. J. Heller, *J. Chem. Phys.* **68**, 2066 (1978); N. E. Henriksen and E. J. Heller, *J. Chem. Phys.* (,91) 4700 (1989).

[179] D. H. Zhang, O. A. Sharafeddin, and J. Z. H. Zhang, *Chem. Phys.* **167**, 137 (1992).

[180] R. Schinke, *Photodissociation Dynamics: Spectroscopy and Fragmentation of Small Polyatomic Molecules* (Cambridge University Press, Cambridge, 1993).

[181] P. Jensen, *J. Mol. Spectrosc.* **133**, 438 (1989).

[182] V. Engel and R. Schinke, *J. Chem. Phys.* **88**, 129 (1988).

[183] Z. T. Cai, D. H. Zhang, and J. Z. H. Zhang, *J. Chem. Phys.* **100**, 5631 (1994).

[184] P. Villarreal, S. Miret-Artes, O. Roncero, G. Delgado-Barrio, J. A. Beswick, N. Halberstadt, and R. D. Coalson, *J. Chem. Phys.* **94**, 4230 (1991).

[185] D. H. Zhang and J. Z. H. Zhang, *J. Chem. Phys.* **95**,; *J. Phys. Chem.* **96**, 1575 (1992).

[186] J. A. Beswick and J. Jortner, *Chem. Phys. Lett.* **49**, 13 (1977); G. E. *Chem. Phys.* **29**, 253 (1978).

[187] J. I. Cline, B. P. Reid, D. D. Evard, N. Sivakumar, N. Halberstadt, and K. C. Janda, *J. Chem. Phys.* **89**, 3535 (1988).

[188] D. C. Clary, *J. Chem. Phys.* **96**, 90 (1992).

[189] C. M. Lovejoy, D. D. Nelson, Jr., and D. J. Nesbitt, *J. Chem. Phys.* **89**, 7180 (1988).

[190] D. H. Zhang, J. Z. H. Zhang, and Z. Bacic, *J. Chem. Phys.* **97**, 927 (1992); D. H. Zhang, J. Z. H. Zhang, and Z. Bacic, *Chem. Phys. Lett.* **194**, 313 (1992).

[191] D. E. Manolopoulos and M. H. Alexander, *J. Chem. Phys.* **97**, 2527 (1992).

[192] T. Seideman and W. H. Miller, *J. Chem. Phys.* **96**, 4412 (1992).

[193] J. T. Farrell and D. J. Nesbitt (private communication).

[194] D. H. Zhang, Q. Wu, and J. Z. H. Zhang, *J. Chem. Phys.* **102**, 124 (1995).

[195] M. Quack and M. A. Suhm, Mol. Phys. **69**, 791 (1990); *J. Chem. Phys.* **95**, 28 (1991).

[196] A. S. Pine and G. T. Fraser, *J. Chem. Phys.* **89**, 6636 (1988).

[197] D. J. Kouri, D. K. Hoffman, T. Peng, and J. Z. H. Zhang, *Chem. Phys. Lett.* **262**, 519 (1996).

[198] D. Y. Wang, W. Zhu, J. Z. H. Zhang, and D. J. Kouri, *J. Chem. Phys.* **107**, 751 (1997).

[199] O. A. Sharafeddin and J. Z. H. Zhang, *Chem. Phys. Lett.* **204**, 190 (1993).

[200] A. Cruz and B. Jackson, *J. Chem. Phys.* **84**, 5715 (1991).

[201] J. Sheng and J. Z. H. Zhang, *J. Chem. Phys.* **99**, 1373 (1993).

[202] J. Dai, J. Sheng, and J. Z. H. Zhang, *J. Chem. Phys.* **101**, 1555 (1994); *Surf. Sci.* **319**, 193 (1994).

[203] J. Dai and J. Z. H. Zhang, *Surf. Sci.* **319**, 193 (1994).

[204] G. R. Darling and S. Holloway, *J. Chem. Phys.* **101**, 3268 (1994).

[205] J. Sheng and J. Z. H. Zhang, *J. Chem. Phys.* **97**, 6784 (1992).

[206] M. Balooch, M. J. Cardillo, R. R. Miller, and R. E. Stickney, *Surf. Sci.* **46**, 358 (1974); M. Balooch and R. E. Stickney, *Surf. Sci.* **44**, 310 (1974).

[207] G. Anger, A. Winkler, and K. D. Rendulic, *Surf. Sci.* **220**, 1 (1989).

[208] B. E. Hayden and C. L. A. Lamont, *Phys. Rev. Lett.* **63**, 1823 (1089); *Chem. Phys. Lett.* **160**, 331 (1989).

[209] G. Comsa and R. David, *Surf. Sci.* **117**, 77 (1982).

[210] G. D. Kubiak, G. O. Sitz and R. N. Zare, *J. Chem. Phys.* **81**, 6398 (1984); **83**, 2538 (1985).

[211] B. Hammer, M. Scheffler, K. W. Jacobsen, and J. K. Norskov, *Phys. Rev. Lett.* **73**, 1400 (1994); J. A. White, D. M. Bird, M. C. Payne, and I. Stich, *Phys. Rev. Lett.* **73**, 1404 (1994).

[212] R. C. Mowrey, *J. Chem. Phys.* **94**, 7098 (1991).

[213] K. Yang and T. S. Rahman, *J. Chem. Phys.* **93**, 6834 (1990).

[214] (a) H. A. Michelsen, C. T. Rettner, and D. J. Auerbach, *Phys. Rev. Lett.* **69**, 2678 (1992); (b) H. A. Michelsen, C. T. Rettner, D. J. Auerbach, and R. N. Zare, *J. Chem. Phys.* **98**, 8294 (1993).

[215] R. E. Langer, *Phys. Rev.* **51**, 669 (1937).

[216] I. S. Gradshteyn and I. M. Ryzhik, *Tables of Integrals, Series, and Products* (Academic Press, San Diego, CA, 1980).

[217] E. T. Wittaker, G. N. Watson, *Modern Analysis*, 4th ed. (Cambridge University Press, London, 1952).

[218] R. P. Feynman and A. P. Hibbs, *Quantum Mechanics and Path Integrals* ((McGraw-Hill, New York, 1965).

[219] H. Goldstein, *Classical Mechanics* (Addision-Wesley, New York, 1959).

[220] J. H. Van Vleck, *Proc. Natl. Acad. Sci.* **14**, 178 (1928).

[221] P. Pechukas, Phys. Rev. **181**, 166 (1969).

[222] M. F. Herman and E. Fluk, *Chem. Phys.* **91**,, 27 (1984); E. Fluk, M. F. Herman, and H.L. Davis, *J. Chem. Phys.* **84**, 326 (1986); M. F. Herman, *J. Chem. Phys.* **85**, 2069 (1986).

[223] E. J. Heller, *J. Chem. Phys.* **94**,, 2723 (1991).

[224] W. H. Miller, *J. Chem. Phys.* **95**,, 9428 (1991).

[225] K. W. Ford and J. A. Wheeler, Ann, Phys., **7**, 259, 287 (1959).

[226] H. Jeffreys, *Asymptotic Approximations* (Oxford University Press, London, 1962).

[227] J. Heading, *An Introduction to Phase Integral Methods*, (Oxford University Press, London, 1962).

[228] N. Fröman and P. O. Fröman, *JWKB Approximation*, (North-Holland, Amsterdam, 1965).

[229] M. V. Berry and K. E. Mount, Rep. Prog. Phys. **35**, 315 (1972).

[230] W. H. Miller, *Adv. Chem. Phys.,* **30**, 77 (1975)

[231] B. C. Eu, *Semiclassical Theories of Molecular Scattering* (Springer-Verlag, Berlin, 1984).

[232] M. S. Child, *Semiclassical Mechanics with Molecular Applications* (Oxford University Press, Cambridge, 1991).

[233] G. Racah, *Phys. Rev.*, **62**, 438 (1942).

[234] A. R. Edmonds, *Angular Momentum in Quantum Mechanics* (Princeton University Press, 1974).

[235] E. P. Wigner, *Group Theory* (Academic Press, New York, 1959).

[236] O. Klein, *Z. Physik*, **58** 730 (1929).

[237] H. B. G. Casimir, *Rotation of a Rigid Body in Quantum Mechanics* (J. H. Woltjers, The Hague, 1931).

[238] J. H. Van Vleck, *Rev. Mod. Phys.* **23**, 213 (1951).

[239] R. T. Pack and J. O. Hirschfelder, *J. Chem. Phys.* **49**, 4009 (1968).

[240] R. N. Zare, *Angular Momentum* (Wiley, New York, 1988).

Index